PREFACE

No calculus text can contain everything that each individual student, or each individual instructor, would like to have included. First of all, there would not be enough room for all this material. More fundamental, however, is that while some readers need a great deal of detail, that detail would be unnecessary and burdensome for others.

Hence The Calculus Companion. It supplements Anton's Calculus with extras that simply could not go into that book. In particular, Volume II of The Companion contains:

- additional explanations of difficult concepts;
- additional computational examples with very detailed solutions;
- suggested procedures for attacking specific types of complex problems;
- numerous warnings concerning common mistakes and trouble spots;
- answers to the often-raised question "Why is this topic important?";
- numerous helpful figures and diagrams;
- optional sections on supplementary topics such as Kepler's Laws of Planetary Motion, the computation of surface integrals in cylindrical and spherical coordinates, and differential forms.

The Companion stresses conceptual understanding and computational skill and, for that reason, it rarely elaborates on proofs or derivations. The writing style is informal and chatty, both to make the reading more appealing and to make the material less intimidating.

For ease of use, The Companion is organized in modular fashion: each section in Anton's Calculus has a corresponding section in The Companion, and different sections of The Companion rarely refer to each other. Thus readers can refer to those sections which are useful to them, and skip those which are not. Moreover, The Companion has a detailed index (covering both Volumes I and II), so finding a particular topic is easy.

The Companion has been used in classes at Bowdoin College since 1981 and we are indebted to the many students and colleagues who made helpful suggestions about it. In particular we want to thank R. Wells Johnson for the derivation of Kepler's Laws that appears in Section 14.6. We are also grateful to Gary W. Ostedt and Robert W. Pirtle, mathematics editors at John Wiley and Sons, who have been cooperative and encouraging throughout this project, and to Barbara J. Moody of "North Country Technical Typing" who performed the Herculean task of typing the manuscript so beautifully. Finally, we must thank Howard Anton, both for requesting us to write The Calculus Companion, and for his willingness to be spoofed occasionally.

William H. Barker
James E. Ward

Bowdoin College
Brunswick, Maine
March, 1984

TO THE INSTRUCTOR

The Companion is a book students can turn to when in need of help on specific topics; for this reason, students having difficulty with the calculus should find it particularly useful. Stronger (or simply more curious) students will obtain a fuller understanding of the calculus by reading The Companion as a regular supplement to Anton's text. The Companion should also prove to be a useful tool when reviewing for examinations.

In addition,
- the numerous lists of suggested procedures and warnings about common trouble spots should contain new lecture ideas even for the most seasoned calculus instructor;
- the optional supplementary sections can be the basis for interesting additions to the standard calculus course, or as enrichment reading for the better students;
- The Companion should answer many of the simpler questions that students often bring to an instructor. By using it for this purpose, the instructor can spend more time helping students with fundamental problems, and less on routine matters.

The Companion has been tested, and has proved to be very effective, in a self-paced calculus program with no regular classroom lectures. When used in this way, it takes the place of the missing lectures.

The Companion can be offered as an optional, not required, text for a calculus course. This is probably a very sensible procedure in many situations.

TO THE STUDENT

You should use The Companion as you would your own, private tutor. After reading a section in Anton's Calculus, consult The Companion if you need additional explanation or if you would like to know more about a particular topic. The preface gives a detailed listing of the sorts of things you can expect to find in this volume.

There is a section in The Companion for each section in Anton's Calculus and, for the most part, the sections are self-contained. Thus you can skip around, reading only those sections which are important to you. The index should prove to be useful in this regard.

Our students have found The Companion to be very helpful. We hope you will too.

W.H.B. and J.E.W.

THE CALCULUS
COMPANION VOL. 2
TO ACCOMPANY

CALCULUS
HOWARD ANTON
SECOND EDITION

WILLIAM H. BARKER
JAMES E. WARD

BOWDOIN COLLEGE

JOHN WILEY & SONS
NEW YORK/CHICHESTER
BRISBANE/TORONTO
SINGAPORE

ISBN 0-471-88614-9
Printed in the United States of America

10 9 8 7 6 5 4 3 2 1

CONTENTS OF VOLUME II

CHAPTER **18.** TOPICS IN VECTOR CALCULUS

INDEX FOR VOLUMES I AND II

SUPPLEMENTARY TOPICS IN VOLUME II

Chapter 14:   Vectors in the Plane

Section 14. 1:   Vectors.

1.   Geometric vectors.   Consider the following examples of important physical quantities:

length	force
temperature	velocity
mass	acceleration
electric charge	rate of heat flow

The first column contains scalar quantities, i. e., quantities that are completely described by giving one real number which specifies magnitude.

The second column contains vector quantities, i. e., quantities for which both a magnitude and a direction must be specified.   We represent these in space by directed line segments, or arrows, where

direction of arrow = direction of vector quantity

length of arrow = magnitude of vector quantity

We emphasize that

| Geometric equivalence of vectors | Two vectors $\bar{v}$ and $\bar{w}$ are equivalent, written $\bar{v} = \bar{w}$, if and only if they have the same length and the same direction. | (A) |

two equivalent vectors	two non-equivalent vectors

An important consequence: <u>the placement of the initial point of a vector is unimportant</u>! Hence vectors act like little <u>movable</u> arrows.

There are a number of standard operations on vectors which are easily defined geometrically. Anton carefully defines these operations in words; we will just summarize his definitions and include the standard pictures. These definitions and pictures must become second nature to you!!

<u>Vector addition</u>: $\bar{v} + \bar{w}$. Place the initial

point $\bar{w}$ on the terminal point of $\bar{v}$.

Then $\bar{v} + \bar{w}$ starts at the initial point

of $\bar{v}$ and ends at the terminal point of $\bar{w}$.

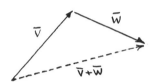

<u>Scalar multiplication</u>: $k\bar{v}$.

    i. If $k \geq 0$, multiply the length of $\bar{v}$

        by k to get $k\bar{v}$. Thus $k\bar{v}$ has

        the <u>same</u> direction as $\bar{v}$.

    ii. If $k < 0$, multiply the length of $\bar{v}$

        by $|k|$ and reverse the direction

        to get $k\bar{v}$. Thus $k\bar{v}$ has the

        <u>opposite</u> direction of $\bar{v}$.

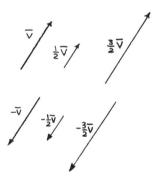

<u>Vector subtraction</u>: $\bar{v} - \bar{w} = \bar{v} + (-\bar{w})$.

    Place the initial points of $\bar{v}$ and $\bar{w}$

    togeher. Then $\bar{v} - \bar{w}$ starts at the terminal

    point of $\bar{w}$ and ends at the terminal point of $\bar{v}$.

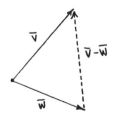

The following statement needs to be emphasized in view of our subsequent work: <u>everything</u> <u>said up until now applies to vectors in space as well as to vectors in the plane.</u>

2. <u>Algebraic vectors in the plane.</u> For the rest of Chapter 14 Anton restricts his attention to vectors <u>in the plane</u>. This is done so that the introduction to vector algebra will be as gradual as possible. Vectors in space will be treated in Chapter 15; there you will see that everything we are about to do in the plane has its straightforward analogue in 3-dimensional space.

In order to deal effectively with vectors we must somehow convert our geometry into algebra (the usual story of analytic geometry ...). To do so we introduce an xy-coordinate system on the plane. Then any "geometric vector" $\overline{v}$ in the plane can be converted into an "algebraic vector" as follows:

| Vector components in the plane | Position $\overline{v}$ so that its initial point is at the origin. Then $\overline{v}$ is completely determined by the coordinates (v_1, v_2) of its terminal point, and we write $$\overline{v} = \langle v_1, v_2 \rangle$$ with v_1 and v_2 called the <u>components</u> of $\overline{v}$. | 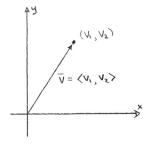 |

Thus every geometric vector $\overline{v}$ in the plane determines an ordered pair of numbers $\langle v_1, v_2 \rangle$ -- an algebraic vector -- and vice versa. In particular, suppose $\overline{v}$ is a vector in the plane with initial point $P_1(x_1, y_1)$ and terminal point $P_2(x_2, y_2)$, written

$\bar{v} = \overrightarrow{P_1 P_2}$. Then the components of

$\bar{v}$ are as follows:

$$\bar{v} = \overrightarrow{P_1 P_2} = \langle x_2 - x_1, y_2 - y_1 \rangle$$

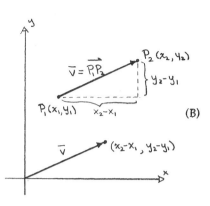

(B)

i.e., the components of any vector

are the coordinates of its terminal

point minus the coordinates of its initial point.

Be sure you understand the meaning of this result: as Figure (B) makes clear, when the initial point of $\bar{v} = \overrightarrow{P_1 P_2}$ is moved to the origin, then the terminal point of $\bar{v}$ becomes $(x_2 - x_1, y_2 - y_1)$.

The geometric equivalence of vectors defined in (A) now has an algebraic formulation which is crucial for our later work: Suppose $\bar{v} = \langle v_1, v_2 \rangle$ and $\bar{w} = \langle w_1, w_2 \rangle$. Then

Algebraic equivalence of vectors

$$\bar{v} = \bar{w} \text{ if and only if } \begin{cases} v_1 = w_1 \\ v_2 = w_2 \end{cases}$$

(C)

i.e., two vectors are equivalent if and only if

their corresponding components are equal.

Thus one vector equation $\bar{v} = \bar{w}$ in the plane is equivalent to two simultaneous scalar equations: $\begin{cases} v_1 = w_1 \\ v_2 = w_2 \end{cases}$

Example A. Is there any value of t which makes the following two vectors equivalent?

$$\bar{v} = \overrightarrow{P_1 P_2}, \quad \text{where} \quad P_1 = (t-1, 6) \quad \text{and} \quad P_2 = (4, 1-t)$$

$$\bar{w} = \overrightarrow{Q_1 Q_2}, \quad \text{where} \quad Q_1 = (t, -t) \quad \text{and} \quad Q_2 = (1, 1)$$

Solution. Using (B) we obtain $\qquad \bar{v} = (5-t, -5-t)$

$$\bar{w} = (1-t, 1+t)$$

Thus, by the algebraic equivalence of vectors (C), we see that $\bar{v} = \bar{w}$ means

$$\begin{cases} 5-t = 1-t \\ -5-t = 1+t \end{cases}$$

or

$$\begin{cases} 4 = 0 \\ -3 = t \end{cases}$$

Oops! The equation $4 = 0$ is nonsense, and hence there is no value of t which makes $\bar{v} = \bar{w}$. [However, if we change P_2 to $(0, 1-t)$, then $t = -3$ will work. The reason? Because the nonsense equation $4 = 0$ will be replaced by $0 = 0$, and thus $t = -3$ will make $\bar{v}$ and $\bar{w}$ equivalent.]

3. Algebraic operations on vectors. Geometric definitions for vector addition, scalar multiplication and vector subtraction were given earlier; we now list the corresponding algebraic formulations in terms of components:

Assume $\bar{v} = \langle v_1, v_2 \rangle$, $\bar{w} = \langle w_1, w_2 \rangle$

and k is any scalar. Then

$$\bar{v} + \bar{w} = \langle v_1 + w_1, v_2 + w_2 \rangle$$

$$\bar{v} - \bar{w} = \langle v_1 - w_1, v_2 - w_2 \rangle$$

$$k\bar{v} = \langle kv_1, kv_2 \rangle$$

Notice that <u>on the left</u> of the equal signs we have the <u>vector operations</u> of addition, subtraction and scalar multiplication, while <u>on the right</u> we have the corresponding <u>scalar operations</u> (i. e. , ordinary addition, subtraction and multiplication of numbers). These definitions are no trouble to remember because they are so natural: <u>operate componentwise.</u> Thus, to add two vectors, simply add their corresponding components. Here is an example to show how easy vectors are to deal with when expressed in component form:

<u>Example B.</u>   Suppose $\bar{u} = \langle 1, 0 \rangle$, $\bar{v} = \langle 2, 1 \rangle$ and $\bar{w} = \langle 1, -1 \rangle$. Find the vector $\bar{x}$ that satisfies $\bar{u} - \bar{x} + 2\bar{v} = 3\bar{x} + \bar{w}$.

<u>Solution.</u>   Let $\bar{x} = \langle x_1, x_2 \rangle$, and convert our desired vector equation into component form:

$$\langle 1, 0 \rangle - \langle x_1, x_2 \rangle + 2 \langle 2, 1 \rangle = 3 \langle x_1, x_2 \rangle + \langle 1, -1 \rangle$$

Performing the designated operations componentwise yields

$$\langle 5 - x_1, 2 - x_2 \rangle = \langle 3x_1 + 1, 3x_2 - 1 \rangle$$

However, two vectors are equivalent if and only if their corresponding components are equal by (C), and thus our vector equation converts into a system of two simultaneous scalar equations:

$$\begin{cases} 5 - x_1 = 3x_1 + 1 \\ 2 - x_2 = 3x_2 - 1 \end{cases}$$

which simplify to

$$\begin{cases} x_1 = 1 \\ x_2 = 3/4 \end{cases}$$

Hence our desired vector is $\bar{x} = \langle 1, 3/4 \rangle$.   □

The basic rules for vector arithmetic are given in Anton's Theorem 14. 1. 5. Since these rules are like those of "ordinary arithmetic," they are very natural and easy to remember.

There is only one common pitfall: in each rule you must carefully distinguish between the vectors and the scalars. Fortunately most mistakes along these lines can be avoided by heeding the following two warnings:

> A scalar can <u>never</u> be added to a vector!
>
> Two vectors can <u>never</u> be multiplied together! *

Examples of such undefined expressions are: $\bar{v} + 3$, $\bar{v}\,\bar{w}$, $\bar{v}^2$.

We give one example to show how easy these rules are to use:

<u>Example C.</u> Simplify $3[6(\bar{u} - 7\bar{v}) - 3(2\bar{u})] + 128\bar{v}$.

<u>Solution.</u>

$$3[6(\bar{u} - 7\bar{v}) - 3(2\bar{u})] + 128\bar{v}$$

$$= 18(\bar{u} - 7\bar{v}) - 9(2\bar{u}) + 128\bar{v}$$

$$= 18\bar{u} - 126\bar{v} - 18\bar{u} + 128\bar{v}$$

$$= 2\bar{v} \qquad\qquad \square$$

As you can see, when we multiplied and collected terms, we treated the vector quantities $\bar{u}$ and $\bar{v}$ in the same way that we would have treated scalar ("ordinary") quantities.

The standard unit vectors $\bar{i}$ and $\bar{j}$ in the plane are defined by $\bar{i} = \langle 1, 0 \rangle$ and $\bar{j} = \langle 0, 1 \rangle$. It is then easy to establish the following useful result:

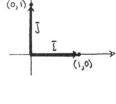

> Suppose $\bar{v} = \langle v_1 , v_2 \rangle$
>
> Then $\bar{v} = v_1\bar{i} + v_2\bar{j}$

_____

* In Chapter 15 we will introduce an operation on pairs of vectors in the plane which we'll call the "dot product." However, the dot product of two vectors will be a <u>scalar</u> quantity, hardly what a "multiplication" of two vector quantities should produce. <u>There is no "usual" form of multiplication for vectors.</u>

Anton establishes this result previous to his Example 5. Thus $\bar{v} = v_1 \bar{i} + v_2 \bar{j}$ is an alternate way to express the vector $\bar{v} = \langle v_1, v_2 \rangle$. We will use both of these expressions in work to follow, in any given instance choosing the one that is the most convenient.

4. <u>The norm of a vector.</u> The <u>length</u> or <u>norm</u> of $\bar{v} = \langle v_1, v_2 \rangle$ is seen from the distance formula to be

$$\|\bar{v}\| = \sqrt{v_1^2 + v_2^2} \qquad \text{(D)}$$

The norm has two important properties:

$$\|k\bar{v}\| = |k| \, \|\bar{v}\| \qquad \text{for any scalar } k$$

$$\|\bar{v} + \bar{w}\| \le \|\bar{v}\| + \|\bar{w}\| \qquad \text{(The Triangle Inequality)} \qquad \text{(E)}$$

An algebraic proof (i. e., using components) is readily obtained for the first property:

$$\|k\bar{v}\| = \|k\langle v_1, v_2 \rangle\|$$

$$= \|\langle kv_1, kv_2 \rangle\|$$

$$= \sqrt{k^2 v_1^2 + k^2 v_2^2} \qquad \text{by (D)}$$

$$= \sqrt{k^2 \left(v_1^2 + v_2^2 \right)}$$

$$= \sqrt{k^2} \sqrt{v_1^2 + v_2^2}$$

$$= |k| \, \|\bar{v}\| \qquad \text{from (D) and } \sqrt{k^2} = |k|$$

The <u>triangle inequality</u>

expresses the geometric fact that

the sum of the lengths of two sides

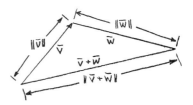

of a triangle is greater than or equal to the length of the third side. However, the <u>algebraic</u> proof

of this result is surprisingly tricky. It can be found in any linear algebra text (see, for example,

Theorem 4. 16 in Anton's <u>Elementary Linear Algebra</u>, Third Edition).

<u>Example D.</u>   If $\|\bar{v}\| = 3$ and $\|\bar{w}\| = 1$, then determine the maximum length of $2\bar{v} - 3\bar{w}$.

<u>Solution.</u> $\|2\bar{v} - 3\bar{w}\| \leq \|2\bar{v}\| + \|-3\bar{w}\|$ by the triangle inequality

$$= |2| \, \|\bar{v}\| + |-3| \, \|\bar{w}\| \quad \text{by (E)}$$

$$= 2(3) + 3(1)$$

$$= \boxed{9}$$ □

Any vector $\bar{u}$ whose length is   1   is called a   <u>unit vector.</u>   There are two very useful

results about unit vectors which Anton gives (in Examples 4 and 6) :

1.   Given any non-zero vector $\bar{v}$, then the

<u>unit</u> vector $\bar{u}$ with the same direction as $\bar{v}$ is

$$\boxed{\bar{u} = \bar{v} / \|\bar{v}\|}$$ (F)

$\bar{u}$ is called the <u>unit direction vector</u> for $\bar{v}$,   or the <u>normalized</u> form of $\bar{v}$.

That $\bar{u}$ and $\bar{v}$ have the same direction is easy to see: multiplication or division by a

positive scalar does not change the direction of a vector. To prove that $\bar{u}$ is a unit vector

is also easy:

$$\|\bar{u}\| = \left\| \frac{\bar{v}}{\|\bar{v}\|} \right\| = \frac{1}{\|\bar{v}\|} \|\bar{v}\| = 1, \quad \text{as desired.}$$

by (E)

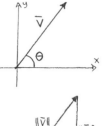

2. Suppose $\bar{v}$ is any nonzero vector in the plane, and θ is the angle from the positive x-axis to $\bar{v}$. Then, as can be seen from the right triangle formed by $\bar{v}$, the components of $\bar{v}$ are $\|\bar{v}\| \cos \theta$ and $\|\bar{v}\| \sin \theta$. Thus any nonzero vector in the plane can be written as

$$\bar{v} = \|\bar{v}\| \cos \theta \, \bar{i} + \|\bar{v}\| \sin \theta \, \bar{j}$$

In particular, any unit vector $\bar{u}$ (i. e., $\|\bar{u}\| = 1$) can be written as

$$\bar{u} = \cos \theta \, \bar{i} + \sin \theta \, \bar{j}$$

and any vector $\bar{v}$ can be expressed as the scalar multiplication of its <u>unit direction vector</u> $\bar{u} = \cos \theta \, \bar{i} + \sin \theta \, \bar{j}$ by its magnitude $\|\bar{v}\|$:

Polar form
of a vector

$$\bar{v} = \|\bar{v}\| \, \bar{u} \, ,$$
$$\text{where} \quad \bar{u} = \cos \theta \, \bar{i} + \sin \theta \, \bar{j}$$

(G)

This is known as the <u>polar form</u> of the vector $\bar{v}$ since, when the initial point of $\bar{v}$ is placed at the origin, the terminal point has <u>polar</u> coordinates $(\|\bar{v}\|, \theta)$.

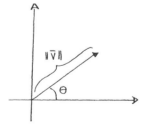

<u>Example E.</u> Find the components of the vector $\bar{v}$ whose length is 3 and whose direction is the same as the vector $\bar{w} = \langle 4, -3 \rangle$.

Solution.   The unit direction vector $\bar{u}$ for $\bar{w}$ is given by (F) as

$$\bar{u} = \bar{w}/\|\bar{w}\| = \langle 4, -3 \rangle / \sqrt{16 + 9}$$

$$= \langle 4, -3 \rangle / 5 = \langle 4/5, -3/5 \rangle$$

Since $\bar{v}$ and $\bar{w}$ have the same direction, then $\bar{u}$ is also the unit direction vector for $\bar{v}$, so by the polar form for $\bar{v}$ we have

$$\bar{v} = \|\bar{v}\|\, \bar{u}$$

$$= 3 \langle 4/5, -3/5 \rangle$$

$$= \langle 12/5, -9/5 \rangle \qquad\qquad \square$$

Example E illustrates an important point:

The unit vector is usually

the most convenient vector

to use in calculations.

When asked to find a vector in a certain direction and of a certain length (as in Example E), for example, it is best first to find a unit vector in that direction and then to multiply it by the desired length.

Section 14.2 :   Vector Calculus in Two Dimensions.

1.   <u>Vector-valued functions.</u>   To this point we have studied the calculus of functions $f : \mathbb{R} \to \mathbb{R}$,

i.e. , functions $y = f(x)$, where f assigns a real number y to any real number x in

the domain of f . In this section we begin the study of functions whose values lie <u>in $\mathbb{R}^2$</u> ,

$\overline{r} : \mathbb{R} \to \mathbb{R}^2$, where $\mathbb{R}^2$ is the set of points in the coordinatized xy-plane :

$$\mathbb{R}^2 = \{ \text{all ordered pairs } (x,y) \text{ of real numbers } x \text{ and } y \}$$

The point (x,y) in $\mathbb{R}^2$ can also be considered as a vector $\langle x,y \rangle = x\,\overline{i} + y\,\overline{j}$. For

this reason functions of the form $\overline{r} : \mathbb{R} \to \mathbb{R}^2$ are called <u>vector-valued functions</u> or

<u>curves in the plane.</u>

The term <u>curve in the plane</u> is used because of the following important interpretation.

Let $(x(t), y(t))$ be the coordinates at <u>time</u> t of the position of an object in the plane. The

vector $\overline{r}(t)$ whose components are $x(t)$ and $y(t)$, i.e.,

$$\overline{r}(t) = x(t)\,\overline{i} + y(t)\,\overline{j}$$

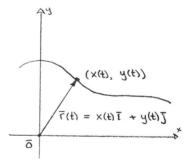

is called the <u>position vector at time t</u> , and

its terminal point $(x(t), y(t))$ traces out

a <u>curve</u> in the plane, the path of movement

of our object.

<u>Example A.</u>   Suppose Howard Ant is crawling along the xy-plane with his position at time $t \geq 0$

being

$$\overline{r}(t) = \sqrt{t}\,\overline{i} + t\,\overline{j}$$

Describe his motion.

<u>Solution.</u>   Howard's movement is governed by the parametric equations $x = \sqrt{t}$ and $y = t$

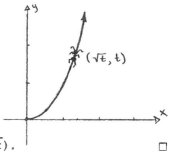

for $t \geq 0$. The curve C traced out is one-half the parabola $y = x^2$. Thus Howard starts at the origin and moves out along the parabola with a steady speed in the y-direction $(y = t)$ but a decreasing speed in the x-direction $(x = \sqrt{t})$.

□

There are many advantages to vector notation for motion in the plane which will be developed in this chapter.

2. Limits of vector-valued functions. Everything that needs to be said about limits of vector-valued functions is summed up in Definition 14. 2. 1:

<table>
<tr><td>Limits
computed
component-
wise</td><td>Limits of vector-valued functions

are computed <u>componetwise</u>, i. e.

if $\bar{r}(t) = x(t)\,\bar{i} + y(t)\,\bar{j}$

then $\lim \bar{r}(t) = (\lim x(t))\,\bar{i} + (\lim y(t))\,\bar{j}$</td></tr>
</table>

In particular, the vector limit $\lim \bar{r}(t)$ exists if and only if the two (scalar) limits $\lim x(t)$ and $\lim y(t)$ exist. This definition reduces vector limits to pairs of "ordinary" limits as illustrated in the next example.

<u>Example B.</u>   Compute $\displaystyle\lim_{t \to 1} \frac{\bar{r}(t) - \bar{r}(1)}{t - 1}$ if $\bar{r}(t) = \sqrt{t}\,\bar{i} + t\,\bar{j}$.

<u>Solution.</u>   We have only to rewrite the vector expression $\dfrac{\bar{r}(t) - \bar{r}(1)}{t - 1}$ in component form and take the ("ordinary" scalar) limit of each term:

$$\frac{\overline{r}(t) - \overline{r}(1)}{t - 1} = \frac{(\sqrt{t}\,\overline{i} + t\overline{j}) - (\overline{i} + \overline{j})}{t - 1}$$

$$= \frac{(\sqrt{t} - 1)\overline{i} + (t - 1)\overline{j}}{t - 1} \qquad \text{by grouping like terms together}$$

$$= \left(\frac{\sqrt{t} - 1}{t - 1}\right)\overline{i} + \overline{j}$$

$$= \left(\frac{1}{\sqrt{t} + 1}\right)\overline{i} + \overline{j} \qquad \text{since } t - 1 = (\sqrt{t} + 1)(\sqrt{t} - 1)$$

Hence $\lim\limits_{t \to 1} \dfrac{\overline{r}(t) - \overline{r}(1)}{t - 1} = \left(\lim\limits_{t \to 1}\left(\dfrac{1}{\sqrt{t} + 1}\right)\right)\overline{i} + \left(\lim\limits_{t \to 1} 1\right)\overline{j}$ by componentwise computation of limits!

$$= \boxed{\frac{1}{2}\,\overline{i} + \overline{j}} \qquad\qquad\qquad \square$$

3.   <u>Derivatives of vector-valued functions.</u>   As you have probably noticed, the definition of the derivative for a vector-valued function $\overline{r}(t)$ is very similar to the definition for a real-valued function $f(x)$:

| real-valued derivative | $f'(x) = \lim\limits_{h \to 0} \dfrac{f(x + h) - f(x)}{h}$ | Definition 3.1.2 |

| vector-valued derivative | $\overline{r}'(t) = \lim\limits_{\Delta t \to 0} \dfrac{\overline{r}(t + \Delta t) - \overline{r}(t)}{\Delta t}$ | Definition 14.2.2. |

The crucially important geometric fact about this definition is:

$\overline{r}'(t)$ is <u>tangent</u> to the graph
of $\overline{r}$ at the terminal point of $\overline{r}(t)$,
and points in the direction of
increasing values of the parameter t.

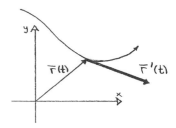

It takes some careful thought to see why this important fact is true. From the geometric

definition of vector subtraction we see that the vector $\bar{r}(t + \Delta t) - \bar{r}(t)$ is parallel to

the secant line through the terminal points of the position

vectors $\bar{r}(t + \Delta t)$ and $\bar{r}(t)$. Thus, as

Δt approaches zero, the directions of

these "secant vectors" approach the direction

of the tangent line to the graph of $\bar{r}$ at the

terminal point of $\bar{r}(t)$. However, the lengths

of the secant vectors approach zero; to have

any hope of obtaining a non-zero tangent vector

as Δt approaches zero we must divide

$\bar{r}(t + \Delta t) - \bar{r}(t)$ by Δt. Then

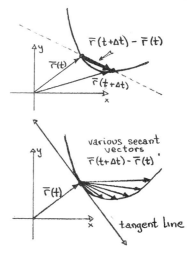

$$\frac{\bar{r}(t + \Delta t) - \bar{r}(t)}{\Delta t} \qquad (A)$$

points from $\bar{r}(t)$ to $\bar{r}(t + \Delta t)$ when $\Delta t > 0$,

or from $\bar{r}(t + \Delta t)$ to $\bar{r}(t)$ when $\Delta t < 0$.

Thus in either case ($\Delta t > 0$ or $\Delta t < 0$) the secant

vectors of the form given in (A) will point in the direction

of increasing parameter t. The limits of these vectors

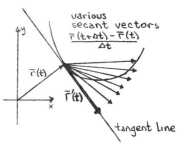

will be a (generally nonzero) vector $\bar{r}'(t)$ which is parallel to the tangent line.

So far we have elaborated on the geometry of vector derivatives; fortunately there is

little need to elaborate on the algebra of vector derivatives in view of Theorem 14.2.3:

Derivatives
computed
component-
wise

derivatives of vector valued function

are computed <u>component-wise</u>, i. e.,

if $\quad \overline{r}(t) = x(t)\,\overline{i} + y(t)\,\overline{j}$

then $\quad \overline{r}'(t) = x'(t)\,\overline{i} + y'(t)\,\overline{j}$

This theorem reduces vector derivatives to pairs of "ordinary" derivatives, and thus reduces our work tremendously.

<u>Example C.</u> Suppose $\overline{r}(t) = \sin 2t\,\overline{i} + t^2\,\overline{j}$. Compute $\overline{r}'(\pi/3)$.

<u>Solution.</u> $\qquad \overline{r}'(t) = \dfrac{d}{dt}[\sin 2t]\,\overline{i} + \dfrac{d}{dt}[t^2]\,\overline{j}$

$$= 2\cos 2t\,\overline{i} + 2t\,\overline{j}$$

Hence $\qquad \overline{r}'(\pi/3) = 2\cos(2\pi/3)\,\overline{i} + (2\pi/3)\,\overline{j}$

$$= \boxed{-\,i + (2\pi/3)\,\overline{j}} \qquad \text{since} \quad \cos(2\pi/3) = -\dfrac{1}{2}. \qquad \square$$

<u>Example D.</u> Redo Example B using derivative rules.

<u>Solution.</u> $\qquad \lim\limits_{t \to 1} \dfrac{\overline{r}(t) - \overline{r}(1)}{t - 1} = \overline{r}'(1) \qquad$ by definition of the derivative.

Thus $\qquad \overline{r}'(t) = \dfrac{d}{dt}[\sqrt{t}\,\overline{i} + t\,j] = (1/2\sqrt{t})\,\overline{i} + \overline{j}$

$$\overline{r}'(1) = \boxed{(1/2)\,\overline{i} + \overline{j}} \qquad \text{as obtained previously.} \qquad \square$$

The ability to compute vector derivatives in a component-wise fashion leads immediately to the following powerful result:

> all the usual algebraic rules for
>
> "ordinary" derivatives carry over
>
> to vector derivatives.

These rules are summarized in Theorem 14.2.4. For example,

$$\frac{d}{dt} \left[\overline{r}_1(t) + \overline{r}_2(t) \right] = \frac{d}{dt} \left[\overline{r}_1(t) \right] + \frac{d}{dt} \left[\overline{r}_2(t) \right]$$

is just the usual "derivative of a sum is the sum of the derivatives" rule (now written for vector-valued functions). There is only one pitfall to watch out for: notice that the product rule involves a <u>vector</u>-valued function $\overline{r}(t)$ multiplied by a <u>scalar</u>-valued function $f(t)$:

$$\frac{d}{dt} \left[f(t) \, \overline{r}(t) \right] = f(t) \, \frac{d}{dt} \left[\overline{r}(t) \right] + \overline{r}(t) \, \frac{d}{dt} \left[f(t) \right]$$

<u>Example E.</u> Suppose $\overline{r}_1$ and $\overline{r}_2$ are vector-valued functions for which

$$\overline{r}_1(1) = 3\,\overline{i} + \overline{j} \quad , \quad \overline{r}_1'(1) = 2\,\overline{i} + \sqrt{2}\,\overline{j}$$

$$\overline{r}_2(1) = \overline{i} + \overline{j} \quad , \quad \overline{r}_2'(1) = -3\,\overline{i}$$

Compute $\overline{r}'(1)$ where $\overline{r}(t) = t^2 \overline{r}_1(t) - 3\,\overline{r}_2(t)$.

<u>Solution.</u> We first work out an expression for $\overline{r}'(t)$ using the usual derivatives rules (Theorem 14.2.4) and then plug in $t = 1$.

$$\overline{r}'(t) = \frac{d}{dt} \left[t^2 \overline{r}_1(t) - 3\,\overline{r}_2(t) \right]$$

$$= \frac{d}{dt} \left[t^2 \overline{r}_1(t) \right] - \frac{d}{dt} \left[3\,\overline{r}_2(t) \right]$$

$$= \left[2t\,\overline{r}_1(t) + t^2 \overline{r}_1'(t) \right] - 3\,\overline{r}_2'(t)$$

Putting in $t = 1$ yields

$$\bar{r}'(1) = 2\,\bar{r}_1(1) + \bar{r}_1'(1) - 3\,\bar{r}_2'(1)$$

$$= [6\,\bar{i} + 2\,\bar{j}] + [2\,\bar{i} + \sqrt{2}\,\bar{j}] + [9\,\bar{i}]$$

$$= 17\,\bar{i} + (2 + \sqrt{2})\,\bar{j} \qquad\qquad \square$$

Anton also gives a version of the Chain Rule for vector-valued functions (Theorem 14.2.5) which we'll restate in a slightly expanded form:

If $\bar{u} = \bar{r}(t)$ is a differentiable <u>vector</u>-valued function of t, and

$t = g(w)$ is a differentiable <u>scalar</u>-valued function of w, then

The
Chain
Rule

$$\frac{d}{dw}\,\bar{r}(g(w)) = \bar{r}'(g(w))\,g'(w)$$

$$\text{or} \quad \frac{d\bar{u}}{dw} = \frac{d\bar{u}}{dt}\,\frac{dt}{dw}$$

This is the same Chain Rule that you know and love (Theorem 3.5.2) except that now the outer function, $\bar{u} = \bar{r}(t)$, is vector-valued. That is the <u>only</u> difference. Here are two typical uses of this Chain Rule:

<u>Example F.</u> Suppose $\bar{u} = e^{2t}\,\bar{i} + t^3\,\bar{j}$ and $t = \ln(w+1)$. Calculate $d\bar{u}/dw$ first by the Chain Rule, and then by expressing $\bar{u}$ in terms of w and differentiating.

<u>First Solution.</u> $\quad \dfrac{d\bar{u}}{dw} = \dfrac{d\bar{u}}{dt}\,\dfrac{dt}{dw}$ the Chain Rule

$$= (2e^{2t}\,\bar{i} + 3t^2\,\bar{j})\left(\frac{1}{w+1}\right) \qquad \text{by differentiation}$$

$$= (2(w+1)^2\,\bar{i} + 3(\ln(w+1))^2\,\bar{j})\left(\frac{1}{w+1}\right)$$

by substituting $t = \ln(w+1)$ and

using $e^{2\ln(w+1)} = (w+1)^2$

$$= 2(w+1)\,\bar{i} + 3(\ln(w+1))^2\,(w+1)^{-1}\,\bar{j}$$

Second Solution. $\overline{u} = e^{2t}\overline{i} + t^3\overline{j}$

$\qquad\qquad = (w+1)^2\overline{i} + (\ln(w+1))^3\overline{j}$

$\qquad\qquad\qquad$ by substituting $\; t = \ln(w+1)\;$ and

$\qquad\qquad\qquad$ using $\; e^{2\ln(w+1)} = (w+1)^2$

Thus $\; \dfrac{d\overline{u}}{dw} = 2(w+1)\overline{i} + 3(\ln(w+1))^2 (w+1)^{-1}\overline{j}$

which agrees (as it must) with the answer in the first solution.   □

Example G.   Suppose $\; \overline{u} = \overline{r}(t)\;$ is a differentiable vector-valued function of $\; t \;$, and $\; s = s(t) \;$ is a differentiable scalar-valued function of $\; t \;$ with a differentiable inverse function. Suppose further that $\; ds/dt = t^2 + 1 \;$. Determine a formula for $\; d\overline{u}/ds \;$ in terms of $\; d\overline{u}/dt \;$.

Solution. $\dfrac{d\overline{u}}{ds} = \dfrac{d\overline{u}}{dt} \cdot \dfrac{dt}{ds}$ by the Chain Rule

$\qquad\qquad = \dfrac{d\overline{u}}{dt}\left(\dfrac{1}{ds/dt}\right)$ by Theorem 8.1.5, Equation (15)

$\qquad\qquad = \left(\dfrac{1}{t^2+1}\right)\dfrac{d\overline{u}}{dt}$ since $\; ds/dt = t^2 + 1$ □

In Example G we were able to express a vector derivative with respect to one parameter (s) in terms of the vector derivative with respect to another parameter (t). This might not appear to be a big deal ... but you'll see this very procedure used over and over again in the next few sections, with $\; s \;$ being the "arc length" parameter.

4. Integrals of vector-valued functions.   As with limits, the important properties of the integral of a vector-valued function are all immediate from its definition (Definition 14.2.6):

integrals of vector-valued functions

are computed component-wise, i.e.,

if $\bar{r}(t) = x(t)\bar{i} + y(t)\bar{j}$

then $\int \bar{r}(t)\,dt = \left(\int x(t)\,dt\right)\bar{i} + \left(\int y(t)\,dt\right)\bar{j}$

$\int_a^b \bar{r}(t)\,dt = \left(\int_a^b x(t)\,dt\right)\bar{i} + \left(\int_a^b y(t)\,dt\right)\bar{j}$

Integrals
computed
component-
wise

Anton's Example 6 illustrates the computation of such integrals. In particular, observe that for an indefinite vector-valued integral, the constant of integration is a vector, i.e.,

$$\int \bar{r}(t)\,dt = \bar{R}(t) + \bar{C}$$

where $\bar{C}$ is a vector constant, $\bar{C} = C_1\bar{i} + C_2\bar{j}.$

Here is one further example, this one a definite integral:

Example H. Evaluate the integral

$$\int_0^{\pi/2} [(t\cos t)\bar{i} + t\bar{j}]\,dt$$

Solution. $\int_0^{\pi/2} [(t\cos t)\bar{i} + t\bar{j}]\,dt$

$$= \left(\int_0^{\pi/2} t\cos t\,dt\right)\bar{i} + \left(\int_0^{\pi/2} t\,dt\right)\bar{j}$$

└── Component-wise computation of vector integrals.
    To evaluate the first integral we use integration
    by parts with $u = t$ $dv = \cos t\,dt$
                       $du = dt$ $v = \sin t$

$$= \left(t \sin t \Big|_0^{\pi/2} - \int_0^{\pi/2} \sin t \, dt \right) \overline{i} + \left(\int_0^{\pi/2} t \, dt \right) \overline{j}$$

$$= \left(t \sin t \Big|_0^{\pi/2} + \cos t \Big|_0^{\pi/2} \right) \overline{i} + \left(\frac{1}{2} t^2 \Big|_0^{\pi/2} \right) \overline{j}$$

$$= \left(\frac{\pi}{2} \sin \frac{\pi}{2} - 0 \sin 0 + \cos \frac{\pi}{2} - \cos 0 \right) \overline{i} + \frac{1}{2} \left(\frac{\pi}{2} \right)^2 \overline{j}$$

$$= \left(\frac{\pi}{2} - 1 \right) \overline{i} + \frac{\pi^2}{8} \overline{j} \qquad \qquad \Box$$

There are geometric interpretations which can be given for the (definite) vector integral, but such interpretations are not particularly useful. The important fact in applications is an analytic one: <u>vector integration and vector differentiation are inverse operations to one another, i. e. , the two operations "undo" each other.</u> To make this clear, let $\overline{r}(t) = x(t) \overline{i} + y(t) \overline{j}$ be a vector-valued function with continuous components $x(t)$ and $y(t)$.

i.   If $\overline{r}(t) = \overline{R}'(t)$, then $\int_a^b \overline{r}(t) \, dt = \overline{R}(b) - \overline{R}(a)$, i. e. ,

FTC I

$$\boxed{\int_a^b \frac{d}{dt} \overline{R}(t) \, dt = \overline{R}(b) - \overline{R}(a)}$$

ii.   If $\overline{R}(t) = \int_a^t \overline{r}(t) \, dt$, then $\overline{R}'(t) = \overline{r}(t)$, i. e. ,

FTC II

$$\boxed{\frac{d}{dt} \int_a^t \overline{r}(t) \, dt = \overline{r}(t)}$$

Look very carefully at these results: they are the vector-valued function versions of the First and Second Fundamental Theorems of Calculus, respectively (Theorems 5. 8. 1 and 5. 10. 1). They are proven by applying the ordinary FTC's to the component functions $x(t)$ and $y(t)$ of $\overline{r}(t)$. These results will play a critical role in our application of vector calculus to motion problems in §14. 5. In particular, the FTC II tells us how to obtain an antiderivative $\overline{R}(t)$ for any continuous vector-valued function $\overline{r}(t)$.

Section 14. 3: Unit Tangent and Normal Vectors; Arc Length as a Parameter.

In the first two subsections below we will summarize the computational techniques associated with unit tangent vectors, unit normal vectors, and the use of arc length as a parameter of a curve. In the final two subsections we will discuss the importance and uses of these concepts.

1. Definition and computation of $\overline{T}$ and $\overline{N}$. Suppose C is the graph of the vector-valued function $\overline{r}(t)$ in the plane.

Unit tangent vector:
(Definition 14. 3. 1)

$$\overline{T}(t) = \overline{r}'(t) / \|\overline{r}'(t)\|$$

when $\overline{r}'(t) \neq \overline{0}$

Unit normal vector:
(Definition 14. 3. 2)

$$\overline{N}(t) = \text{the } 90^{o} \text{ counterclockwise}$$

rotation of $\overline{T}(t)$

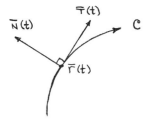

Thus, given any point $\bar{r}(t)$ on C, we obtain a set of two <u>mutually orthogonal unit</u> *

vectors one of which is <u>tangent</u> to the curve C at $\bar{r}(t)$ and the other <u>normal</u> to C at

$\bar{r}(t)$. Two different values of t, say t_1 and t_2, will produce two different sets of vectors:

For this reason the set $\{\bar{T}(t), \bar{N}(t)\}$ is referred to as a <u>moving frame of reference</u> along

the curve C.

Computing $\bar{T}$ is straightforward from its definition. Once you have $\bar{T}$ it is easy

to obtain $\bar{N}$, as Anton shows prior to Example 2:

Calculating $\bar{N}$ from $\bar{T}$	if $\bar{T} = u_1 \bar{i} + u_2 \bar{j}$ then $\bar{N} = -u_2 \bar{i} + u_1 \bar{j}$	(A)

<u>Example A.</u>   Find the unit tangent and unit normal vectors to the graph of $\bar{r}(t) = e^{3t}\bar{i} + (1 - 4e^t)\bar{j}$

at the point t = 0. Then sketch the vectors and a portion of the curve containing the point

corresponding to t = 0.

<u>Solution.</u>   We will compute $\bar{T}(0)$ from its definition, which means first computing $\bar{r}'(0)$

and $\|\bar{r}'(0)\|$.

_____

* Recall from § 14.1 that a <u>unit</u> vector is a vector of length one.

$$\bar{r}'(t) = 3e^{3t}\bar{i} - 4e^t\bar{j} \qquad \text{(componentwise differentiation)}$$

$$\bar{r}'(0) = 3\bar{i} - 4\bar{j}$$

$$\|\bar{r}'(0)\| = \sqrt{3^2 + 4^2} = 5$$

Thus $\qquad \bar{T}(0) = \bar{r}'(0)/\|\bar{r}'(0)\| = \boxed{(3/5)\bar{i} - (4/5)\bar{j}}$

From (A) we then obtain $\qquad \bar{N}(0) = \boxed{(4/5)\bar{i} + (3/5)\bar{j}}$

To obtain the graph of $\bar{r}(t)$ we must use the techniques of §13.4 for sketching the graphs of parametric equations:

$$x(t) = e^{3t} \qquad (\text{so } x > 0)$$

$$y(t) = 1 - 4e^t$$

In this case we can eliminate the parameter t as follows. Since $e^{3t} = (e^t)^3$, we have $e^t = (e^{3t})^{1/3} = x^{1/3}$ so that

$$y = 1 - 4x^{1/3}$$

The graph of this function can be obtained by the curve sketching techniques of §4.5 or by using the Translation and Expansion/Contraction Principles of Appendices E §1 and G §5 on the simpler graph of $y = x^{1/3}$.

The final result is shown to the right. □

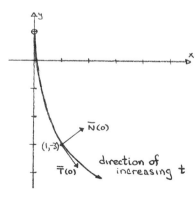

2.   Computations with the arc length parameter.   Be sure that you carefully study Anton's three-
     step method (and the corresponding Figure 14.3.5) for the definition of parameterizing a
     curve by arc length.   Notice that there are two arbitrary choices to make:

     a.   The reference point, i.e. , the point whose arc length parameter
          s   will equal zero.

     b.   The positive/negative orientation of the curve from the reference
          point.

     However, when given a curve already parameterized by a parameter t, the positive/negative
     orientation will be the same as for the parameter t, i.e.,   s > 0 where t > 0 and
     s < 0 where t < 0.

          Theorem 14.3.3 gives the main result connecting an arc length parameter   s   with an
     arbitrary parameter t   for the graph of $\overline{r}(t) = x(t)\overline{i} + y(t)\overline{j}$:

Formulas for
arc length
parameters

$$s = \int_{t_0}^{t} \sqrt{\left(\frac{dx}{du}\right)^2 + \left(\frac{dy}{du}\right)^2} \, du \tag{B}$$

$$\frac{ds}{dt} = \sqrt{\left(\frac{dx}{dt}\right)^2 + \left(\frac{dy}{dt}\right)^2} \tag{C}$$

     Here the arc length reference point (s = 0) is the point for which $t = t_0$. Do not be
     confused by the use of the dummy variable u in the integral for s. The variable t is
     needed as the upper limit of integration and hence a different variable, u, is chosen for use
     inside the integral.

<u>Example B.</u>   Find parametric equations for the graph of

$$\overline{r}(t) = t^3 \overline{i} + t^2 \overline{j}, \qquad t > 0$$

using arc length s as a parameter and the point corresponding to $t = 0$ as the reference

point.

<u>Solution.</u>   The procedure is as follows: (i) compute s as a function of t from (B),

(ii) invert the relationship found in (i) to express t as a function of s, and (iii)

substitute your expression for t as a function of s into the original equation for $\overline{r}(t)$.

(i.)   Since $\overline{r}(t) = t^3 \overline{i} + t^2 \overline{j}$, then

$$x = t^3 \qquad \text{and} \qquad y = t^2$$

Thus

$$\frac{dx}{dt} = 3t^2 \qquad \text{and} \qquad \frac{dy}{dt} = 2t$$

which when placed into (B) will yield

$$s = \int_0^t \sqrt{\left(\frac{dx}{du}\right)^2 + \left(\frac{dy}{du}\right)^2}\, du = \int_0^t \sqrt{(3u^2)^2 + (2u)^2}\, du$$

$$= \int_0^t \sqrt{9u^4 + 4u^2}\, du \qquad = \int_0^t u\sqrt{9u^2 + 4}\, du$$

$$\boxed{\text{since} \quad u \geq 0}$$

$$= \int_4^{9t^2+4} v^{1/2}\, dv/18 \qquad = \frac{1}{18}\left[\frac{v^{3/2}}{3/2}\right]\Bigg|_4^{9t^2+4}$$

$$\boxed{\begin{array}{l} v = 9u^2 + 4 \\ dv = 18u\,du \\ dv/18 = u\,du \end{array}}$$

$$= \frac{1}{27}\left[(9t^2 + 4)^{3/2} - 8\right]$$

(ii.)   Express t as a function of s : We have

$$s = \frac{1}{27}\left[(9t^2 + 4)^{3/2} - 8\right]$$

To invert this to express t as a function of s , we proceed as follows:

$$27s + 8 = (9t^2 + 4)^{3/2}$$

$$(27s + 8)^{2/3} = 9t^2 + 4$$

$$9t^2 = (27s + 8)^{2/3} - 4$$

$$t = \frac{1}{3}\left[(27s + 8)^{2/3} - 4\right]^{1/2} \qquad \text{since } t > 0$$

(iii.)   Plugging into $\bar{r}(t) = t^3\bar{i} + t^2\bar{j}$ will yield

$$\bar{r} = \frac{[\,(27s + 8)^{2/3} - 4\,]^{3/2}}{27}\,\bar{i} + \frac{[\,(27s + 8)^{2/3} - 4\,]}{9}\,\bar{j} \qquad \Box$$

Yes, that <u>is</u> a horrible answer... but except for very simple examples, this is the best that can be hoped for. At least we were able to evaluate the integral involved, which is more than you can say for most arc length problems!

Fortunately, as you will see in the next section, we rarely need to write out <u>explicitly</u> a parameterization in terms of arc length! (You do it in this section to make sure you understand the basic principles which are at work.) Instead we will need to compute ds/dt as a function of t. Using Equation (C) we see that in Example B we would have

$$ds/dt = \sqrt{\left(\frac{dx}{dt}\right)^2 + \left(\frac{dy}{dt}\right)^2} = t\sqrt{9t^2 + 4}$$

a computation which should not cause anyone much trouble.

3. <u>The significance of the vectors $\overline{T}$ and $\overline{N}$.</u> From §§14.1 and 14.2 you should under-

stand the computational value of writing a vector $\overline{v}$ in component form: $\overline{v} = v_1 \overline{i} + v_2 \overline{j}$.

Look more closely at what has been done.

The vector $\overline{v}$ has been expressed as a

<u>linear combination</u> of the vectors $\overline{i}$ and

$\overline{j}$, where $\overline{i}$ and $\overline{j}$ are mutually

orthogonal unit vectors.

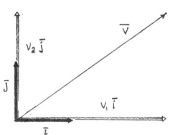

However, when dealing with a vector

quantity $\overline{w}$ <u>along a curve</u>, it is often inconvenient to express $\overline{w}$ as a linear combination of

$\overline{i}$ and $\overline{j}$. Moreover, $\overline{i}$ and $\overline{j}$ depend on how we choose the x- and y-axes, which

may be a very arbitrary choice. Instead we can express $\overline{w}$ as a linear combination of $\overline{T}$

and $\overline{N}$, vectors which depend on the curve but <u>not</u> on the choice of x- and y-axes:

$$\overline{w} = w_1 \overline{T} + w_2 \overline{N}$$

Thus $\overline{w}$ is written as a sum of a

<u>tangential component</u> $w_1 \overline{T}$ plus a

<u>normal component</u> $w_2 \overline{N}$.

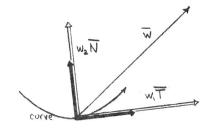

An important use for this representation

of a vector is given in Anton's Equation (21) in §14.5. There we find the "acceleration vector"

$\overline{a}$ of a particle moving along a curve expressed as a linear combination of $\overline{T}$ and $\overline{N}$. This

is an important equation in the physics of motion.

The choice of an appropriate <u>frame of reference</u> in the plane (i.e., two mutually

orthogonal unit vectors such as $\overline{i}, \overline{j}$ or $\overline{T}, \overline{N}$) can be crucial to the solution of many

vector calculus problems. We will see this principle again in our derivation of Kepler's Laws of Planetary Motion in <u>The Companion's</u> optional §14.6.

4.   <u>The significance of arc length parameterization.</u>   We have described the meaning of
"parameterization by arc length" and have shown (in special cases) how to reparameterize a
curve by arc length.   Now we will try to explain why anyone would want to perform such a
bizarre operation!   Here are three reasons:

Reason 1:   <u>Arc length is the natural geometric parameter.</u>   Arc length is the only <u>natural</u>,
<u>geometric</u> parameter that an arbitrary curve can have.   For example, if a person meets you
on a road and asks where a certain location is on that road, then (in the absence of major land-
marks) your probable answer will be the <u>distance</u> to the desired location, i.e., the arc length
of the road from the starting location ("reference point") to the desired terminal location.

Reason 2:   <u>The arc length parameter defines geometric quantities.</u>   Arc length can be used to
define most other geometric quantities associated with the curve.   For example, in Theorem
14.3.4 Anton shows

$$\overline{T} = \frac{d\overline{r}}{ds}$$

i.e., the unit tangent vector $\overline{T}$ is simply the derivative <u>with respect to arc length</u> of any
parameterizing function $\overline{r}(t)$.   As a second illustration we cite the <u>curvature</u> κ to be defined
in the next section by a derivative with respect to arc length.   Further definitions of this sort
will be seen in §15.5.

Reason 3:   <u>The arc length parameter gives simple formulas.</u>   For example, we now have two
formulas for $\overline{T}$:

$$\overline{T} = \frac{d\overline{r}}{ds} \quad \text{and} \quad \overline{T} = \frac{\overline{r}'(t)}{\|\overline{r}'(t)\|}$$

The first equation, expressed solely in terms of the arc length parameter s, is the simpler and _more easily remembered_ of the two, and should be considered as the definition of $\overline{T}$. Unfortunately this is not a very useful _computational_ formula because we rarely have curves explicitly parameterized by arc length. However, it is easy to compute $\overline{T}$ from the second equation, as we did in Example A.

Section 14.4: Curvature.

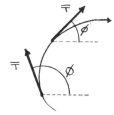

Notice that the definition of _curvature_ κ at a point on a smooth curve C makes use of two quantities which were first introduced in the previous section: the _arc length_ parameter s along the curve, and the _angle of inclination_ ϕ of the unit tangent vector $\overline{T}$ with the positive x-axis, measured counterclockwise:

$$\kappa = \frac{d\phi}{ds}$$

This is a perfect illustration of Reason 2 from the previous section of _The Companion_: the arc length parameter can be used to define geometric quantities along a curve.

Unfortunately, this defining equation is not of much use in computing the curvature since we rarely have any formula for the angle ϕ, let alone a formula for ϕ as a function of arc length. To fill this gap, Anton derives more useful formulas for κ. We will summarize and illustrate these methods for computing κ in the next two subsections. Then we'll elaborate on the geometric meaning of curvature.

1. Computing curvature from $\overline{T}$. First consider a curve which is parameterized by arc length, $\overline{r}(s) = x(s)\overline{i} + y(s)\overline{j}$. Then the absolute value of the curvature κ can be computed from the unit tangent vector $\overline{T}$ by Anton's Formula (3):

$$\left| \kappa \right| = \left\| \frac{d\overline{T}}{ds} \right\| \tag{3}$$

Here $\overline{T}$ is computed by the methods of §14.3:

$$\overline{T} = \frac{d\overline{r}}{ds} \qquad \text{Theorem 14.3.4(b)}$$

To determine the sign of κ we must examine the concavity of the curve and use Theorem 14.4.2.

Anton illustrates the use of this method to compute κ in Example 2. The success of this example depends, however, on being able to obtain an explicit reparameterization by arc length of the curve under consideration. Since such a reparameterization is generally unpleasant (and often impossible), we would prefer a method for computing $\left| \kappa \right|$ which is valid for arbitrary parameterizations. Such a method is given by Anton's Formula (4):

$$\left| \kappa \right| = \frac{\left\| d\overline{T}/dt \right\|}{ds/dt} \tag{4}$$

Here $\overline{T}$ and ds/dt are computed by the methods of §14.3:

14.4.3

$$\overline{T} = \frac{\overline{r}'(t)}{\|\overline{r}'(t)\|} \qquad \text{Definition 14.3.1}$$

$$ds/dt = \|\overline{r}'(t)\| \qquad \text{Theorem 14.3.4(a)}$$

Here's an illustration of this latter method:

<u>Example A.</u> Compute κ for the curve parameterized by $\overline{r}(t) = t^3\overline{i} + t^2\overline{j}$, $t > 0$.

<u>Solution.</u> We will use the equations for ds/dt, $\overline{T}$ and $|\kappa|$ given above.

$$ds/dt = \|\overline{r}'(t)\| = \|3t^2\overline{i} + 2t\overline{j}\| = \sqrt{9t^4 + 4t^2}$$

Thus

$$\boxed{ds/dt = t\sqrt{9t^2 + 4}}$$

$$\overline{T} = \frac{\overline{r}'(t)}{\|\overline{r}'(t)\|} = \frac{3t^2\overline{i} + 2t\overline{j}}{t\sqrt{9t^2 + 4}}$$

Thus

$$\boxed{\overline{T} = (3t\overline{i} + 2\overline{j})/\sqrt{9t^2 + 4}}$$

To determine $|\kappa|$ we must differentiate $\overline{T}$:

$$\frac{d\overline{T}}{dt} = \frac{(9t^2+4)^{1/2}(3\overline{i}) - (3t\overline{i}+2\overline{j})(\frac{1}{2})(9t^2+4)^{-1/2}(18t)}{9t^2+4}$$

$$= \frac{(9t^2+4)(3\overline{i}) - (3t\overline{i}+2\overline{j})(9t)}{(9t^2+4)^{3/2}}$$

$$= \frac{27t^2\overline{i} + 12\overline{i} - 27t^2\overline{i} - 18t\overline{j}}{(9t^2 + 4)^{3/2}}$$

$$= (12\overline{i} - 18t\overline{j})/(9t^2 + 4)^{3/2}$$

Therefore

$$\| d\overline{T}/dt \| = \sqrt{12^2 + 18^2 t^2}/(9t^2 + 4)^{3/2} \qquad \text{using} \qquad \left\| \frac{x\overline{i} + y\overline{j}}{a} \right\| = \frac{\sqrt{x^2 + y^2}}{|a|}$$

$$= 6\sqrt{4 + 9t^2}/(9t^2 + 4)^{3/2}$$

$$= 6/(9t^2 + 4)$$

Thus $\quad |\kappa| = \dfrac{\| d\overline{T}/dt \|}{ds/dt} = \dfrac{6/(9t^2 + 4)}{t\sqrt{9t^2 + 4}} = \dfrac{6}{t(9t^2 + 4)^{3/2}}$

To determine the sign of κ we sketch the graph of $\overline{r}$; this is most easily done by converting

$x(t) = t^3$ and $y(t) = t^2$ into $y = x^{2/3}$, $x > 0$,

whose graph is shown to the right. Since the increasing

t direction is in the positive x direction, the graph

shows that the concave side of the curve is on the right,

and hence $\kappa < 0$ by Theorem 14.4.2. Thus

$$\boxed{\kappa = - \frac{6}{t(9t^2 + 4)^{3/2}}} \qquad \square$$

2. Computing curvature directly from x(t) and y(t). Anton's Formula (5) can be used

to compute the curvature κ directly from the parameterization $\overline{r}(t) = x(t)\overline{i} + y(t)\overline{j}$:

$$\boxed{\kappa = \frac{x' y'' - y' x''}{[x'^2 + y'^2]^{3/2}}} \qquad (5)$$

This formula has one major advantage over our previous procedures: it's a straightforward, plug-in formula which yields the exact value of κ, not just the <u>absolute value</u> of κ. Moreover, computing $|\kappa|$ from the vector $\overline{T}$ can be a lot more unpleasant than it looks; the new equation represents a significant savings in time.

When (5) is combined with earlier equations we obtain a highly efficient method for computing $\overline{T}, \overline{N}$ and κ, as is illustrated in the next example:

<u>Example B.</u>   Obtain $\overline{T}, \overline{N}$ and κ at $t = 2$ for the curve

$$\overline{r}(t) = t^3 \overline{i} + t^2 \overline{j} \quad.$$

<u>Solution.</u>   We have

$$x = t^3 \qquad\qquad y = t^2$$

$$x' = 3t^2 \qquad\qquad y' = 2t$$

$$x'' = 6t \qquad\qquad y'' = 2$$

At $t = 2$ this gives

$$x'(2) = 12 \qquad\qquad y'(2) = 4$$

$$x''(2) = 12 \qquad\qquad y''(2) = 2$$

Hence, at $t = 2$ we have

$$\| \overline{r}'(2) \| = \sqrt{x'^2 + y'^2} = \sqrt{144 + 16} = \sqrt{160} = 4\sqrt{10}$$

This yields

$$\overline{T}(2) = \frac{\overline{r}'(2)}{\| \overline{r}'(2) \|} = \frac{12\overline{i} + 4\overline{j}}{4\sqrt{10}} = \frac{3\sqrt{10}}{10}\overline{i} + \frac{\sqrt{10}}{10}\overline{j}$$

$$\overline{N}(2) = -\frac{\sqrt{10}}{10}\overline{i} + \frac{3\sqrt{10}}{10}\overline{j} \quad \text{since} \quad \overline{N} = -u_2\overline{i} + u_1\overline{j} \text{ if } \overline{T} = u_1\overline{i} + u_2\overline{j}$$

$$\kappa(2) = \frac{x'y'' - x''y'}{[x'^2 + y'^2]^{3/2}} = \frac{24 - 48}{[160]^{3/2}} = -\frac{3\sqrt{10}}{800} \approx \boxed{-.012} \qquad \square$$

Equation (5)

Anton's last major formula is an equation for finding the curvature on the graph of a

function   y = f(x)   (Formula 10):

$$\kappa = \frac{\dfrac{d^2 y}{dx^2}}{\left[1 + \left(\dfrac{dy}{dx} \right)^2 \right]^{3/2}}$$ (10)

and it is derived from (5) simply by parameterizing the graph of   y = f(x)   by $\bar{r}(t) = t\,\bar{i} +$

+ f(t) $\bar{j}$ (i.e., by letting   t = x). Notice that the increasing direction of the curve is taken

to be in the positive   x-direction. Anton's

Example 5   illustrates the use of this formula.

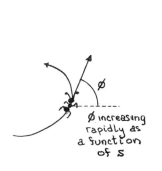

3.   Geometric aspects of curvature.   Suppose we place Howard Ant on a

smooth parameterized curve and have him crawl along the

curve at a constant rate of one inch per second in the

positive direction, i.e., in the direction of increasing

parameter. How would Howard detect the curvature κ

of the curve at any given point? Well, suppose at a point   P

along the curve Howard is in the process of turning sharply

to his left. He would say there is a large curvature at   P,

and that the concave side of the curve is to his left. This

corresponds perfectly to our definition of κ: since ϕ

is increasing rapidly as a function of   s,   then   dϕ/ds = κ will be a large positive number.

Similar reasoning will justify the statements in the following table:

Howard Ant experiences...	Curvature κ is...	Concave side of curve is on the...
a sharp turn to his <u>left</u>	large <u>positive</u>	left
a slight turn to his <u>left</u>	small <u>positive</u>	
no turning sensation	zero	? ... there <u>might</u> not be a concave side of the curve at this point
a slight turn to his <u>right</u>	small <u>negative</u>	right
a sharp turn to his <u>right</u>	large <u>negative</u>	

<u>Warning</u>: the sign of the curvature κ depends on the direction in which the curve is parameterized! In other words, turn Howard around on the curve (i.e., reparameterize the curve in the opposite direction) and all "lefts/rights" become reversed, as do all "positive/negatives."

Let's also discuss Howard Ant's relationship with the unit normal vector $\overline{N}$. Since Howard is always heading in the direction of the unit tangent vector $\overline{T}$ (i.e., in the direction of increasing parameter), and since $\overline{N}$ is obtained from $\overline{T}$ by a 90° rotation of $\overline{T}$ counterclockwise, then it is clear that

$$\boxed{\overline{N} \text{ points off to Howard Ant's left side}}$$

Now consider the direction of $\kappa\overline{N}$. If $\kappa > 0$, then $\kappa\overline{N}$ points in the same direction as $\overline{N}$, i.e., to Howard's left side. Moreover, the table above tells us that Howard's left side is the concave side of the curve. Thus $\kappa\overline{N}$ points to the concave side of the curve if $\kappa > 0$.

We claim this same fact holds true even when $\kappa < 0$! For in that case, $\kappa\overline{N}$ will point to Howard's right side, which from our table is now the concave side of the curve. In summary:

$$\boxed{\kappa\overline{N} \text{ points into the concave side of the curve whenever } \kappa \neq 0}$$

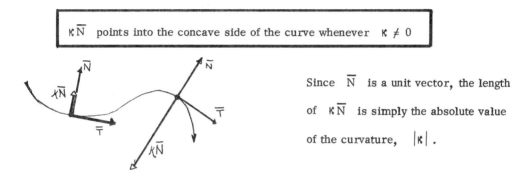

Since $\overline{N}$ is a unit vector, the length of $\kappa\overline{N}$ is simply the absolute value of the curvature, $|\kappa|$.

Section 14.5 :   Motion in a Plane.

Much of calculus was originally formulated as a result of attempts to solve problems of motion, especially problems involving the motion of the planets. We have discussed rectilinear motion, i.e., motion along a straight line, three times (§§2.3, 5.4, 5.8). Our vector methods now give us the ability to handle motion problems in a plane. The effort required to expand from the plane into space is then quite minimal, and will be done in §15.5.

1. __The basic definitions.__   There is a geometric motivation for the definition of the __velocity vector__ for motion in a plane. Compare what follows with the discussion given in §2.3 for rectilinear velocity.

Consider $\bar{r}(t)$ to be the __position__ vector for an object P in the plane at time t , and suppose we consider a time interval from t to $t + \Delta t$, where Δt is some small time increment. How should we define the __average velocity__ of P during this time interval? The natural definition is

$$\bar{v}_{av}(t, t + \Delta t) = \frac{\text{change in position}}{\text{change in time}}$$

$$= \frac{\bar{r}(t + \Delta t) - \bar{r}(t)}{(t + h) - t}$$

$$= \frac{\bar{r}(t + \Delta t) - \bar{r}(t)}{\Delta t}$$

How can we use this __average velocity__ on intervals $[t, t + \Delta t]$ to define the (instantaneous) velocity at t ? Again, our intuition should say "take the limit of $\bar{v}_{av}(t, t + \Delta t)$ as Δt goes to zero," i.e.,

$$v(t) = \lim_{\Delta t \to 0} \bar{v}_{av}(t, t + \Delta t) = \lim_{\Delta t \to 0} \frac{\bar{r}(t + \Delta t) - \bar{r}(t)}{\Delta t}$$

This last limit should be recognized as the definition of $\bar{r}'(t)$, the derivative of $\bar{r}(t)$ (Definition 14. 2. 2). Thus

$$\boxed{\text{velocity} = \bar{v}(t) = \bar{r}'(t)}$$

which is indeed Definition 14. 5. 1 for the velocity vector. Since we know that $\bar{r}'(t)$ is tangent to the curve at $\bar{r}(t)$, then we have the expected result that the velocity vector $\bar{v}(t)$ is tangent to the curve at $\bar{r}(t)$.

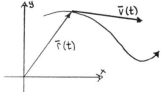

Much more mysterious than velocity is <u>acceleration,</u> denoted by $\bar{a}(t)$. The definition

is simple enough:

$$\boxed{\text{acceleration} = \bar{a}(t) = \bar{v}'(t) \; .}$$

<u>Example A.</u>   Compute the velocity and acceleration for the circular motion

$$\bar{r}(t) = \cos\left(\frac{t}{2}\right)\bar{i} + \sin\left(\frac{t}{2}\right)\bar{j}$$

<u>Solution.</u>   We have $\bar{v}(t) = -\frac{1}{2}\sin\left(\frac{t}{2}\right)\bar{i} + \frac{1}{2}\cos\left(\frac{t}{2}\right)\bar{j}$

and $\bar{a}(t) = -\frac{1}{4}\cos\left(\frac{t}{2}\right)\bar{i} - \frac{1}{4}\sin\left(\frac{t}{2}\right)\bar{j} = -\frac{1}{4}\bar{r}(t)$ □

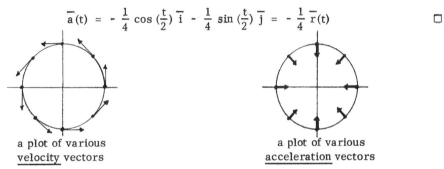

a plot of various   a plot of various
velocity vectors   acceleration vectors

The velocity vectors look perfectly reasonable, but what is the meaning to be found in the

acceleration vectors? Well, just as

$$\text{velocity} = \text{rate of change of } \underline{\text{position}}$$

we also have

$$\text{acceleration} = \text{rate of change of } \underline{\text{velocity.}}$$

Here's an illustration. Suppose at $\bar{r}(t_0)$ the velocity and acceleration vectors look as

follows:

Figure A

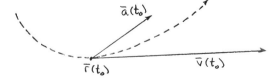

Then the subsequent velocity vectors for $t > t_0$ should approximately be as shown in the next picture:

Figure B

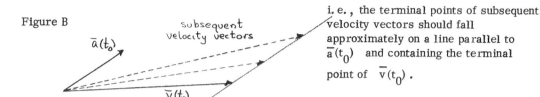

i. e. , the terminal points of subsequent velocity vectors should fall approximately on a line parallel to $\bar{a}(t_0)$ and containing the terminal point of $\bar{v}(t_0)$.

Here's a proof of our claim. Any subsequent velocity vector is of the form $\bar{v}(t_0 + \Delta t)$ for some $\Delta t > 0$. Then

$$\bar{v}(t_0 + \Delta t) = \bar{v}(t_0) + \Delta t \left(\frac{\bar{v}(t_0 + \Delta t) - \bar{v}(t_0)}{\Delta t} \right) \qquad \text{by simple algebra}$$

$$\underset{\boxed{\text{for } \Delta t \text{ small}}}{\cong} \bar{v}(t_0) + \Delta t \lim_{\Delta t \to 0} \left(\frac{\bar{v}(t_0 + \Delta t) - \bar{v}(t_0)}{\Delta t} \right)$$

$$= \bar{v}(t_0) + \Delta t \, \bar{v}'(t_0) \qquad \text{by definition of derivatives}$$

$$= \bar{v}(t_0) + \Delta t \, \bar{a}(t_0) \qquad \text{by definition of } \bar{a}(t_0)$$

Hence the terminal point of $\bar{v}(t_0 + \Delta t)$ lies approximately on the line parallel to $\bar{a}(t_0)$ and containing the terminal point of $\bar{v}(t_0)$.

Hence, given $\bar{v}(t_0)$ and $\bar{a}(t_0)$ as shown in Figures A & B, our object should be turning upward (since the $\bar{v}(t)$ vectors are turning upward) and its speed should be increasing (since the $\bar{v}(t)$ vectors are increasing in length).

In general, the <u>direction</u> of $\bar{a}(t)$ indicates the direction in which the object is attempting to turn, the <u>magnitude</u> of $\bar{a}(t)$ indicates how quickly the attempted change in direction is occurring, and the <u>angle</u> θ between $\bar{a}(t)$ and $\bar{v}(t)$ indicates whether the object is speeding up or slowing down:

$0 \le \theta < \frac{\pi}{2}$ implies speeding up

$\frac{\pi}{2} < \theta \le \pi$ implies slowing down

$\theta = \frac{\pi}{2}$ implies no change in speed

So far we have three vector quantities for studying motion in a plane:

$$\bar{r}(t) = \underline{\text{position}} \text{ at time } t$$

$$\bar{v}(t) = \bar{r}'(t) = \underline{\text{velocity}} \text{ at time } t$$

$$\bar{a}(t) = \bar{v}'(t) = \underline{\text{acceleration}} \text{ at time } t$$

We now bring in two scalar quantities:

$$s(t) = \underline{\text{distance travelled}} \text{ from time } t_0 \text{ to time } t$$

$$s'(t) = \underline{\text{speed}} \text{ at time } t$$

Clearly $s(t)$ is the arc length of

our curve from $\bar{r}(t_0)$ to $\bar{r}(t)$,

and thus, from Theorem 14.3.3,

we have

$$s(t) = \int_{t_0}^{t} \sqrt{\left(\frac{dx}{du}\right)^2 + \left(\frac{dy}{du}\right)^2} \; du = \int_{t_0}^{t} \| \bar{v}(u) \| \; du$$

$$s'(t) = \sqrt{\left(\frac{dx}{dt}\right)^2 + \left(\frac{dy}{dt}\right)^2} = \| \bar{v}(t) \|$$

Hence

$$\text{distance} = s(t) = \int_{t_0}^{t} \|\overline{v}(u)\| \, du$$

$$\text{speed} = s'(t) = \|\overline{v}(t)\|$$

Don't be discouraged by all these formulas! They are summarized in a very handy way in our <u>Five-Box Diagram</u>, similar to the Four-Box Diagram introduced in §5.4 of <u>The Companion</u>, only now with three of the boxes representing vector quantities:

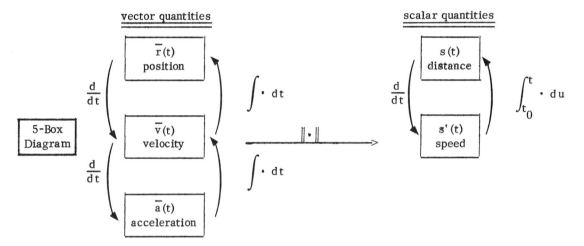

Notice that differentiation moves down while integration (undoing differentiation) moves up. Here we have used both the ordinary and the vector-valued versions of the Fundamental Theorems of Calculus. (See §14.2.4 in <u>The Companion.</u>)

<u>Example B.</u> Determine all the equations of motion if we know $\overline{v}(t) = 2t\,\overline{i} + t^2\,\overline{j}$ when $t \geq 0$, and $\overline{r}(1) = \overline{i} - \overline{j}$.

<u>Solution.</u> We are given the velocity, which is the middle box on the left in our 5-Box Diagram.

So we just follow the arrows to get everything else...

1. $\bar{a}(t) = d\bar{v}/dt = 2\bar{i} + 2t\bar{j}$

2. $\bar{r}(t) = \int \bar{v}(t)dt = \int (2t\bar{i} + t^2\bar{j})dt$

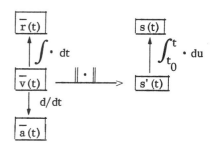

$= \left(\int 2t\,dt\right)\bar{i} + \left(\int t^2\,dt\right)\bar{j}$

$= (t^2 + c_1)\bar{i} + (t^3/3 + c_2)\bar{j}$

where c_1 and c_2 are constants of integration which must be determined by using $\bar{r}(1) = \bar{i} - \bar{j}$, i.e.,

$\bar{i} - \bar{j} = \bar{r}(1) = (1 + c_1)\bar{i} + (1/3 + c_2)\bar{j}$

which, when equating coefficients of like terms, yields

$1 = 1 + c_1$ and $-1 = 1/3 + c_2$

or $c_1 = 0$ and $c_2 = -4/3$.

Hence $\bar{r}(t) = t^2\bar{i} + (t^3/3 - 4/3)\bar{j}$

3. $s'(t) = \|\bar{v}(t)\| = \sqrt{4t^2 + t^4} = t\sqrt{4 + t^2}$ since $t \geq 0$.

4. $s(t) = \int_0^t s'(u)\,du$,   where we have taken $t_0 = 0$.

$= \int_0^t u\sqrt{4 + u^2}\,du = \int_4^{4+t^2} w^{1/2}\,(dw/2)$

$$
\boxed{\begin{array}{l} w = 4 + u^2 \\[4pt] dw = 2u\,du \\[4pt] \dfrac{dw}{2} = u\,du \end{array}}
$$

$= \frac{1}{3}\,w^{3/2}\,\Big|_4^{4+t^2}$

$= \frac{1}{3}\,(4 + t^2)^{3/2} - \frac{8}{3}$ □

2. Motion due to a force. Probably the most important reason for a detailed consideration of

acceleration is Newton's Second Law of Motion:

$$\boxed{\overline{F}(t) \; = \; m \, \overline{a}(t)}$$

where   m = the mass of an object

$\overline{F}(t)$ = the force exerted on the object at time t , and

$\overline{a}(t)$ = the resulting acceleration of the object at time t

Anton uses this law, along with the equations of Subsection 1, to obtain motion equations for

objects in free flight (i. e. , the only force is gravity) near the surface of the earth.   Let's see

how Anton's derivations fit into the 5-Box Diagram.

Example C.   Suppose an object in free flight starts initially at a height s_0 above the surface

of the earth and with an initial velocity vector $\overline{v}_0$. Find a vector equation for $\overline{r}(t)$, the

position of the object at any time   t .

Solution.   Consider the set-up to the

right.   Anton shows, by using the force

of gravity $\overline{F} = - m g \, \overline{j}$ in Newton's Law

$\overline{F} = m \overline{a}$,   that

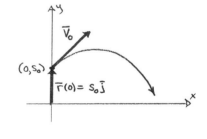

$$\boxed{\overline{a}(t) \; = \; - g \, \overline{j}} \qquad\qquad (A)$$

Ahh... we now have the acceleration, and our 5-Box Diagram shows that two integrations

(carefully handling the constants of integration) will give $\overline{r}(t)$. Anton carries out these

computations (study them!) to obtain

$$\overline{v}(t) = -gt\,\overline{j} + \overline{v}_0 \tag{B}$$

$$\overline{r}(t) = \left(-\frac{1}{2}gt^2 + s_0\right)\overline{j} + \overline{v}_0\,t \tag{C} \quad \square$$

If we wish to see what the graph of this equation looks like we need to eliminate the parameter t. To do so, suppose $\overline{v}_0 = v_1\overline{i} + v_2\overline{j}$. Then

$$\overline{r}(t) = v_1\,t\,\overline{i} + \left(-\frac{1}{2}gt^2 + v_2 t + s_0\right)\overline{j}$$

so that $x = v_1 t$ and $y = -\frac{1}{2}gt^2 + v_2 t + s_0$. Assuming $v_1 \neq 0$ (otherwise we have an object moving in a purely vertical path), then $t = x/v_1$ which gives

$$y = \left(-\frac{g}{2v_1^2}\right)x^2 + \left(\frac{v_2}{v_1}\right)x + s_0 \tag{D}$$

so that we have a parabolic trajectory.

initial velocity vector

trajectory of object

Motion problems involving free flight occur so often that it might be valuable to memorize Equations (A) through (C). On the other hand, Equations (B) and (C) are so easily derived from (A) via the 5-Box Diagram (especially in specific cases) that memorization could be more trouble than it is worth. It's a matter of individual preference, and we'll leave it entirely to you.

Anton's Example 3 gives an excellent illustration of the use of these equations. The example starts off as many such problems do: rather than explicitly giving the initial velocity vector $\overline{v}_0$, we are given the initial angle of flight α and the initial speed v_0. (This would be realistic information in a missile launch, for example.) To begin the problem, we must express $\overline{v}_0$ in terms of ν_0 and α. This can be done by using the

formula in Example 6 of §14.1:

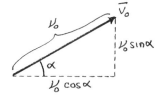

$$\overline{v}_0 = \|\overline{v}_0\| (\cos \alpha \overline{i} + \sin \alpha \overline{j})$$

Thus, since $\|\overline{v}_0\| = \nu_0$, we obtain

$$\boxed{\overline{v}_0 = \nu_0 \cos \alpha \overline{i} + \nu_0 \sin \alpha \overline{j}} \qquad \text{(E)}$$

Example D. A shell is fired from the top of a 100 foot cliff with an initial speed of 500 feet/sec

and at an initial angle of 30°. Where does the

shell strike the ground?

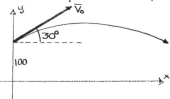

Solution. We wish to find that x value for which

$y = 0$. It is thus convenient to have y expressed as a function of x, i.e., as in Equation

(D).

Our given information is $s_0 = 100$ feet, $\nu_0 = 500$ ft/sec and $\alpha = 30^\circ$. Thus,

from (E), our initial velocity vector is

$$\overline{v}_0 = v_1 \overline{i} + v_2 \overline{j} = 500 \cos 30^\circ \overline{i} + 500 \sin 30^\circ \overline{j}$$

$$= 250 \sqrt{3}\ \overline{i} + 250 \overline{j}$$

or $v_1 = 250\sqrt{3}$ and $v_2 = 250$. Hence, using (D), we have

$$0 = \left(-\frac{32}{2(250\sqrt{3})^2}\right) x^2 + \left(\frac{250}{250\sqrt{3}}\right) x + 100$$

which is a quadratic equation in x. Using the quadratic formula (and a hand calculator) will

yield $x \cong 6935$. $\qquad\qquad \square$

3.   <u>Components of acceleration.</u>   As we discussed in §14.3.3, the orthogonal unit vectors $\overline{T}$

and $\overline{N}$ form what is called the <u>moving frame of reference</u> along a smooth curve. In terms

of this frame Anton shows that the velocity and acceleration vectors for an object moving

along this curve can be expressed as follows:

$$\overline{v} = \frac{ds}{dt} \overline{T}$$

$$\overline{a} = \frac{d^2 s}{dt^2} \overline{T} + \kappa \left(\frac{ds}{dt}\right)^2 \overline{N} \qquad\qquad (F)$$

The second equation is often written as

$$\overline{a} = a_T \overline{T} + a_N \overline{N}, \text{ where}$$

$$a_T = \frac{d^2 s}{dt^2} \text{ is the scalar } \underline{\text{tangential component}} \text{ of acceleration}$$

$$a_N = \kappa \left(\frac{ds}{dt}\right)^2 \text{ is the scalar } \underline{\text{normal component}} \text{ of acceleration.}$$

These components (unlike the "usual" $\overline{i}$ and $\overline{j}$ components) have important physical meaning,

mostly in connection with Newton's Second Law of Motion, $\overline{F} = m\overline{a}$.

An an example, consider an auto travelling along a road. If $\overline{r}(t)$ represents the path

of the auto, then (again from $\overline{F} = m\overline{a}$) the force which must

be exerted to keep the car on the road is

$$\overline{F}_N = m a_N \overline{N} = m\kappa \left(\frac{ds}{dt}\right)^2 \overline{N}$$

(The tangential component of the force or acceleration is irrelevant in this regard since it is in

the direction of travel.) Look at our equation closely, for it contains some interesting information.

If the curvature in the road doubles (i.e., the car reaches a turn which is "twice as sharp" as the one presently on), then the "road maintaining force" must also double. However, if the

speed doubles... well, the road maintaining force must quadruple. Thus, taking a turn at

30 mph requires 4 times the force as taking it at 15 mph. Take the same turn at 60 mph requires

16 times the force as taking it at 15 mph!

Here is another illustration of the use of Equation (F).

Example E. Suppose that Howard Ant is crawling along a smooth curve in such a way that his

velocity and acceleration vectors are always perpendicular. Then what can you say about

Howard's speed?

Solution. Since $\bar{v}$ is a non-zero multiple of $\bar{T}$,

then $\bar{a}$ is perpendicular to $\bar{v}$ if and only if $\bar{a}$

is a multiple of $\bar{N}$. Thus the tangential component

of $\bar{a}$ must be zero, proving

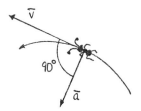

$$0 = a_T = \frac{d^2 s}{d t^2} = \frac{d}{dt}\left(\frac{d s}{d t}\right)$$

Thus the derivative of the speed ds/dt is constantly zero, which means that Howard's

speed is a constant. □

Explicit computations of

$$a_T = \frac{d^2 s}{d t^2} \quad \text{and} \quad a_N = \kappa \left(\frac{d s}{d t}\right)^2$$

for a given curve $\bar{r}(t)$ present nothing new to us since we have already learned how to compute

ds/dt and κ , and $d^2 s/dt^2$ is merely the derivative of ds/dt. However, since κ can be a pain to compute, there is a worthwhile shortcut to be aware of. As Anton shows,

$$\|\bar{a}\|^2 = a_T^2 + a_N^2$$

Thus our computations can proceed as follows:

---

1. From $\bar{r}$ compute $\bar{a}$ and ds/dt as in the 5-Box Diagram.

2. Compute a_T by taking the derivative of ds/dt, i.e., $a_T = \dfrac{d}{dt}\left(\dfrac{ds}{dt}\right)$

3. Compute $|a_N|$ by $|a_N| = \left(\|\bar{a}\|^2 - a_T^2\right)^{1/2}$

Generally we only need the magnitudes of a_N and a_T so that this method is usually

sufficient. A good illustration of this technique is found in Anton's Example 4.

Section 14.6: Polar Coordinate Motion and Kepler's Laws (Optional)

In this additional section to Anton's Chapter 14 our goal is the proof of Johannes

Kepler's (1571-1630) three laws of planetary motion. There are several reasons for wishing

to develop these laws. The first is to give a non-trivial and important application of the vector

calculus we have studied in this chapter. Another is simply to present a beautiful piece of

mathematical and scientific work purely for its own value. The analytic derivation of Kepler's

Laws was the first truly monumental success of calculus, and represents an important turning

point in the history of science.

Planetary motion in the solar system is governed by a central force, i.e., a force which

is always directed toward one fixed point, in this case the sun. In such a situation, polar

coordinates are extremely useful. Hence, before stating and proving Kepler's Laws, we will develop some necessary machinery concerning motion problems in polar coordinates.

[It might be wise at this point to review

the material of Chapter 13, in

particular §13.3.]

1. The polar coordinate frame of reference.

Recall the polar coordinates r, θ of a point $P(x, y)$

$$r = \|\overline{P}\| = \sqrt{x^2 + y^2}$$

θ = the angle from the

    x-axis to P measured

    counterclockwise.

Then $x = r \cos \theta$ and $y = r \sin \theta$.

A curve of motion in the plane can be represented as follows:

$$\overline{r}(t) = x(t) \overline{i} + y(t) \overline{j} = r(t) \cos \theta(t) \overline{i} + r(t) \sin \theta(t) \overline{j}$$

where $r(t)$ and $\theta(t)$ are both functions of time t, i.e., the polar coordinates of our travelling object vary with time. For notational convenience we will simply write

$$\boxed{\overline{r}(t) = r \cos \theta \, \overline{i} + r \sin \theta \, \overline{j}} \qquad (A)$$

where it is understood that r and θ vary with t ($r = r(t)$ and $\theta = \theta(t)$).

Here is an important observation:

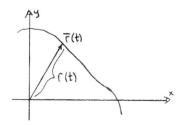

$\overline{r}(t)$ is the position vector at time t, while

$r = r(t)$ is just the scalar distance from the

origin to the moving object, i.e.,

$$r = r(t) = \|\overline{r}(t)\|$$

We now rewrite Equation (A) in a useful way. Let

$$\overline{u} = \langle \cos\theta \, , \, \sin\theta \rangle$$

a unit vector pointing in the <u>radial</u> direction, i. e. , in the direction of $\overline{r}(t)$. From Equation (A)

we see

$$\overline{r}(t) = r\overline{u}$$

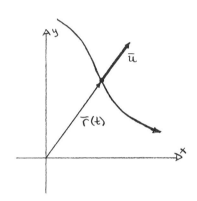

This simple equation will allow us to obtain

an important acceleration equation which

is crucial to Kepler's Laws. To obtain

the acceleration we will need to take two

derivatives of $\overline{r}(t)$ with respect to time t .

Hence we will need derivatives of the vector $\overline{u}$.

The prime (') notation will always

denote differentiation <u>with respect to t</u> .

(This is an important convention because we will have to take derivatives with respect to **many**

different variables.)

Thus $\overline{u}' = \dfrac{d\overline{u}}{dt} = \dfrac{d\overline{u}}{d\theta}\dfrac{d\theta}{dt}$ (the Chain Rule)

$$= \langle -\sin\theta \, , \, \cos\theta \rangle \, \theta'$$

We therefore define

$$\overline{w} = \langle -\sin\theta \, , \, \cos\theta \rangle$$

to obtain

$$\overline{u}' = \theta' \overline{w}$$

What is the geometric significance of the $\overline{w}$ vector? Like $\overline{u}$, it is easy to see that $\overline{w}$ is a **unit** vector. (Check for youself that $\|\overline{w}\| = 1$.) However, we further claim that $\overline{w}$ is obtained by a $90°$ counterclockwise roation of $\overline{u}$. To prove this, simply note that from the trigonometric identities

$$- \sin \theta = \cos (\theta + 90°)$$

and

$$\cos \theta = \sin (\theta + 90°)$$

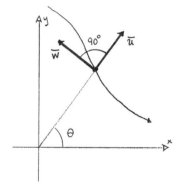

we have

$$\overline{w} = \langle - \sin \theta , \cos \theta \rangle$$

$$= \langle \cos (\theta + 90°) , \sin (\theta + 90°) \rangle$$

which is clearly a $90°$ counterclockwise rotation of

$$\overline{u} = \langle \cos \theta , \sin \theta \rangle$$

Hence $\overline{u}$ and $\overline{w}$ are **orthogonal.** We say that $\overline{w}$ points in the **angular** direction, i.e., the direction of "increasing θ." Now take the derivative of $\overline{w}$:

$$\overline{w}' = \frac{d\overline{w}}{dt} = \frac{d\overline{w}}{d\theta} \frac{d\theta}{dt} = \langle - \cos \theta , - \sin \theta \rangle \theta'$$

or

$$\overline{w}' = - \theta' \overline{u}$$

Let's summarize:

<u>Definition.</u>   The orthogonal unit vectors $\bar{u} = \bar{u}(t)$ and $\bar{w} = \bar{w}(t)$ form the moving

polar coordinate frame of reference for the curve

$\bar{r}(t) = r \cos \theta \, \bar{i} + r \sin \theta \, \bar{j}$.

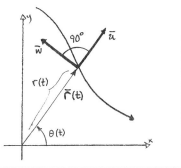

The important equations are

$\bar{u} = \langle \cos \theta , \sin \theta \rangle , \qquad \bar{u}' = \theta ' \bar{w}$

$\bar{w} = \langle - \sin \theta , \cos \theta \rangle , \qquad \bar{w}' = - \theta ' \bar{u}$

$\bar{r}(t) = r \bar{u}$

We thus have developed three frames of reference for motion in a coordinate plane: $\{\bar{i}, \bar{j}\}$,

$\{\bar{T}, \bar{N}\}$ and $\{\bar{u}, \bar{w}\}$.   It is this third frame which will be crucial for establishing Kepler's

Laws.

<u>Example A.</u>   Suppose $\bar{r}(t) = r \cos \theta \, \bar{i} + r \sin \theta \, \bar{j}$ is given by

$$r(t) = 1 + t \qquad \text{and} \qquad \theta(t) = t^2$$

(a)   Calculate $\bar{u}'$ and $\bar{w}'$ in terms of t .

(b)   Compute $\bar{u}'$ and $\bar{w}'$ at $t = \sqrt{\pi}$.

<u>Solution (a).</u> $\bar{u}' = \theta ' \bar{w} = (t^2)' \langle - \sin (t^2) , \cos (t^2) \rangle$

$= 2t \langle - \sin (t^2) , \cos (t^2) \rangle$

$\bar{w}' = - \theta ' \bar{u} = - (t^2)' \langle \cos (t^2) , \sin (t^2) \rangle$

$= - 2t \langle \cos (t^2) , \sin (t^2) \rangle$

<u>Solution (b).</u>

$\bar{u}' = 2 \sqrt{\pi} \langle - \sin \pi , \cos \pi \rangle = 2 \sqrt{\pi} \langle 0 , - 1 \rangle = \langle 0 , - 2 \sqrt{\pi} \rangle$

$\bar{w}' = - 2 \sqrt{\pi} \langle \cos \pi , \sin \pi \rangle = - 2 \sqrt{\pi} \langle - 1 , 0 \rangle = \langle 2 \sqrt{\pi} , 0 \rangle$ $\square$

The formula $\bar{r}(t) = r\bar{u}$ expresses the position vector in terms of $\bar{u}$; we now obtain formulas for $\bar{v}(t)$ and $\bar{a}(t)$ in terms of $\bar{u}$ and $\bar{w}$.

$$\bar{v}(t) = \bar{r}'(t) = (r\bar{u})' \qquad \text{since} \quad \bar{r}(t) = r\bar{u}$$

$$= r'\bar{u} + r\bar{u}' \qquad \text{by the product rule}$$

$$= r'\bar{u} + r(\theta'\bar{w}) \qquad \text{since} \quad \bar{u}' = \theta'\bar{w}$$

$$= r'\bar{u} + r\theta'\bar{w}$$

$$\bar{a}(t) = \bar{v}'(t) = (r'\bar{u} + r\theta'\bar{w})' \qquad \text{by the formula for} \quad \bar{v}(t)$$

$$= (r'\bar{u})' + (r\theta'\bar{w})'$$

$$= (r''\bar{u} + r'\bar{u}') + ((r\theta')'\bar{w} + r\theta'\bar{w}') \qquad \text{by the product rule}$$

$$= r''\bar{u} + r'(\theta'\bar{w}) + r'\theta'\bar{w} + r\theta''\bar{w} + r\theta'(-\theta'\bar{u})$$

$$\qquad \text{since} \quad \bar{u}' = \theta'\bar{w} \quad \text{and} \quad \bar{w}' = -\theta'\bar{u}$$

$$= (r'' - r(\theta')^2)\bar{u} + (2r'\theta' + r\theta'')\bar{w}$$

where we have gathered together the terms involving $\bar{u}$ and the terms involving $\bar{w}$.

Here's a summary of our <u>polar coordinate motion equations</u> (not to be memorized!):

$$\boxed{\begin{aligned} \bar{r}(t) &= r\bar{u} \\ \bar{v}(t) &= r'\bar{u} + r\theta'\bar{w} \\ \bar{a}(t) &= (r'' - r(\theta')^2)\bar{u} + (2r'\theta' + r\theta'')\bar{w} \end{aligned}} \qquad (B)$$

These equations must look pretty weird to you, and their importance is certainly not evident at first glance. The value is:

the velocity and acceleration vectors are both

written as sums of two component vectors,

one of which is <u>radial</u> (i. e. , a multiple of $\bar{u}$)

the other of which is <u>angular</u> (i. e. , a

multiple of $\bar{w}$) .

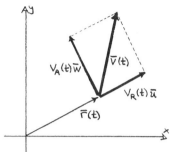

Hence $\bar{v}(t) = v_R(t) \bar{u} + v_A(t) \bar{w}$ where

$v_R(t) = r'$

$v_A(t) = r\theta'$

a similar equation can be written for $\bar{a}(t)$. Thus the polar coordinate motion equations play

the same role for the frame $\{\bar{u}, \bar{w}\}$ that the equations in Anton's Theorem 14.5.3 play for

the frame $\{\bar{T}, \bar{N}\}$. Compare these results!

The importance of separating the "radial" and "angular" components of $\bar{v}(t)$ and $\bar{a}(t)$

is still probably mysterious, but the value will become clear when proving Kepler's Laws.

<u>Example B.</u>   Consider again the curve of motion $\bar{r}(t) = r \cos \theta \, \bar{i} + r \sin \theta \, \bar{j}$ defined by

$r(t) = 1 + t$ and $\theta(t) = t^2$.

(a)   Calculate the angular components of $\bar{v}(t)$ and $\bar{a}(t)$.

(b)   Compute the angular components of $\bar{v}(t)$ and $\bar{a}(t)$ at $t = \sqrt{\pi}$.

<u>Solution (a).</u> $v_A(t) \bar{w} = r\theta' \bar{w} = (1 + t)(2t) \langle -\sin(t^2), \cos(t^2) \rangle$

$= (2t + 2t^2) \langle -\sin(t^2), \cos(t^2) \rangle$

$a_A(t) \bar{w} = (2 r'\theta + r\theta'') \bar{w}$

$= (2(1)(2t) + (1 + t)(2)) \langle -\sin(t^2), \cos(t^2) \rangle$

$= (6t + 2) \langle -\sin(t^2), \cos(t^2) \rangle$

Solution (b). $v_A(\sqrt{\pi})\,\overline{w} = (2\sqrt{\pi} + 2\pi)\,\langle -\sin\pi,\ \cos\pi\rangle$

$$= (2\sqrt{\pi} + 2\pi)\,\langle 0,\ -1\rangle = \langle 0,\ -2\sqrt{\pi} - 2\pi\rangle$$

$a_A(\sqrt{\pi})\,\overline{w} = (6\sqrt{\pi} + 2)\,\langle -\sin\pi,\ \cos\pi\rangle$

$$= (6\sqrt{\pi} + 2)\,\langle 0,\ -1\rangle = \langle 0,\ -6\sqrt{\pi} - 2\rangle \qquad\qquad \square$$

Exercises.

1.   Using our polar coordinate motion equations, derive an expression for $\nu(t)$, the speed of our moving particle, in terms of r, r' and θ'.

2.   Obtain an expression for $s(t)$, the <u>distance</u> the particle has travelled from time 0 to time t. This formula will of course also measure the <u>arc length</u> of the curve from $\overline{r}(0)$ to $\overline{r}(t)$.

   For the remaining problems suppose $\overline{r}(t)$ is defined by the polar coordinates

$$\begin{cases} r(t) = 1 + \cos t \\ \theta(t) = t \end{cases}$$

3.   Sketch the image of this curve for $0 \le t \le 2\pi$.

4.   Calculate the following quantities in terms of t:

$$r',\ r'',\ \theta',\ \theta''$$

5.   Express $\overline{r}$, $\overline{v}$ and $\overline{a}$ in terms of t, $\overline{u}$ and $\overline{w}$.

6.   Compute $\overline{u}$ and $\overline{w}$ at $t = 0$ and graph the resulting vectors with their initial points placed at $\overline{r}(0)$. Repeat this for $t = \dfrac{\pi}{2}$.

7.   Compute the radial and the angular components of velocity at t = 0 . Graph these

two component vectors along with the total velocity vector. The initial points of all

three vectors should be placed at $\bar{r}(0)$. What should these vectors tell us about the

particle motion at t = 0 ? Repeat the procedure for t = $\frac{\pi}{2}$.

8.   Repeat Exercise 7 for the components of acceleration.

Answers.   (1) $\nu(t) = \| r' \bar{u} + r\theta' \bar{w} \|$ (2) $s(t) = \int_0^t \nu(u)\, du$

$= ((r')^2 + r^2(\theta')^2)^{\frac{1}{2}}$

$$= \int_0^t ((r')^2 + r^2(\theta')^2)^{\frac{1}{2}}\, du$$

(3)   A cardioid   (4) $r' = -\sin t,\quad \theta' = 1$

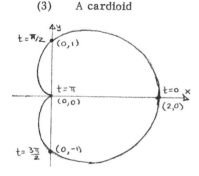

$r'' = -\cos t,\quad \theta'' = 0$

(5) $\bar{r} = (1 + \cos t)\bar{u}$

$\bar{v} = -\sin t\,\bar{u} + (1 + \cos t)\bar{w}$

$\bar{a} = -(2\cos t + 1)\bar{u} - 2(\sin t)\bar{w}$

(6) $\bar{u}(0) = \langle 1, 0 \rangle = \bar{i}$ $\bar{u}(\frac{\pi}{2}) = \langle 0, 1 \rangle = \bar{j}$

$\bar{w}(0) = \langle 0, 1 \rangle = \bar{j}$ $\bar{w}(\frac{\pi}{2}) = \langle -1, 0 \rangle = -\bar{i}$

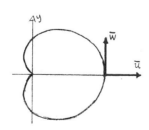
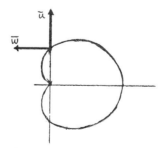

(7) $v_R(0)\bar{u} = \langle 0,0 \rangle = \bar{0}$

$v_A(0)\bar{w} = \langle 0,2 \rangle = 2\bar{j}$

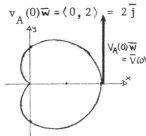

Particle's distance to

$(0,0)$ is not changing

(instantaneously) at

$\bar{r}(0) = \langle 2,0 \rangle$

$v_R(\frac{\pi}{2})\bar{u} = \langle 0,-1 \rangle = -\bar{j}$

$v_A(\frac{\pi}{2})\bar{w} = \langle -1,0 \rangle = -\bar{i}$

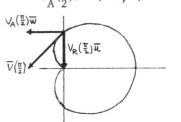

Particle's distance to $(0,0)$

is decreasing at the same rate

as θ is changing at $\bar{r}(\frac{\pi}{2})$

(8) $a_R(0)\bar{u} = \langle -3,0 \rangle = -3\bar{i}$

$a_R(0)\bar{w} = \langle 0,0 \rangle = \bar{0}$

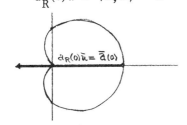

Particle's velocity is

changing only in the

radial direction at

$\bar{r}(0) = \langle 2,0 \rangle$

$a_R(\frac{\pi}{2})\bar{u} = \langle 0,-1 \rangle = -\bar{j}$

$a_A(\frac{\pi}{2})\bar{w} = \langle 2,0 \rangle = 2\bar{i}$

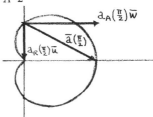

Particle's velocity is slowing

down at $\bar{r}(\frac{\pi}{2}) = \langle 0,1 \rangle$, with

most of the change in the

angular direction.

2.   The statement of Kepler's Laws.

In the early 1600's Johannes Kepler postulated the following three laws of planetary

motion:

I.   In relation to the sun, a
      planet travels in an
      elliptical path with the
      sun located at a focus of
      the ellipse.

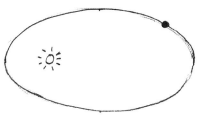

II.   The radial arm (position
      vector from sun to planet)
      sweeps out equal areas
      over equal time intervals.
      Hence a planet's speed is
      faster when it is nearer
      the sun.

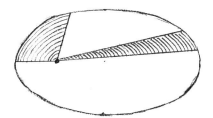

III. If T is the time required
      for one complete orbit
      around the sun, and a is
      the length of the semi-
      major axis of the orbit,
      then $\dfrac{T^2}{a^3}$ has the same

      value for all the planets

      revolving about the sun.

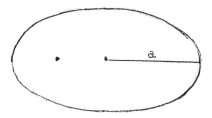

Kepler obtained these three laws strictly from observational data, a truly remarkable feat of scientific reasoning. It was not until the late 1600's that Isaac Newton proved Kepler's Laws from his own more basic laws of motion. Although the proofs we will give are not exactly the same as Newton's, they are based on the same careful and systematic exploitation of the calculus of curves. The derivation is a true tour de force.

3. <u>The general setup and proof of the 2nd Law.</u>

Place the sun at the origin of the xy-plane and let $\overline{r}(t) = r \cos \theta \, \overline{i} + r \sin \theta \, \overline{j}$ be the path of a planet (or, more generally, any object controlled by the gravitational pull of the sun). Then the <u>force</u> $\overline{F}$ (or <u>gravitational pull</u>) of the sun on this planet is a <u>central force.</u> In terms of our polar coordinate frame of reference $\overline{u}, \overline{w}$ we therefore have

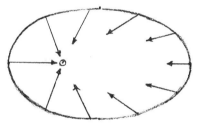

$$\overline{F} = f(t) \, \overline{u}$$

where $f(t)$ is some scalar function of time t. The significance of this equation is the <u>absence of an angular component for the force</u> $\overline{F}$.

We now bring in Newton's Law $\overline{F} = m\overline{a}$, where m = the mass of the planet and $\overline{a}$ = the acceleration of the planet which is caused by the force $\overline{F}$. Thus

$$f(t) \, \overline{u} = m(r'' - r(\theta')^2) \, \overline{u} + m(2r'\theta' + r\theta'') \, \overline{w} \qquad (C)$$

where we used our polar coordinate formula (B) for $\overline{a}$. Notice, however, that the left-hand side of this equation has no angular term $\alpha \, \overline{w}$; we should therefore expect that the angular term on the right-hand side must be zero, i.e.,

$$2r'\theta' + r\theta'' = 0 \qquad (D)$$

If this were not so, then the non-zero multiple of $\bar{w}$ given by

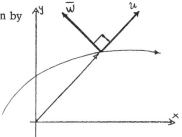

$$m(2r'\theta + r\theta'')\bar{w}$$

would equal the multiple of $\bar{u}$ given by

$$[f(t) - m(r'' - r(\theta')^2)]\bar{u}$$

and hence the multiple of $\bar{u}$ would be parallel to the multiple of $\bar{w}$. Since $\bar{u}$ and $\bar{w}$ are

<u>orthogonal</u> unit vectors, such a relationship cannot be true, proving (D).

Note that Equation (D) will prove that the derivative of $r^2\theta'$ is zero:

$$\frac{d}{dt}(r^2\theta') = 2rr'\theta' + r^2\theta'' \qquad \text{by the product rule}$$

$$= r(2r'\theta' + r\theta'')$$

$$= 0 \qquad \text{by Equation (D)}$$

Hence, since $\frac{d}{dt}(r^2\theta') = 0$, we obtain that $r^2\theta'$ is a constant function of t, so that

$$r^2\theta' = J \qquad (E)$$

for some fixed scalar constant J.

What is the significance of Equation (E)? Answer: it immediately proves Kepler's 2nd

Law! To see this we must recall our formula for the calculation of polar coordinate area

(Anton's Definition 13.3.2):

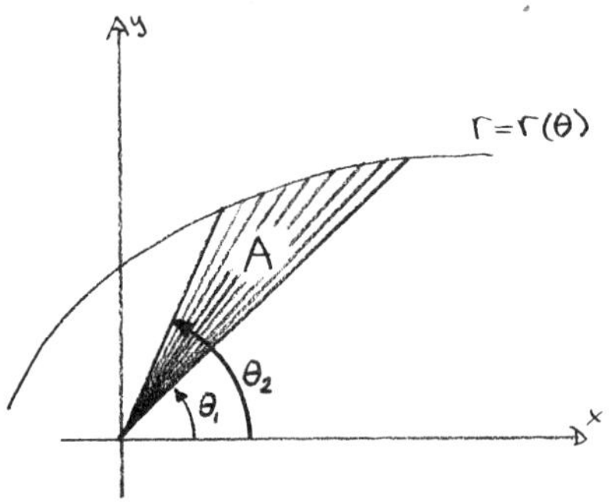

$$\text{Area of A} = \frac{1}{2}\int_{\theta_1}^{\theta_2} r^2\, d\theta \qquad \text{when } r \text{ is a function of } \theta.$$

However, if r and θ are both functions of t, then we can change variables* in this integral by using

$$\theta = \theta(t) \qquad \theta_1 = \theta(t_1)$$

$$d\theta = \theta'\, dt \qquad \theta_2 = \theta(t_2)$$

Hence the area swept out by the radial arm from time t_1 to time t_2 is given by

$$\frac{1}{2}\int_{t_1}^{t_2} \underbrace{r^2\theta'}\, dt$$

ahh, but (E) gives that $r^2\theta'$ is the constant J.

$$= \frac{1}{2}\int_{t_1}^{t_2} J\,dt = \frac{1}{2}J(t_2 - t_1)$$

Hence the amount of area swept out by the radial arm depends only on Δt, the length of time involved, not on the particular time values t_1 and t_2. Hence equal areas are swept out during equal time intervals,

proving Kepler's 2nd Law.

We emphasize two aspects of this derivation. The first is the importance of the formula which resolves $\bar{a}(t)$ into its radial and angular components with respect to the polar coordinate

---

* This is merely a substitution into a definite integral. See Anton's Theorem 5.9.3, or §5.9.2 in The Companion.

frame of reference $\{\bar{u}, \bar{w}\}$.

The second observation is that we only used that the gravitational force was <u>central</u>,

i.e.,

$$\bar{F} = f(t)\,\bar{u}$$

We did not need the actual value of $f(t)$! Hence Kepler's second law is valid for any object

which travels under the influence of a central force.

3.   <u>The proof of the 1st Law.</u>   The 1st and 3rd Laws are not in general valid for <u>all</u> central forces.

Instead we need to use the actual value for the gravitational force,

$$\boxed{\bar{F}(t) = -\frac{GMm}{r^2}\,\bar{u}}$$

where   M = mass of the sun

m = mass of the planet

G = gravitational constant

r = r(t) = distance from the planet to the sun at time t.

This is Newton's Law of Gravitation, a fundamental principle of mechanics which Newton

formulated in the latter part of the seventeenth century. The important aspect of this formula

is the <u>inverse square dependence on the quantity   r</u>. Our equation (C) now becomes

$$-\frac{GMm}{r^2}\,\bar{u} = m(r'' - r(\theta')^2)\,\bar{u} + m(2r'\theta' + r\theta'')\,\bar{w} \qquad (F)$$

However, we already determined that $2r'\theta' + r\theta'' = 0$,   and coupling this with what remains of (F) yields the two basic equations

$$2\,r'\,\theta' + r\theta'' = 0 \tag{F a}$$

$$r'' - r(\theta')^2 = -(G\,M/r^2) \tag{F b}$$

We will use these two equations to prove the 1st Law, i.e., that the planets travel in elliptical orbits. Actually we prove more. We will obtain that the path of any object moving under the influence of the sun must be a <u>conic</u> -- a circle, ellipse, parabola or hyperbola -- with the sun at a foci. We do this by manipulating Equations (F a) and (F b) in some very strange ways to show that r as a function of θ has the following form:

$$r = \frac{\alpha}{1 - \epsilon\,\cos\,(\theta - \theta_0)} \tag{G}$$

for some constants $\alpha > 0$, $\epsilon \geq 0$ and $0 \leq \theta_0 \leq 2\pi$. [Fact: this is the general polar coordinate equation for a conic with one focus at the origin. This claim is fairly easily demonstrated by converting (G) into xy coordinates (when $\theta_0 = 0$). However, in order not to interrupt our development of Kepler's Laws we will prove the stated properties of (G) in an appendix to this section.]

To prove the 1st Law we must therefore take Equations (F a) and (F b), eliminate the parameter t from them, and end up with r as a function of θ in the form of (G). So here we go... this IS a complicated derivation, so don't be discouraged if it seems confusing.

As shown in the discussion leading up to Equation (E), Equation (F a) is equivalent to

$$\theta' = \frac{J}{r^2} \tag{H a}$$

We can also change (F b) by observing $r(\theta')^2 = \dfrac{r^4\,(\theta')^2}{r^3} = \dfrac{J^2}{r^3}$, so

$$r'' - \frac{J^2}{r^3} = -\frac{GM}{r^2} \qquad \text{(H b)}$$

The (H) equations are better than the (F) equations in that no θ'' appears and (H a) gives a way to eliminate θ' whenever it appears. We will do this shortly. First we play with (H b) to get rid of the second derivative which appears there. To eliminate r'', first multiply (H b) by $2r'$ to obtain

$$2r' r'' + J^2(-2r^{-3}r') - 2GM(-r^{-2}r') = 0$$

Can you see that the left-hand side of this equation is the derivative of another, simpler expression? The equation is

$$\frac{d}{dt}[(r')^2 + J^2 r^{-2} - 2GMr^{-1}] = 0$$

Hence

$$(r')^2 + \frac{J^2}{r^2} - \frac{2GM}{r} = C \qquad \text{(I)}$$

where C is a constant independent of t.

We now wish to change (I) from a differential equation in $\frac{d}{dt}$ to a differential equation in $\frac{d}{d\theta}$ (this step corresponds to our desire to eliminate the parameter t). We use a clever trick to accomplish our purpose:

$$\text{define} \quad s = \frac{1}{r} \qquad \text{(K)}$$

Then $r = \frac{1}{s}$ and $r' = -\frac{s'}{s^2}$, so (I) becomes

$$\boxed{\left(\frac{s'}{s^2}\right)^2 + J^2 s^2 - 2GMs = C} \tag{L}$$

Now here's how we convert from $\dfrac{d}{dt}$ to $\dfrac{d}{d\theta}$:

$$s' = \frac{ds}{dt} = \frac{ds}{d\theta}\frac{d\theta}{dt} \qquad \text{by the Chain Rule,}$$

$$= \frac{ds}{d\theta}\left(\frac{J}{r^2}\right) \qquad \text{by (Ha), as promised earlier,}$$

$$= Js^2\frac{ds}{d\theta} \qquad \text{since } \frac{1}{r^2} = s^2$$

Hence $\quad \dfrac{s'}{s^2} = J\dfrac{ds}{d\theta}$, so (L) becomes

$$\boxed{J^2\left(\frac{ds}{d\theta}\right)^2 + J^2 s^2 - 2GMs = C} \tag{M}$$

As promised, we've gotten rid of $\dfrac{d}{dt}$ altogether, and now have a differential equation in $\dfrac{d}{d\theta}$. We will proceed to solve this differential equation for s as a function of θ.

We start by dividing (M) by J^2 :

$$\left(\frac{ds}{d\theta}\right)^2 + s^2 - \frac{2GM}{J^2}s = \frac{C}{J^2}$$

Completing the square in the s variable yields

$$\left(\frac{ds}{d\theta}\right)^2 + \left(s - \frac{GM}{J^2}\right)^2 = \frac{C}{J^2} + \frac{G^2M^2}{J^4} \tag{N}$$

Notice that the left side of this equation is a sum of squares, and hence is non-negative; thus the right side of the equation is also non-negative, and we can therefore express it as the square of some constant E. For convenience we will let

$$D = \frac{GM}{J^2} \qquad \text{and} \qquad E^2 = \frac{C}{J^2} + \frac{G^2 M^2}{J^4} \qquad (O)$$

Then (N) can be written as

$$\left(\frac{ds}{d\theta}\right)^2 + (D - s)^2 = E^2$$

or

$$\left(\frac{1}{E}\frac{ds}{d\theta}\right)^2 + \left(\frac{D - s}{E}\right)^2 = 1 \qquad (P)$$

Now recall a basic trigonometry fact:

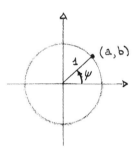

$\Big\|$ If $a^2 + b^2 = 1$, then there exists

$\Big\|$ an angle ψ such that

$\Big\|$ $a = \sin\psi$ and $b = \cos\psi$

Applying this fact to (P) shows that for each θ there exists an angle ψ such that

$$\frac{D - s}{E} = \cos\psi \qquad (Q)$$

or

$$s = D - E\cos\psi \qquad (R)$$

The angle ψ is, by definition, a function of θ; we can show it to be a very simple function as follows. Differentiate (R) with respect to θ to obtain

$$\frac{ds}{d\theta} = E\sin\psi\left(\frac{d\psi}{d\theta}\right)$$

Plugging this equation and (Q) back into (P) yields

$$\sin^2 \psi \left(\frac{d\psi}{d\theta}\right)^2 + \cos^2 \psi = 1$$

$$\left(\frac{d\psi}{d\theta}\right)^2 = \frac{1 - \cos^2 \psi}{\sin^2 \psi} = \frac{\sin^2 \psi}{\sin^2 \psi} = 1$$

$$\frac{d\psi}{d\theta} = \pm 1$$

Thus

$$\psi = \pm (\theta - \theta_0) \quad \text{for some constant} \quad \theta_0$$

Placing this equation for ψ into (R), and using $\cos(-\psi) = \cos \psi$, gives

$$\boxed{s = D - E \cos (\theta - \theta_0)}$$

Thus, using (K), we obtain

$$r = \frac{1}{s} = \frac{1}{D - E \cos (\theta - \theta_0)}$$

$$= \frac{D^{-1}}{1 - (E/D) \cos (\theta - \theta_0)}$$

Finally, using the definition of D in (O), we obtain

$$\boxed{r = \frac{\alpha}{1 - \varepsilon \cos (\theta - \theta_0)} \quad \text{where} \quad \alpha = \frac{J^2}{GM} \quad \text{and} \quad \varepsilon = \frac{E}{D}} \tag{S}$$

... WHICH IS THE DESIRED EQUATION (G)! Thus, if you are still alive and breathing, you have seen the proof of Kepler's 1st Law. The fact that this equation is indeed a conic is proven in the Appendix which follows.

The proof of Kepler's 3rd Law, which is relatively simple and based on facts we have already established, is outlined in the exercises.

Appendix.   The conic equation $r = \dfrac{\alpha}{1 - \epsilon \cos (\theta - \theta_0)}$.

We first analyze the polar equation

$$r = \frac{1}{1 - \epsilon \cos \theta}$$

What is the graph of this equation? This is most easily obtained by conversion to xy-coordinates:

$$r - \epsilon r \cos \theta = 1$$

$$r = 1 + \epsilon r \cos \theta$$

$$r^2 = (1 + \epsilon r \cos \theta)^2$$

$$x^2 + y^2 = (1 + \epsilon x)^2 = 1 + 2\epsilon x + \epsilon^2 x^2$$

$$\boxed{(1 - \epsilon^2)x^2 - 2\epsilon x + y^2 = 1}$$

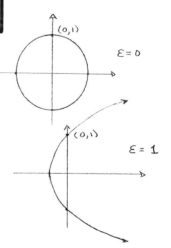

Notice: $\epsilon = 0$ gives $x^2 + y^2 = 1$

a circle

$\epsilon = 1$ gives $x = \frac{1}{2} y^2 - \frac{1}{2}$

a parabola.

Suppose $\epsilon \neq 1$. Then dividing by $1 - \epsilon^2$ gives

$$x^2 - \frac{2\epsilon}{1 - \epsilon^2} x + \frac{y^2}{1 - \epsilon^2} = \frac{1}{1 - \epsilon^2}$$

and completing the square in the x terms gives

$$\left(x - \frac{\epsilon}{1 - \epsilon^2}\right)^2 - \frac{\epsilon^2}{(1 - \epsilon^2)^2} + \frac{y^2}{1 - \epsilon^2} = \frac{1}{1 - \epsilon^2}$$

$$\left(x - \frac{\epsilon}{1 - \epsilon^2}\right)^2 + \frac{y^2}{1 - \epsilon^2} = \frac{1}{1 - \epsilon^2} + \frac{\epsilon^2}{(1 - \epsilon^2)^2}$$

$$= \frac{1 - \epsilon^2 + \epsilon^2}{(1 - \epsilon^2)^2} = \frac{1}{(1 - \epsilon^2)^2}$$

Division by $\dfrac{1}{(1 - \epsilon^2)^2}$ now yields

$$\boxed{\frac{\left(x - \dfrac{\epsilon}{1 - \epsilon^2}\right)^2}{\dfrac{1}{(1 - \epsilon^2)^2}} + \frac{y^2}{\dfrac{1}{1 - \epsilon^2}} = 1} \qquad \text{(T)}$$

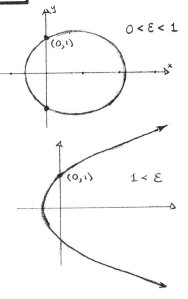

Hence, $\underline{0 < \epsilon < 1}$ gives $1 - \epsilon^2 > 0$, and

(T) is an <u>ellipse</u>;

$\underline{1 < \epsilon}$ gives $1 - \epsilon^2 < 0$, and

(T) is a <u>hyperbola</u> (one branch).

In all four of the above cases, ϵ is called

the <u>eccentricity</u> of the conic, and we can

show $(0, 0)$ is a <u>focus</u> of the conic (see

Anton's Chapter 12).

So what happens when we generalize from

$$r = \frac{1}{1 - \epsilon \cos} \qquad \text{to} \qquad r = \frac{\alpha}{1 - \epsilon \cos(\theta - \theta_0)} \quad ?$$

Not much! Take the graph of the first equation,

rotate it by the angle θ_0, and expand it by

the multiplicative factor α. The result will

be the graph of the second equation... still a

conic with focus at $(0,0)$.

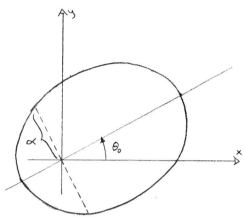

<u>Exercises.</u>   These problems make use of Kepler's 2nd Law as well as Equations (Ha), (Hb),

(S) and (T). Exercise 1 from the previous set of exercises will also be helpful.

9.   Suppose an object is moving through the gravitational force field of the sun in a <u>circular</u>

path (with the sun necessarily at the center of the circle).

a.   Use (Hb) to compute the radius of the circular orbit in terms of J, G and M.

(Hint: What can you say about r' and r''?)

b.   Compute θ' in terms of J, G and M.

c.   Prove that the object has <u>constant speed</u> ν by computing the speed in terms of J,

G and M. Is the <u>velocity</u> constant?

10.   Recall that $\dfrac{(x - x_0)^2}{a^2} + \dfrac{(y - y_0)^2}{b^2} = 1$ is the equation

of an ellipse with center (x_0, y_0) and semi-major and

semi-minor axes of lengths a and b (assuming

$a \geq b > 0$). The <u>area</u> of such an ellipse is πab.

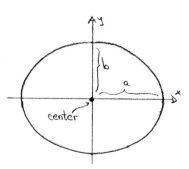

a.   Consider the ellipse given by the polar equation

$r = \dfrac{1}{1 - \epsilon \cos \theta}$, $0 < \epsilon < 1$. Express a and b in terms of ϵ.

14. 6. 24

b.   Do the same for the ellipse $r = \dfrac{\alpha}{1 - \varepsilon \cos (\theta - \theta_0)}$.

(Hint: (b) is easy if you use geometric arguments similar to those given at the end of the appendix.)

11.   <u>A proof of Kepler's 3rd Law.</u>   Let   T   denote the <u>period</u> of the orbit of a planet about the sun, i. e. , the time necessary to complete one revolution.

a.   Use the derivation of Kepler's 2nd Law to prove

$$\frac{1}{2} JT = \pi ab$$

where   a   and   b   are the lengths of the semi-major and semi-minor axes of the path of motion.

b.   Making use of Exercise (10), express   a   and   b   in terms of ε, J, G   and   M.

c.   Conclude from the above that T^2/a^3 is a constant which is the same for all the planets revolving about the sun.   This establishes Kepler's 3rd Law.

<u>Solutions.</u>

9.   Since we have a circular path, then the radius   r   is a constant.   Hence   r' = r" = 0.
This is the key to this problem.

a.   Using   r" = 0   changes   (Hb)   into

$$J^2/r^3 = GM/r^2 , \quad \text{so that} \quad \boxed{r = J^2/GM}$$

b.   Placing our formula for   r   into   (Ha)   will yield

$$\boxed{\theta' = G^2M^2/J^3}$$

c.   From Exercise 1 we know that

$$\text{speed} = \left((r')^2 + r^2(\theta')^2 \right)^{1/2}$$

$$= \left(0^2 + \frac{J^4}{G^2 M^2} \cdot \frac{G^4 M^4}{J^6} \right)^{1/2} \qquad \text{by (a) and (b)}$$

$$= \left(\frac{G^2 M^2}{J^2} \right)^{1/2} = \frac{GM}{J}$$

Thus $\boxed{\text{speed} = GM/J}$, which is a constant. However, from the polar

coordinate motion equation for velocity $\overline{v}(t)$, we have

$$\overline{v}(t) = r' \overline{u} + r\theta' \overline{w} = 0 \cdot \overline{u} + \frac{J^2}{GM} \cdot \frac{G^2 M^2}{J^3} \overline{w}$$

$$= \frac{GM}{J} \langle - \sin \theta , \cos \theta \rangle$$

which is NOT a constant vector.

10.   a.   Reading off a and b from (T) we obtain

$$a = \frac{1}{1 - \epsilon^2} \qquad \text{and} \qquad b = \frac{1}{\sqrt{1 - \epsilon^2}}$$

b.   The ellipse $r = \dfrac{\alpha}{1 - \epsilon \cos (\theta - \theta_0)}$ is merely the ellipse $r = 1/(1 - \epsilon \cos \theta)$

rotated by the angle θ_0 and magnified or contracted by the factor α (see the

last diagram in the Appendix). Thus a and b are merely the same as in

part (a), but multiplied by α, i.e.,

$$a = \frac{\alpha}{1 - e^2} \quad \text{and} \quad b = \frac{\alpha}{\sqrt{1 - e^2}}$$

11.   a.   In the proof of Kepler's 2nd Law we saw that the area swept out from time t_1 to time t_2 is

$$\frac{1}{2} J(t_2 - t_1)$$

Hence, if we go around one complete revolution, then the area swept out is the full area of the ellipse ($\pi a b$ from the given information in Exercise 10), and $t_2 - t_1 = T$. Thus

$$\pi a b = \begin{bmatrix} \text{area swept} \\ \text{out in one} \\ \text{revolution} \end{bmatrix} = \frac{1}{2} J T$$

as desired.

b.   From (S) we have $\alpha = J^2/GM$; when combined with the results of Exercise 10b we obtain

$$a = \frac{J^2}{GM}\left(\frac{1}{1 - e^2}\right) \quad \text{and} \quad b = \frac{J^2}{GM}\left(\frac{1}{\sqrt{1 - e^2}}\right)$$

c.   To show T^2/a^3 is a constant which is the same for all the planets, we start with $JT/2 = \pi a b$ from part (a). Then

$$T/a = 2\pi b/J$$

$$T^2/a^2 = 4\pi^2 b^2/J^2$$

Multiplying by   1/a   gives

$$\frac{T^2}{a^3} = \left(\frac{4\pi^2}{J^2}\right)\left(\frac{b^2}{a}\right)$$

$$= \left(\frac{4\pi^2}{J^2}\right)\left(\frac{J^4/G^2M^2(1-e^2)}{J^2/GM(1-e^2)}\right) \qquad \text{from part (b)}$$

$$= \left(\frac{4\pi^2}{J^2}\right)\left(\frac{J^2}{GM}\right)$$

$$= \frac{4\pi^2}{GM}$$

Ahh! This expression involves only the constants $4\pi^2$, G and M, all of which are independent of the planet under consideration.

This proves Kepler's 3rd Law.

Chapter 15 : Three-dimensional Space

Section 15. 1 : Rectangular Coordinates in 3-Space; Cylindrical Surfaces.

1. $\mathbb{R}^3$. We live in a three-dimensional world, and hence many important problems require

mathematical techniques which deal with three dimensions. The starting place for such techniques

is in coordinatizing three dimensional space. You are already familiar with the coordinatized

xy-plane; to coordinatize 3-space we simply add a third axis - the z axis - which is

perpendicular to both the x and y axes. We orient the x, y and z axes by the right-

hand rule, as shown in Anton's Figure 15. 1. 3(a).   The y - and z-axes
                                                                      lie on the page

Coordinatizing space sets up the following

correspondence: The x-axis points
                                                                      out of the page

> every point   P   in space is assigned a triple of
>
> numbers (x, y, z)   called the coordinates of P .

Conversely,

> every triple of numbers   (x, y, z)   gives the
>
> coordinates of some point   P   in space.

In this way 3-space is identified with the set of all ordered triples of real numbers, denoted

by $\mathbb{R}^3$, i.e.,

$$\mathbb{R}^3 = \{(x, y, z) , \text{ where } x, y \text{ and } z \text{ are real numbers}\}$$

This is just a straightforward extension of the familiar identification of the two dimensional plane

with the set of all ordered pairs of real numbers, denoted by

$$\mathbb{R}^2 = \{(x, y), \text{ where } x \text{ and } y \text{ are real numbers}\}$$

The coordinatizing of space has profound implications for mathematics: it enables three dimensional "geometry" problems to be converted into "algebraic" problems, and vice versa. You have already seen this happen with the coordinate plane; * you will now see this happen with coordinatized space.

2. <u>The distance formula.</u> The first formula which Anton gives is that for computing the distance d between two points $P_1(x_1, y_1, z_1)$ and $P_2(x_2, y_2, z_2)$ in $\mathbb{R}^3$:

Distance
Formula

$$d = \sqrt{(x_2 - x_1)^2 + (y_2 - y_1)^2 + (z_2 - z_1)^2} \qquad \text{(A)}$$

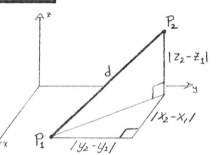

Anton's derivation is easily understood: it consists of two applications of the Pythagorean Theorem to the right triangles shown in our picture (note that the absolute value signs are necessary because we do not know if $x_1 \leq x_2$ or $x_2 \leq x_1$, and similarly for the y and z variables). Moreover, (A) is easily remembered as the obvious generalization of the two dimensional formula: the distance in $\mathbb{R}^2$ between $P_1(x_1, y_1)$ and $P_2(x_2, y_2)$ is

$$\sqrt{(x_2 - x_1)^2 + (y_2 - y_1)^2}$$

<u>Don't</u>, however, let the simplicity of its derivation and form lead you to underestimate

---

* The introductory comments in §1.3 of the <u>Companion</u> are relevant here.

the importance of the distance formula! Calculus is based on limits, a concept which in turn is based on "distances getting very small." Thus, besides its practical value, the distance formula is at the heart of the theory of three dimensional calculus.

3.   **Spheres and cylinders.**   In later sections we will discuss much more complicated shapes in 3-space; for now we content ourselves with spheres and cylinders.

It is an easy application of the distance formula to show that any sphere with center (x_0, y_0, z_0) and radius $r > 0$ consists of those points (x, y, z) which satisfy the equation

| Equation of a Sphere |

$$(x - x_0)^2 + (y - y_0)^2 + (z - z_0)^2 = r^2 \qquad \text{(B)}$$

Anton further shows (Theorem 15.1.1) that any equation of the form

$$x^2 + y^2 + z^2 + Gx + Hy + Iz + J = 0 \qquad \text{(C)}$$

(in which the coefficients of all the squared terms are 1) represents a sphere or a point, or else has no graph. Given a specific equation of form (C), to determine the graph type you must convert your equation to form (B); this step always requires you to complete the squares in the three separate variables x, y and z. (See Appendix D §7 if you need a review of completing the square.) Carefully read Anton's Example 3 and the discussion which follows it.

Graphs of equations which contain only two of the three variables x, y and z are called cylindrical surfaces. Graphing such surfaces, although very easy once you catch on to the concept, tends to give people some trouble. Here is an example showing how it is done:

Example A.   Sketch the surface $x^2 + 4y^2 = 4$.

Solution.   Dividing by 4 gives the equation

$$\frac{x^2}{4} + y^2 = 1$$

which in the two dimensional xy-plane is the

equation for an ellipse C as shown in

Figure 1 (see Appendix E §3). Place this

sketch onto the xy-plane in xyz-space as

in Figure 2 (the drawing trick here is to achieve

the proper perspective - you are looking at the

ellipse from an angle).

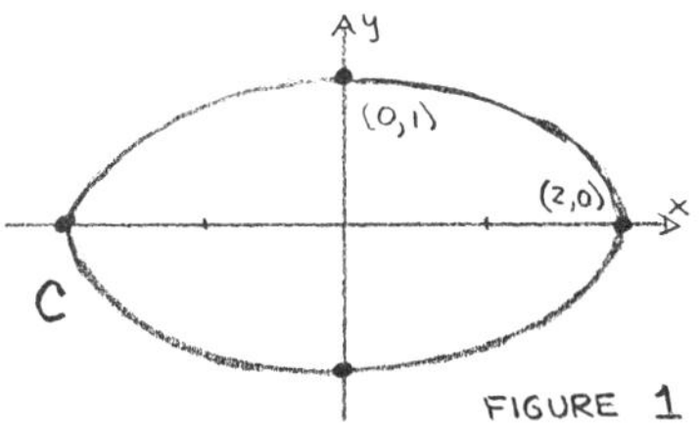

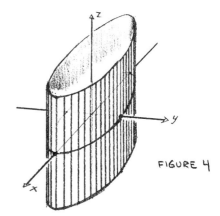

FIGURE 1

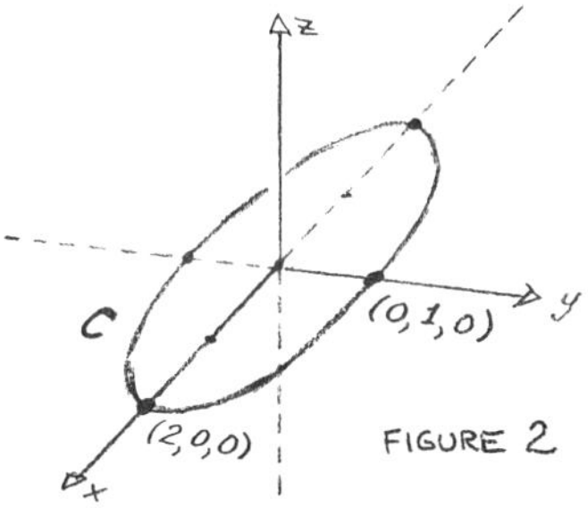

FIGURE 2

Clearly all the points shown in Figure 2 are on our surface:

they are all of the form $(x_0, y_0, 0)$ where x_0 and y_0

satisfy Equation (D). However, if $(x_0, y_0, 0)$ satisfies (D), then (x_0, y_0, z) satisfies

<u>(D) for any value of z</u>! The set of all these points, i.e.,

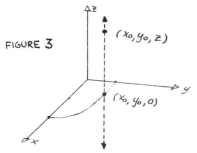

FIGURE 3

$\{(x_0, y_0, z)$, where z is any real number$\}$

is a straight line parallel to the z-axis and

passing through $(x_0, y_0, 0)$, as shown in Figure

3. Thus, each point on ellipse C in Figure 2

determines a whole vertical line of points on our

graph. Placing all these lines into Figure 2

like pieces of spagetti (uncooked!) produces

a picture of our graph as shown in Figure 4.

In Anton's terminology, the line in Figure 3

is a <u>generator</u> for our surface. As this line

traverses the ellipse C in Figure 2, it

generates the cylindrical surface shown in Figure 4.

FIGURE 4

Another way to view our cylindrical surface is as follows: Take the ellipse C in

Figure 2 and translate it up and down along the z-axis. The surface so generated is again

the one shown in Figure 4.   □

Section 15.2.   Vectors and Lines in 3-Space.

1.   Geometric vectors.   Except for the discussion on representing lines in 3-space, all the

material in Anton's §15.2 is a straightforward generalization of that presented in §14.1 for

2-space: instead of two components we now need three. In fact, the initial geometric discussion

of vectors and the operations of vector addition, vector subtraction and scalar multiplication

given in §14.1 applies equally well to 2-space or 3-space, and should be re-read at this

stage. We emphasize that

Geometric equivalence of vectors	two vectors $\bar{v}$ and $\bar{w}$ are equivalent, written $\bar{v} = \bar{w}$, if and only if they have the same length and the same direction.	(A)

Thus, as with vectors in the plane, the placement of the initial point of a 3-space vector is

unimportant.

2.   Algebraic vectors in 3-space.   Suppose we now introduce an xyz-coordinate system on

3-space. We can then convert any "geometric vector" $\bar{v}$ into an "algebraic vector" as follows:

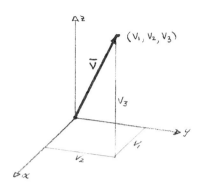

Position $\bar{v}$ so that its initial point is at the origin. Then $\bar{v}$ is completely determined by the coordinates (v_1, v_2, v_3) of its terminal point, and we write

$$\bar{v} = \langle v_1, v_2, v_3 \rangle$$

with v_1, v_2 and v_3 called the components of $\bar{v}$.

Vector components in space

Thus every geometric vector $\bar{v}$ in 3-space determines an ordered triple of numbers $\langle v_1, v_2, v_3 \rangle$ -- an algebraic vector -- and vice versa. In particular, suppose $\bar{v}$ is a vector in 3-space with initial point $P_1(x_1, y_1, z_1)$ and terminal point $P_2(x_2, y_2, z_2)$, written $\bar{v} = \overrightarrow{P_1 P_2}$. Then the components of $\bar{v}$ are as follows

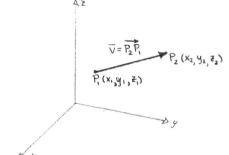

$$\boxed{\bar{v} = \overrightarrow{P_1 P_2} = \langle x_2 - x_1, y_2 - y_1, z_2 - z_1 \rangle} \tag{B}$$

i.e., the components of any vector are the coordinates of its terminal point minus the coordinates of its initial point.

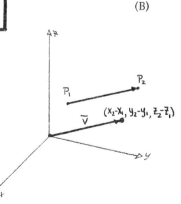

The geometric equivalence of vectors defined in (A) now has an algebraic formulation: Suppose $\bar{v} = \langle v_1, v_2, v_3 \rangle$ and $\bar{w} = \langle w_1, w_2, w_3 \rangle$. Then

Algebraic
equivalence
of vectors

$$\overline{v} = \overline{w} \quad \text{if and only if} \quad \begin{cases} v_1 = w_1 \\ v_2 = w_2 \\ v_3 = w_3 \end{cases}$$ (C)

i. e., <u>two vectors are equivalent if and only</u>

<u>if their corresponding components are equal.</u>

Thus one vector equation $\overline{v} = \overline{w}$ in 3-space is equivalent to three simultaneous scalar

equation: $\begin{cases} v_1 = w_1 \\ v_2 = w_2 \\ v_3 = w_3 \end{cases}$

3. <u>Algebraic operations on vectors.</u>   The algebraic formulations in 3-space for vector addition,

vector subtraction and scalar multiplication are the same as in 2-space -- the plane --

except now we have <u>three</u> components, not <u>two.</u> For example:

2-space : $\overline{v} + \overline{w} = \langle v_1, v_2 \rangle + \langle w_1, w_2 \rangle = \langle v_1 + w_1, v_2 + w_2 \rangle$

3-space : $\overline{v} + \overline{w} = \langle v_1, v_2, v_3 \rangle + \langle w_1, w_2, w_3 \rangle = \langle v_1 + w_1, v_2 + w_2, v_3 + w_3 \rangle$

Thus you have nothing new to learn in this regard! Moreover, all the basic rules for vector

arithmetic as given in Theorem 14.1.5 for 2-space are also valid for 3-space. So again,

nothing new to learn! We do wish to recall, however, the two warnings we voiced in §14.1:

> A scalar can <u>never</u> be added to a vector!
>
> Two vectors can <u>never</u> be multiplied together! *

Just as 2-space has its standard unit vectors $\bar{i}$ and $\bar{j}$, so does 3-space ---

although there are 3 standard unit vectors in 3-space:

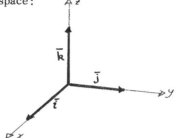

$$\bar{i} = \langle 1, 0, 0 \rangle$$

$$\bar{j} = \langle 0, 1, 0 \rangle$$

$$\bar{k} = \langle 0, 0, 1 \rangle$$

It is then easy to establish the following useful result:

> Suppose $\bar{v} = \langle v_1, v_2, v_3 \rangle$
>
> Then $\bar{v} = v_1 \bar{i} + v_2 \bar{j} + v_3 \bar{k}$

Thus $\bar{v} = \langle v_1, v_2, v_3 \rangle$ and $\bar{v} = v_1 \bar{i} + v_2 \bar{j} + v_3 \bar{k}$ are alternate ways of saying the same

thing; in specific situations we will use whichever form is most convenient.

4. <u>The norm of a vector.</u> The <u>length</u> or <u>norm</u> of $\bar{v} = \langle v_1, v_2, v_3 \rangle$ is seen from the

distance formula in $\mathbb{R}^3$ to be

$$\|\bar{v}\| = \sqrt{v_1^2 + v_2^2 + v_3^2} \tag{D}$$

---

* Later in this Chapter we will introduce two operations on pairs of vectors in 3-space which
we call the "dot product" and the "cross product." However, the dot product of two vectors is
a <u>scalar</u> quantity (hardly what a "multiplication" should produce), while the cross product has
all sorts of "non-multiplicative" properties. There is no "usual" form of multiplication for
vectors!!

As in 2-space, the important properties of the norm in 3-space are:

$$\|k\overline{v}\| \;=\; |k|\,\|\overline{v}\| \quad \text{for any scalar } k,$$

$$\|\overline{v} + \overline{w}\| \;\leq\; \|\overline{v}\| + \|\overline{w}\| \quad \text{(The Triangle Inequality)}$$

(E)

These are used in the same way in 3-space as in 2-space -- see Example D in §14.1 of The Companion.

Any vector $\overline{u}$ whose length is 1 is termed a **unit vector**. Here are two important results concerning unit vectors:

1. Given any non-zero vector $\overline{v}$, then the unit vector $\overline{u}$ with the same direction as $\overline{v}$ is

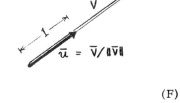

$$\overline{u} \;=\; \overline{v}/\|\overline{v}\|$$

(F)

$\overline{u}$ is called the **unit direction vector** for $\overline{v}$.

2. Suppose $\overline{v}$ is any non-zero vector in 3-space. Then from Equation (F) we obtain

$$\overline{v} \;=\; \|\overline{v}\|\,\overline{u},$$

$$\text{where } \overline{u} \text{ is the unit direction vector for } \overline{v}$$

(G)

This is the **polar form** of the vector $\overline{v}$. The major difference between this expression and the corresponding one in the plane is: unit vectors in the plane all have the form

$$\overline{u} \;=\; \cos\theta\,\overline{i} + \sin\theta\,\overline{j}$$

while there is no equally simple analogous expression for unit vectors in 3-space.

5. <u>Lines in 3-space</u>. This is an extremely important section. Multidimensional calculus makes heavy use of lines in space, and we will need convenient algebraic ways of handling them. You are well advised to go over all of this material in Anton and <u>The Companion</u> as many times as necessary in order to grasp fully the material. In addition to being important, this material also proves to be quite difficult for many people.

Deriving the equations for lines in 3-space rests on the following simple observation:

> two vectors $\bar{v}$ and $\bar{w}$ are parallel if and only if one is a scalar multiple of the other.

(H)

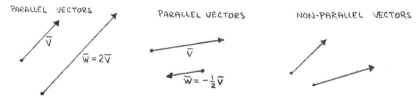

The truth of (H) can be seen by considering all vectors as having their initial points at the origin.

Anton derives what are known as <u>parametric scalar equations</u> for a line. We will discuss these as well as two other methods for describing a line. Our major purpose is to clarify the geometric meaning of the parameter t.

Suppose L is the line in $\mathbb{R}^3$ passing through the <u>point</u> P_0 and parallel to the <u>direction vector</u> $\bar{v} = \langle a, b, c \rangle \neq \bar{0}$. <u>Be careful here</u>:

(i) The terminal <u>point</u> of $\bar{v}$, when starting at $\bar{0}$, will <u>not</u> in general lie on L. (See diagram below.)

(ii) By contrast, the <u>vector</u> $\bar{P}_0$ determined by P_0 will <u>not</u> necessarily be parallel to L. (See diagram below.)

(1) A parametric vector equation for L .

Suppose $\overline{X}$ is any point on the line L .

Then the vector $\overline{X} - \overline{P}_0$ is certainly

parallel to $\overline{v}$, and hence, by (H), is a scalar

multiple of $\overline{v}$, i. e.,

$$\overline{X} - \overline{P}_0 = t\overline{v} \quad \text{for some} \quad -\infty < t < \infty$$

so

$$\boxed{\overline{X} = \overline{P}_0 + t\overline{v} \quad \text{for} \quad -\infty < t < \infty} \qquad \text{(I)}$$

We have shown that every point $\overline{X}$ on the line L satisfies this equation for some t .

Conversely, it is not hard to see that any $\overline{X}$ of the form $\overline{P}_0 + t\overline{v}$ is on L. Our equation is

called the parametric vector equation for the line L through the point P_0 and parallel to

the vector $\overline{v}$.

(2) Parametric scalar equations for L .

We can rewrite the preceeding parametric vector equation into a system of simultaneous scalar

equations by substituting in components:

$$\overline{v} = \langle a, b, c \rangle$$
$$\overline{P}_0 = \langle x_0, y_0, z_0 \rangle$$
$$\overline{X} = \langle x, y, z \rangle$$

Then

$$\langle x, y, z \rangle = \langle x_0, y_0, z_0 \rangle + t \langle a, b, c \rangle \qquad \text{from (I)}$$
$$\langle x, y, z \rangle = \langle x_0 + at, y_0 + bt, z_0 + ct \rangle$$

But by (C), two vectors are equal if and only if their corresponding components are equal, so

$$\begin{cases} x = x_0 + at \\ y = y_0 + bt \\ z = z_0 + ct \end{cases} \qquad \text{for} \quad -\infty < t < \infty \qquad\qquad (J)$$

These are <u>parametric scalar equations</u> for L , and form the heart of Anton's Theorem 15. 2. 1.

Conversely: any such set of three scalar equations

determines a line L in $\mathbb{R}^3$ provided a , b and

c are not all simultaneously equal to zero.

(3) <u>Symmetric equations</u> for L .

We can eliminate the parameter "t" from our parametric scalar equations in the following way. Take the three parametric scalar equations and solve each one for t :

$$t = \frac{x - x_0}{a} \qquad t = \frac{y - y_0}{b} \qquad t = \frac{z - z_0}{c}$$

Then L consists of all points X = (x, y, z) satisfying

$$\boxed{\frac{x - x_0}{a} = \frac{y - y_0}{b} = \frac{z - z_0}{c}} \qquad\qquad (K)$$

This is a set of <u>symmetric equations</u> for L . Notice that the equation is meaningful in the usual algebraic sense only if $a \neq 0$, $b \neq 0$ and $c \neq 0$. However, if one of these constants is zero,

say $a = 0$, then we will allow $\frac{x - x_0}{0}$ to exist as a (formal) expression which is defined to

mean $x = x_0$ for <u>all</u> points on L .

Note that in both the parametric scalar equations and the symmetric equations for L ,
the point $P_0 = \langle x_0, y_0, z_0 \rangle$ is <u>on</u> L and the vector $\bar{v} = \langle a, b, c \rangle$ is <u>parallel</u> to L ,
just as in the parametric vector equation for L . Hence:

*<u>General Principle</u>:   When determining any form of equation for a line

in space, always focus on finding a <u>point</u> P_0 on the line and a   (L)

<u>direction vector</u> $\bar{v}$ parallel to the line.

<u>Example A.</u>   Find various equations for the line L   passing through the two points $P_0 = (1, 2, 0)$
and $P_1 = (1, 3, 1)$.

<u>Solution.</u>   Let $\bar{v} = \bar{P}_1 - \bar{P}_0 = \langle 0, 1, 1 \rangle$.

This is clearly a direction vector for the

line L . Hence one parametric vector

equation for L is given by

$$\bar{X} = \bar{P}_0 + t\bar{v}, \quad -\infty < t < \infty$$

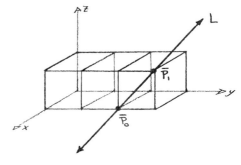

i. e. ,

$$\boxed{\begin{aligned} \bar{X} &= \langle 1, 2, 0 \rangle + t \langle 0, 1, 1 \rangle \\ &\quad -\infty < t < \infty \end{aligned}}$$

The corresponding set of parametric scalar equations are

$$\boxed{\begin{cases} x = 1 \\ y = 2 + t \\ z = t \end{cases} \quad -\infty < t < \infty}$$

15. 2. 10

In terms of symmetric equations we have

$$\frac{x-1}{0} = \frac{y-2}{1} = \frac{z-0}{1}$$

where $\frac{x-1}{0}$ simply means $x = 1$ for __all__ points on L. ☐

What is the geometric meaning for the parameter t? Well, L becomes parameterized by t in that each point of L corresponds to a certain value of t, and each value of t corresponds to a particular point of L. The line L in some sense becomes a copy of the t-axis, as illustrated in the picture to the right.

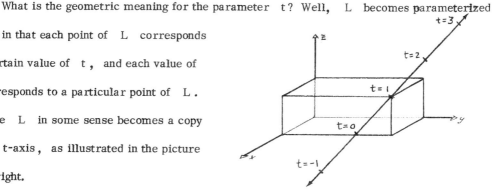

Important observation: The equations for a line L given above are NOT unique, i. e., the same line can be expressed by many different equations corresponding to different parameterizations.

__Example B.__ Consider again the line L passing through the two points $P_0 = (1, 2, 0)$ and $P_1 = (1, 3, 1)$. Clearly the vector

$$\bar{u} = 2(\bar{P}_0 - \bar{P}_1) = \langle 0, -2, -2 \rangle$$

is a direction vector for L, and P_1 is a fixed point point on L. Hence another parametric equation for L is given by

$$\bar{X} = \bar{P}_1 + s\bar{u}, \qquad -\infty < s < \infty$$

i. e.,

$$\bar{X} = \langle 1, 3, 1 \rangle + s \langle 0, -2, -2 \rangle$$

$$-\infty < s < \infty$$

The corresponding set of parametric scalar

equations are

$$\begin{cases} x = 1 \\ y = 3 - 2s \\ z = 1 - 2s \end{cases} \quad -\infty < s < \infty$$

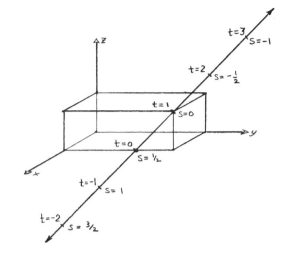

and the symmetric equations are

$$\frac{x - 1}{0} = \frac{y - 3}{-2} = \frac{z - 1}{-2}$$

In the picture to the right you can see how L is parameterized by both t and s, but in different ways. □

As stated in §13.4 and §13.5, it is useful to think of the parameter in a parameterized curve as representing time. Then the parametric equations give the position (as a function of time) of an object moving along the curve.

Example C. Suppose Howard Ant and his wife Pat Ant are running toward each other (i. e. , in opposite directions) along the line L of Examples A and B and that Howard's position at time t is given by the parametric scalar equations of Example A while Pat's position at time s is given by the parametric scalar equations of Example B. If Howard starts at time $t = 0$ and Pat starts at time $s = 0$, then at what point will the Ants collide?

Solution. We have only to find that point $X_0 = (x_0, y_0, z_0)$ for which the two time parameters t and s are equal. (Then Howard and Pat will be in the same place at the same time.) Thus, equating the parametric scalar equations from Examples A and B (using $s = t$) we obtain

$$\begin{cases} 1 = x_0 = 1 \\ 2 + t = y_0 = 3 - 2t \\ t = z_0 = 1 - 2t \end{cases}$$

which quickly yields $t = 1/3$. Substituting this value for t into the parametric equations

yields $\boxed{X_0 = (1, 7/3, 1/3)}$. [This is a reasonable answer given the diagram accompaning

Example B.] $\square$

Anton's Examples 3 - 6 are typical line problems. Here are two more:

Example D. Determine parametric scalar equations for the line L through $(1, -3, 2/3)$ and

parallel to the line L' given by $x = -2 + t$, $y = 1$, and $z = 4 - 2t$.

Solution. By General Principle (L) we need a

point P_0 on L and a direction vector $\bar{v}$.

The point P_0 can be taken to be $(1, -3, 2/3)$,

and since L and L' are parallel, we can use

a direction vector for L' to be a direction vector for L.

A direction vector for L' can be read off from the coefficients of t in the parametric

scalar equations (see J):

$$\begin{cases} x = -2 + t \\ y = 1 \\ z = 4 - 2t \end{cases} \text{gives} \begin{cases} a = 1 \\ b = 0 \\ c = -2 \end{cases} \text{or} \quad \bar{v} = (1, 0, -2)$$

Thus one set of parametric scalar equations for L are

$$\boxed{\begin{aligned} x &= 1 + t \\ y &= -3 \\ z &= 2/3 - 2t \end{aligned}}$$ □

Example E. Let L_1 and L_2 be the lines

$$L_1 : \frac{x-3}{2} = \frac{y+3/2}{-1} = \frac{z-2}{1/4} \quad \text{and} \quad L_2 : \frac{x}{4} = \frac{y+1}{-6} = \frac{z-1}{-2}$$

Do they intersect, and if so, where?

Solution. By (K) we can transform our two sets of symmetric equations into parametric scalar equations:

$$L_1 : \begin{cases} x = 3 + 2t \\ y = -3/2 - t \\ z = 2 + t/4 \end{cases} \qquad L_2 : \begin{cases} x = 4s \\ y = -1 - 6s \\ z = 1 - 2s \end{cases}$$

Notice how we have used _different_ parameters for each line, for if the lines intersect, there is no reason to assume that the parameters will be equal * at the point of intersection $\overline{X}_0 = (x_0, y_0, z_0)$. We now attempt to find values for t and s which give a common point, i. e. ,

$$3 + 2t = x_0 = 4s$$
$$-3/2 - t = y_0 = -1 - 6s$$
$$2 + t/4 = z_0 = 1 - 2s$$

Adding the first equation to twice the second yields $0 = -2 - 8s$, or $s = -1/4$. The first equation then gives $3 + 2t = -1$, or $t = -2$. Since these values for s and t do satisfy the third equation, then the two lines intersect, and do so at the point

$$(x_0, y_0, z_0) = (4s, -1 - 6s, 1 - 2s)$$
$$= (-1, 1/2, 3/2)$$ □

---

* Recall that an intersection point of the two lines need not be a collision point of two objects travelling along the lines. See §13.3.2 of The Companion.

Section 15.3 : Dot Product; Projections.

1. The dot product. Yes, the definition of the dot product of two vectors is a bit strange.

Wouldn't life have been easier if we had just defined the "product" of $\bar{v} = \langle v_1, v_2, v_3 \rangle$ and

$\bar{w} = \langle w_1, w_2, w_3 \rangle$ to be "$\langle v_1 w_1, v_2 w_2, v_3 w_3 \rangle$?" Easier to remember perhaps, but not

useful! The dot product is defined in its strange way for two reasons:

(1) Many quantities in the sciences are most easily described in terms of the

dot product. As an example, suppose a constant force $\bar{F}$ moves an

object from point P to point Q. Then the work done by $\bar{F}$ on the

object is given by

$$W = \bar{F} \cdot \overrightarrow{PQ}$$

(2) Many computations with vectors are most conveniently carried out (and

most easily remembered) in terms of the dot product. One goal of §15.3

is to develop the most basic of these operations.

It is important to distinguish between the defining geometric formula for the dot product,

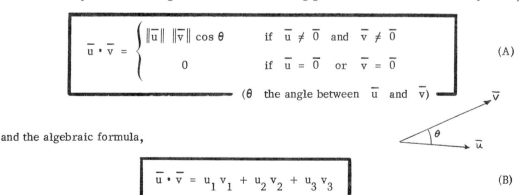

$$\bar{u} \cdot \bar{v} = \begin{cases} \|\bar{u}\| \, \|\bar{v}\| \cos\theta & \text{if } \bar{u} \neq \bar{0} \text{ and } \bar{v} \neq \bar{0} \\ 0 & \text{if } \bar{u} = \bar{0} \text{ or } \bar{v} = \bar{0} \end{cases} \qquad (A)$$

(θ the angle between $\bar{u}$ and $\bar{v}$)

and the algebraic formula,

$$\bar{u} \cdot \bar{v} = u_1 v_1 + u_2 v_2 + u_3 v_3 \qquad (B)$$

The second (very simple) formula is what makes the dot product so useful: we can compute dot products from the vector components using (B) and then (A) allows us to compute lengths of vectors and the angles between them.   Specifically,

vector norms $\|\bar{v}\| = \sqrt{\bar{v} \cdot \bar{v}}$ (C)

angles $\cos \theta = \begin{cases} \dfrac{\bar{u} \cdot \bar{v}}{\|\bar{u}\|\ \|\bar{v}\|} & \text{if } \bar{u} \neq \bar{0} \text{ and } \bar{v} \neq \bar{0} \\[2ex] 0 & \text{if } \bar{u} = \bar{0} \text{ or } \bar{v} = \bar{0} \end{cases}$ (D)

where θ is the angle between $\bar{u}$ and $\bar{v}$.

Example A.   What is the angle between

$$\bar{u} = \langle 4, -2, 5 \rangle \quad \text{and} \quad \bar{v} = \langle \sqrt{2}, 7, -1 \rangle \ ?$$

Solution.

$$\bar{u} \cdot \bar{v} = 4\sqrt{2} - 14 - 5 \cong -13.34315$$

$$\|\bar{u}\| = \sqrt{16 + 4 + 25} \cong 6.70820$$

$$\|\bar{v}\| = \sqrt{2 + 49 + 1} \cong 7.21110$$

Hence

$$\cos \theta \cong -.27584$$

$$\theta \cong 106^{\circ}$$

(Calculators make short work of such problems.)   □

The most important use of the angle formula occurs in determining when two vectors are perpendicular:

two vectors $\bar{u}$ and $\bar{v}$ are <u>perpendicular</u> or <u>orthogonal</u>

if and only if $\bar{u} \cdot \bar{v} = 0$ (E)

Compare this with (H) from §15.2 of the Companion:

> two vectors $\bar{u}$ and $\bar{v}$ are parallel if and
>
> only if one is a scalar multiple of the other.

Notice how two important geometric relationships between vectors have been translated into simple algebraic relationships.

As with vector addition and scalar multiplication, the basic algebraic properties of the dot product must become second nature to you. These properties are nicely summarized in Theorem 15.3.4. Fortunately there are no surprises, for in general any equation which is true for ordinary multiplication is also true for the dot product, so long as all the expressions involved make sense! For example, "$\bar{u} \cdot (\bar{v} \cdot \bar{w}) = (\bar{u} \cdot \bar{v}) \cdot \bar{w}$" is nonsense since $\bar{v} \cdot \bar{w}$ is a scalar quantity, and therefore cannot be "dotted" with vector $\bar{u}$.

> Remember: The dot product of two vectors
>
> gives you a scalar, not a vector.

As you can see, much geometry has been translated into algebra via the dot product; this makes the solution of some otherwise difficult geometry problems quite easy. Anton's Example 3 is one illustration of this principle, and we now give two others:

Example B. Find two unit vectors that make an angle of 60° with $\bar{v} = 4\bar{i} + 3\bar{j}$.

Solution. We are looking for values of x and y such that the vector $\bar{u} = x\bar{i} + y\bar{j}$ has the following properties:

(i) $\bar{u}$ is a unit vector, i.e.,

$$1 = \|\bar{u}\|$$

$$1 = \|\bar{u}\|^2 = \bar{u} \cdot \bar{u} \qquad \text{by (C)}$$

$$\boxed{1 = x^2 + y^2} \tag{F}$$

(ii)   the angle between $\bar{u}$ and $\bar{v}$ is 60°, i.e.,

$$\frac{\bar{u} \cdot \bar{v}}{\|\bar{u}\| \, \|\bar{v}\|} = \cos 60^\circ \qquad \text{by (D)}$$

$$\bar{u} \cdot \bar{v} = (1/2) \|\bar{u}\| \, \|\bar{v}\| \qquad \text{since} \quad \cos 60^\circ = 1/2$$

$$4x + 3y = (1/2) \sqrt{x^2 + y^2} \, \sqrt{4^2 + 3^2}$$

$$\boxed{4x + 3y = 5/2} \qquad \text{by (F)} \tag{G}$$

We thus have only to solve the simultaneous equations (F) and (G). Solving (G) for   y   in terms of   x   yields

$$y = 5/6 - (4/3)x \tag{H}$$

Plugging this into (F) and simplifying yields

$$100 x^2 - 80x - 11 = 0$$

which can be solved via the quadratic formula (Appendix D) to give

$$x = 2/5 \pm \sqrt{27}/10$$

$$\cong .92 , -.12$$

From (H) we get $\qquad\qquad y = 3/10 \mp 2\sqrt{27}/15$

$$\cong -.39 , .99$$

so our two vectors are $\qquad\qquad \boxed{.92\,\bar{i} - .39\,\bar{j} , \; -.12\,\bar{i} + .99\,\bar{j}}$ $\qquad\qquad$ □

<u>Example C.</u> Let $\bar{u}$ and $\bar{v}$ be nonzero vectors. Determine a vector $\bar{w}$ which bisects the

angle between $\bar{u}$ and $\bar{v}$, i.e., angle between

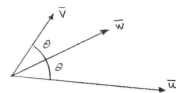

$\bar{u}$ and $\bar{w}$

     = the angle between $\bar{w}$ and $\bar{v}$.

[Hint: Assume $\bar{w}$ is of the form $\bar{w} = \ell\bar{u} + k\bar{v}$, for constants ℓ and k which must

     be determined.]

<u>Solution.</u> Let θ = angle between $\bar{w}$ and $\bar{v}$, so that

                     θ = angle between $\bar{u}$ and $\bar{w}$

From (D) we then can compute the cosine of θ in two ways:

$$\cos\theta = \frac{\bar{w}\cdot\bar{v}}{\|\bar{w}\|\ \|\bar{v}\|} = \frac{(\ell\bar{u}+k\bar{v})\cdot\bar{v}}{\|\bar{w}\|\ \|\bar{v}\|} = \frac{\ell\bar{u}\cdot\bar{v}+k\|\bar{v}\|^2}{\|\bar{w}\|\ \|\bar{v}\|}$$

and

$$\cos\theta = \frac{\bar{u}\cdot\bar{w}}{\|\bar{u}\|\ \|\bar{w}\|} = \frac{\bar{u}\cdot(\ell\bar{u}+k\bar{v})}{\|\bar{u}\|\ \|\bar{w}\|} = \frac{\ell\|\bar{u}\|^2+k\bar{u}\cdot\bar{v}}{\|\bar{u}\|\ \|\bar{w}\|}$$

Equating these two expressions for $\cos\theta$ yields

$$\frac{\ell\bar{u}\cdot\bar{v}+k\|\bar{v}\|^2}{\|\bar{w}\|\ \|\bar{v}\|} = \frac{\ell\|\bar{u}\|^2+k\bar{u}\cdot\bar{v}}{\|\bar{u}\|\ \|\bar{w}\|}$$

Cancelling $\|\bar{w}\|$ from both sides and cross multiplying gives

$$\ell\|\bar{u}\|\ \bar{u}\cdot\bar{v} + k\|\bar{u}\|\ \|\bar{v}\|^2 = \ell\|\bar{v}\|\ \|\bar{u}\|^2 + k\|\bar{v}\|\ \bar{u}\cdot\bar{v}$$

Separating the ℓ and the k terms yields

$$\ell(\|\bar{u}\|\ \bar{u}\cdot\bar{v} - \|\bar{v}\|\ \|\bar{u}\|^2) = k(\|\bar{v}\|\ \bar{u}\cdot\bar{v} - \|\bar{u}\|\ \|\bar{v}\|^2)$$

$$\frac{\ell}{k} = \frac{\|\bar{v}\|\ \bar{u}\cdot\bar{v} - \|\bar{u}\|\ \|\bar{v}\|^2}{\|\bar{u}\|\ \bar{u}\cdot\bar{v} - \|\bar{v}\|\ \|\bar{u}\|^2}$$

$$= \frac{\|\bar{v}\|\ (\bar{u}\cdot\bar{v} - \|\bar{u}\|\ \|\bar{v}\|)}{\|\bar{u}\|\ (\bar{u}\cdot\bar{v} - \|\bar{v}\|\ \|\bar{u}\|)} = \frac{\|\bar{v}\|}{\|\bar{u}\|}$$

Thus ℓ and k can be any numbers which are in the proportion $\|\bar{v}\| : \|\bar{u}\|$. In particular we could take $\ell = \|\bar{v}\|$ and $k = \|\bar{u}\|$ so that

$$\boxed{\bar{w} = \|\bar{v}\|\ \bar{u} + \|\bar{u}\|\ \bar{v}}$$
□

2. **Projections.** Although it may seem pretty esoteric at first, the decomposition of a vector $\bar{u}$ into a sum

$$\boxed{\bar{u} = \bar{w}_1 + \bar{w}_2}$$

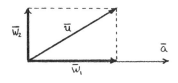

where $\bar{w}_1$ is <u>parallel</u> to a specified nonzero vector $\bar{a}$ and $\bar{w}_2$ is <u>perpendicular</u> to $\bar{a}$, is an important operation. You have already encountered one such decomposition in §14.5 on motion in a plane. There the acceleration vector of a particle moving along a plane curve was decomposed into a sum of two pieces (Theorem 14.5.3), the piece parallel to the curve's tangent vector, and the other perpendicular to the tangent vector. A similar procedure was employed in <u>The Companion</u>'s derivation of Kepler's Laws in §14.6. Such decomposions are essential in many applications of multidimensional calculus.

The vector $\bar{w}_1$ (the piece of $\bar{u}$ which is parallel to $\bar{a}$) is called the <u>orthogonal projection of $\bar{u}$ on $\bar{a}$</u>, and is denoted by

$$\overline{w}_1 = \text{proj}_{\overline{a}} \, \overline{u}$$

In Theorem 15.3.5 the following important formula is given:

Formula for
orthogonal
projection

$$\text{proj}_{\overline{a}} \, \overline{u} = \left(\frac{\overline{u} \cdot \overline{a}}{\|\overline{a}\|^2} \right) \overline{a} \qquad \qquad \text{(I)}$$

Unfortunately the proof given for (I) is fairly abstract, and thus does not help in clarifying this rather mysterious looking formula. There is, however, a more intuitive and geometric derivation for (I), which aids in both understanding and remembering the formula. It is based on the simple fact that the triangle formed by the vectors $\overline{u}$ and $\text{proj}_{\overline{a}} \, \overline{u}$ is a <u>right</u> triangle (see the

accompaning diagram). From this we can compute the <u>magnitude</u> of the projection:

$$\text{From} \quad \cos\theta = \frac{\text{adjacent}}{\text{hypotenuse}} = \frac{\|\text{proj}_{\overline{a}} \, \overline{u}\|}{\|\overline{u}\|} \quad \text{we obtain}$$

$$\|\text{proj}_{\overline{a}} \, \overline{u}\| = \|\overline{u}\| \cos\theta = \frac{\|\overline{u}\| \, \|\overline{a}\| \cos\theta}{\|\overline{a}\|} = \frac{\overline{u} \cdot \overline{a}}{\|\overline{a}\|}$$

multiplication by $\|\overline{a}\| / \|a\|$ \qquad\qquad definition of $\overline{u} \cdot \overline{a}$

This proves

magnitude
of projection

$$\|\text{proj}_{\overline{a}} \, \overline{u}\| = \|\overline{u}\| \cos\theta = \frac{\overline{u} \cdot \overline{a}}{\|\overline{a}\|} \qquad\qquad \text{(J)}$$

(which happen to be Anton's formulas (9) and (10)).

The <u>direction</u> of the projection can be specified by the <u>unit</u> vector

unit direction
of projection

$$\overline{h} = \frac{\overline{a}}{\|\overline{a}\|} \qquad \text{(see (F) in §15.2)}$$

Hence, by (G) in §15.2, the projection is the scalar multiplication of its unit direction by its magnitude, i.e.,

$$\text{proj}_{\bar{a}}\,\bar{u} = \|\text{proj}_{\bar{a}}\,\bar{u}\|\,\bar{h} = \left(\frac{\bar{u}\cdot\bar{a}}{\|\bar{a}\|^2}\right)\bar{a}\ ,$$

which verifies (I) as desired.

Theorem 15.3.5, in addition to giving a formula for $\text{proj}_{\bar{a}}\,\bar{u}$ (the vector component of $\bar{u}$ __along__ $\bar{a}$), also gives a formula for $\bar{w}_2$, the vector component of $\bar{u}$ __orthogonal__ to $\bar{a}$. There is no need to memorize this! Simply remember

$$\bar{w}_2 = \bar{u} - \text{proj}_{\bar{a}}\,\bar{u}$$

along with the formula for $\text{proj}_{\bar{a}}\,\bar{u}$.

Anton's Examples 7 and 8 illustrate the use of the projection formulas. Here is another example:

__Example D.__ Express the vector $\bar{u} = 8\,\bar{i} + \bar{j} - \bar{k}$ as a sum $\bar{u} = \bar{w}_1 + \bar{w}_2$, where $\bar{w}_1$ is parallel to the vector $\bar{a} = 3\,\bar{i} - 2\,\bar{k}$ and $\bar{w}_2$ is perpendicular to $\bar{a}$.

__Solution.__ The vector $\bar{w}_1$ is the orthogonal projection of $\bar{u}$ on $\bar{a}$, so

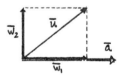

$$\bar{w}_1 = \text{proj}_{\bar{a}}\,\bar{u} = \left(\frac{\bar{u}\cdot\bar{a}}{\|\bar{a}\|^2}\right)\bar{a}$$

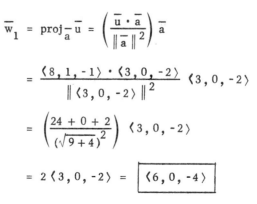

$$= \frac{\langle 8,1,-1\rangle\cdot\langle 3,0,-2\rangle}{\|\langle 3,0,-2\rangle\|^2}\,\langle 3,0,-2\rangle$$

$$= \left(\frac{24+0+2}{(\sqrt{9+4})^2}\right)\langle 3,0,-2\rangle$$

$$= 2\,\langle 3,0,-2\rangle = \boxed{\langle 6,0,-4\rangle}$$

$$\overline{w}_2 = \overline{u} - \overline{w}_1 = \langle 8, 1, -1 \rangle - \langle 6, 0, -4 \rangle$$

$$= \boxed{\langle 2, 1, 3 \rangle}$$

Check:

$$\overline{w}_1 + \overline{w}_2 = \langle 6, 0, -4 \rangle + \langle 2, 1, 3 \rangle$$

$$= \langle 8, 1, -1 \rangle = \overline{u}$$

Thus $\overline{u} = \overline{w}_1 + \overline{w}_2$, as desired.

$$\overline{w}_1 = \langle 6, 0, -4 \rangle = 2 \langle 3, 0, -2 \rangle = 2\overline{a}$$

Thus $\overline{w}_1$ is parallel to $\overline{a}$, as desired.

$$\overline{w}_2 \cdot \overline{a} = \langle 2, 1, 3 \rangle \cdot \langle 3, 0, -2 \rangle = 6 + 0 - 6 = 0$$

Thus $\overline{w}_2$ is perpendicular to $\overline{a}$, as desired. $\square$

3. Work. Suppose we have a constant force F acting on an object and moving it in a straight line path a distance d. If the force is <u>acting in the direction of motion,</u> then W the work done by the force, is given by

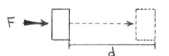

$$W = Fd$$

i. e., work equals force times distance. Now suppose that the force is NOT acting in the direction of motion. For example:

Suppose Atlas Anton is pushing a hundred ton concrete block from point P to point Q by exerting a constant force $\overline{F}$ with an angle θ to the direction of motion. How much work is done by Atlas (i. e., by the force he exerts)?

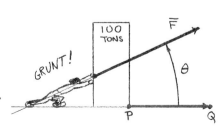

Unfortunately Atlas has a basic law of experimental physics making things harder for him. From experimentation it has been found that the only effective part of a force (i. e. , the only part which does any work) is the component of force along the direction of motion, i. e. ,

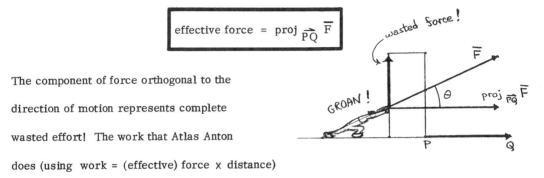

$$\text{effective force} = \text{proj}_{\overrightarrow{PQ}} \overline{F}$$

The component of force orthogonal to the direction of motion represents complete wasted effort! The work that Atlas Anton does (using work = (effective) force × distance) is therefore

$$W = \|\text{proj}_{\overrightarrow{PQ}} \overline{F}\| \, \|\overrightarrow{PQ}\|$$

$$= (\|\overline{F}\| \cos \theta) \, \|\overrightarrow{PQ}\| \qquad \text{by (J)}$$

This yields

$$\boxed{W = (\|\overline{F}\| \cos \theta) \, \|\overrightarrow{PQ}\| = F \cdot \overrightarrow{PQ}}$$

which is Anton's definition (13).

Example E.   Joanne Ant runs the 200 yard dash into a

side wind coming at her from 30^0 from her

direction of sprint. If the wind exerts 20 pounds

of force on Joanne, how much work does the wind do?

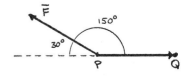

Solution.   The distance vector $\overrightarrow{PQ}$ and the force vector

$\overline{F}$ are related as shown to the right. Thus the angle

between them is $\theta = 180^0 - 30^0 = 150^0$. The work

done by the wind is therefore

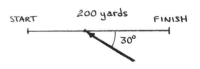

$$W = \|\overline{F}\| \, \|\overrightarrow{PQ}\| \cos \theta$$

$$= (20 \text{ lbs}) (600 \text{ feet}) \cos 150^{\circ}$$

$$= (12000 \text{ ft-lbs}) (- \sqrt{3}/2) = \boxed{- 6000 \sqrt{3} \text{ ft-lbs}} \qquad \square$$

Do not be confused by the negative sign in this answer. The work done by the wind is negative $6000 \sqrt{3}$ ft-lbs, which means the work done against the wind is (positive) $6000 \sqrt{3}$ ft-lbs. Thus Joanne, in running into the wind, must exert an additional $6000 \sqrt{3}$ ft-lbs of work.

Section 15.4.   Cross Product.

All the previously derived results hold equally well in $\mathbb{R}^2$ and $\mathbb{R}^3$ (and, in fact, in "any dimensional" space). However, we now introduce another vector product which exists only for vectors in $\mathbb{R}^3$. This cross product is of crucial importance in the sciences as well as in our development of three dimensional vector algebra.

1.   **Determinants.**   Anton first introduces the   2 x 2 determinant

$$\begin{vmatrix} a & b \\ c & d \end{vmatrix} = ad - bc$$

i.e., the product of the terms along the major diagonal (⬊)   minus the product of the term along the minor diagaonal (⬋).

A   3 x 3 determinant   is defined from   2 x 2 determinants   by the formula

$$\begin{vmatrix} a_1 & a_2 & a_3 \\ b_1 & b_2 & b_3 \\ c_1 & c_2 & c_3 \end{vmatrix} = a_1 \begin{vmatrix} & & \\ & b_2 & b_3 \\ & c_2 & c_3 \end{vmatrix} - a_2 \begin{vmatrix} b_1 & & b_3 \\ c_1 & & c_3 \end{vmatrix} + a_3 \begin{vmatrix} b_1 & b_2 & \\ c_1 & c_2 & \end{vmatrix}$$

$$= a_1(b_2 c_3 - b_3 c_2) - a_2(b_1 c_3 - b_3 c_1) + a_3(b_1 c_2 - b_2 c_1)$$

$$= a_1 b_2 c_3 - a_1 b_3 c_2 - a_2 b_1 c_3 + a_2 b_3 c_1 + a_3 b_1 c_2 - a_3 b_2 c_1 \quad .$$

We have written the first formula in as graphic a way as possible:

> each element a_j in the first row is multiplied by the 2×2 determinant
>
> which remains after both the row and column containing a_j have been eliminated.
>
> These terms are added (with the a_2 term receiving a minus sign) to obtain the
>
> full determinant.

This is how you should remember and compute 3×3 determinants.

Example A.
$$\begin{vmatrix} 1 & -2 & -1 \\ 0 & 2 & 3 \\ -1 & 0 & 4 \end{vmatrix} = 1 \begin{vmatrix} 2 & 3 \\ 0 & 4 \end{vmatrix} - (-2) \begin{vmatrix} 0 & 3 \\ -1 & 4 \end{vmatrix} + (-1) \begin{vmatrix} 0 & 2 \\ -1 & 0 \end{vmatrix}$$

$$= (8 - 0) + 2(0 + 3) - (0 + 2)$$

$$= 8 + 6 - 2 = \boxed{12} \qquad \qquad \square$$

Warning: A very common mistake made when computing determinants is forgetting the minus sign in front of the a_2 term. Beware!

The importance of the determinant at this point is as a means of defining the cross product of two vectors. However, after introducing the determinant, we would be somewhat remiss if we did not at least state the most important geometric properties of determinants:

(i)   The <u>area of the parallelogram</u>

formed by the two vectors

$\bar{a} = \langle a_1, a_2 \rangle$ and $\bar{b} = \langle b_1, b_2 \rangle$

is given by the absolute

value of $\begin{vmatrix} a_1 & a_2 \\ b_1 & b_2 \end{vmatrix}$

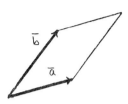

(ii)   The <u>volume of the three dimensional</u>

<u>parallelepiped</u> formed by the three

vectors $\bar{a} = \langle a_1, a_2, a_3 \rangle$,

$\bar{b} = \langle b_1, b_2, b_3 \rangle$ and

$\bar{c} = \langle c_1, c_2, c_3 \rangle$ is given by the

absolute value of

$$\begin{vmatrix} a_1 & a_2 & a_3 \\ b_1 & b_2 & b_3 \\ c_1 & c_2 & c_2 \end{vmatrix}$$

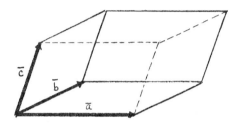

These two facts are crucial for the theory of integration in $\mathbb{R}^2$ and $\mathbb{R}^3$.   Both are conse-·

quences of the cross product properties which Anton develops in this section.

2.   <u>The cross product: built for utility, not beauty.</u>   The determinant notation now allows for an

easily remembered definition for the cross product of two vectors in $\mathbb{R}^3$:

<u>Definition.</u>   Suppose $\bar{u} = \langle u_1, u_2, u_3 \rangle$ and $\bar{v} = \langle v_1, v_2, v_3 \rangle$.   Then the cross product of $\bar{u}$

with $\bar{v}$ is defined by

$$\overline{u} \times \overline{v} = \begin{vmatrix} \overline{i} & \overline{j} & \overline{k} \\ u_1 & u_2 & u_3 \\ v_1 & v_2 & v_3 \end{vmatrix}$$

i. e. ,

$$\overline{u} \times \overline{v} = \overline{i} \begin{vmatrix} u_2 & u_3 \\ v_2 & v_3 \end{vmatrix} - \overline{j} \begin{vmatrix} u_1 & u_3 \\ v_1 & v_3 \end{vmatrix} + \overline{k} \begin{vmatrix} u_1 & u_2 \\ v_1 & v_2 \end{vmatrix}$$

$$= (u_2 v_3 - u_3 v_2)\overline{i} - (u_1 v_3 - u_3 v_1)\overline{j} + (u_1 v_2 - u_2 v_1)\overline{k}$$

$$= \langle u_2 v_3 - u_3 v_2 , u_3 v_1 - u_1 v_3 , u_1 v_2 - u_2 v_1 \rangle$$

The use of the determinant notation in the definition is just a handy way to remember the cross product formula. It is not really a determinant since the first row consists of __vector__ entries -- the $\overline{i}, \overline{j}$ and $\overline{k}$ -- not __scalar__ entries.

__Example B.__   Evaluate $\langle 2, 1, 0 \rangle \times \langle -1, 1, 3 \rangle$.

__Solution.__

$$\langle 2, 1, 0 \rangle \times \langle -1, 1, 3 \rangle = \begin{vmatrix} \overline{i} & \overline{j} & \overline{k} \\ 2 & 1 & 0 \\ -1 & 1 & 3 \end{vmatrix} = \overline{i} \begin{vmatrix} 1 & 0 \\ 1 & 3 \end{vmatrix} - \overline{j} \begin{vmatrix} 2 & 0 \\ -1 & 3 \end{vmatrix} + \overline{k} \begin{vmatrix} 2 & 1 \\ -1 & 1 \end{vmatrix}$$

$$= 3\overline{i} - (6)\overline{j} + (2 - (-1))\overline{k} = 3\overline{i} - 6\overline{j} + 3\overline{k}$$

$$= \langle 3, -6, 3 \rangle \qquad\qquad \square$$

After seeing this definition of cross product, many of you are probably ready to drop calculus! Don't be hasty. The definition of $\overline{u} \times \overline{v}$ does seem quite weird, but in fact it has some very nice geometric properties which make it __extremely__ important in three dimensional calculus. We summarize the most basic of these properties as

The Cross Product Theorem.

(1) $\bar{u} \times \bar{v}$ is <u>perpendicular</u> to both $\bar{u}$ and $\bar{v}$, and is oriented so that

$\bar{u}, \bar{v}, \bar{u} \times \bar{v}$ obey the <u>right-hand rule</u>,

i.e., $\bar{u}, \bar{v}, \bar{u} \times \bar{v}$ are oriented in

the same way as the x, y, z axes.

(2) Let θ be the angle between $\bar{u}$ and $\bar{v}$. Then the <u>norm</u> of $\bar{u} \times \bar{v}$ is

given by

$$\|\bar{u} \times \bar{v}\| = \|\bar{u}\| \, \|\bar{v}\| \sin \theta$$

= the area of the

parallelogram

formed by $\bar{u}$ and $\bar{v}$.

area is
$\|\bar{u}\| \|\bar{v}\| \sin \theta$
$= \|\bar{u} \times \bar{v}\|$

Anton discusses all of these results following his description of the cross product properties of

the standard unit vectors $\bar{i}, \bar{j}$ and $\bar{k}$. Part (1) is essentially Anton's Theorem 15.4.2,

while part (2) is his formula (7).

<u>We cannot overemphasize the importance of the Cross Product Theorem.</u> The first aspect

you should note is that the theorem gives geometric meaning to both the <u>direction</u> and <u>length</u> of

the cross product. In fact, you can think of the Theorem as being simply an alternate geometric

definition for the cross-product (a definition which Anton emphasizes is <u>coordinate free</u> *). The

importance of the cross product lies in the beautiful interplay between the algebraic and geometric

definitions:

---

* The importance of having <u>coordinate free definitions</u> for quantities in $\mathbb{R}^2$ and $\mathbb{R}^3$ can be
difficult to understand at this stage. Suffice it to say that, with rare exceptions, a mathematical
quantity whose definition is not coordinate free cannot be measuring an actual physical property
of the real world.

    i.   the geometric definition of $\overline{u} \times \overline{v}$ (i.e., the Cross Product Theorem) tells us

          that quantities of importance are being measured (more on this shortly...), while

    ii.   the algebraic definition of $\overline{u} \times \overline{v}$ allows us to calculate $\overline{u} \times \overline{v}$ very easily.

These aspects of the cross product now allow us to perform easily the following important

geometric calculations:

Given any two (non-parallel) vectors $\overline{u}$ and $\overline{v}$ in 3-space, we are able to

determine

(1) a non-zero vector orthogonal to both $\overline{u}$ and $\overline{v}$ (this vector is $\overline{u} \times \overline{v}$)

(2) the area of the parallelogram formed by $\overline{u}$ and $\overline{v}$ (it is $\| \overline{u} \times \overline{v} \|$)

The first of these operations is especially crucial to our later work! The important step in

many three dimensional problems is the computation of a non-zero vector which is orthogonal

to two given vectors. The condition that $\overline{u}$ and $\overline{v}$ be non-parallel is necessary for the

following reason: if $\overline{u}$ and $\overline{v}$ are parallel, then $\overline{u} \times \overline{v} = 0$ (Theorem 15.4.3), and thus

we would not have a non-zero orthogonal vector. Also, there is no parallelogram formed by

$\overline{u}$ and $\overline{v}$ if $\overline{u}$ and $\overline{v}$ are parallel!

3. Algebraic properties of the cross product. You should become comfortable with the cross

product relationships between the standard unit vectors $\overline{i}, \overline{j}$ and $\overline{k}$.

Example C. Compute $\overline{i} \times \overline{j}$.

Solution.

$$\bar{i} \times \bar{j} = \begin{vmatrix} \bar{i} & \bar{j} & \bar{k} \\ 1 & 0 & 0 \\ 0 & 1 & 0 \end{vmatrix} = \bar{i} \begin{vmatrix} 0 & 0 \\ 1 & 0 \end{vmatrix} - \bar{j} \begin{vmatrix} 1 & 0 \\ 0 & 0 \end{vmatrix} + \bar{k} \begin{vmatrix} 1 & 0 \\ 0 & 1 \end{vmatrix}$$

$$= 0\bar{i} - 0\bar{j} + 1\bar{k}$$

$$= \bar{k}$$

Thus $\qquad\qquad \bar{i} \times \bar{j} = \bar{k}$ □

Similarly $\qquad\qquad \bar{j} \times \bar{k} = \bar{i}$

$$\bar{k} \times \bar{i} = \bar{j}$$

These formulas can be remembered by using a little triangle:

The cross product of any two of the vectors (taken in the "forward" direction, i. e. , clockwise) is just the third vector in the triangle. Anton discusses these relationships following Theorem 15. 4. 4 .

The general algebraic properties of the cross product are summarized in Anton's Theorem 15. 4. 4. It is useful, however, to divide up these properties into the "expected" and the "surprising. " In the "expected" category the cross product behaves quite well with vector addition and scalar multiplication:

(b) $\quad \bar{u} \times (\bar{v} + \bar{w}) = (\bar{u} \times \bar{v}) + (\bar{u} \times \bar{w})$

(c) $\quad (\bar{u} + \bar{v}) \times \bar{w} = (\bar{u} \times \bar{w}) + (\bar{v} \times \bar{w})$

(d) $\quad \bar{u} \times (k\bar{v}) \quad = (k\bar{u}) \times \bar{v} = k(\bar{u} \times \bar{v})$

Also expected would be the "cross product with zero" rule:

(e) $\bar{u} \times \bar{0} = \bar{0} \times \bar{u} = \bar{0}$

As can be seen, there are "no surprises" on this list. However, the cross product does NOT

have the "expected properties" of multiplication when considered in relation to itself:

(a) $\boxed{\bar{u} \times \bar{v} = -\bar{v} \times \bar{u}}$, i.e., the cross product is <u>anti</u>-commutative, not commutative.

Example: $\bar{k} \times \bar{j} = -(\bar{j} \times \bar{k}) = -\bar{i}$

(f) $\boxed{\bar{u} \times \bar{u} = \bar{0}}$, i.e., the cross product of any vector with itself is the <u>zero</u> vector.

Also: $\boxed{\bar{u} \times (\bar{v} \times \bar{w}) \text{ is NOT equal to } (\bar{u} \times \bar{v}) \times \bar{w}}$, i.e., the cross product is NOT

associative.

Example: $\bar{i} \times (\bar{j} \times \bar{i}) = \bar{i} \times \bar{0} = \bar{0}$

$(\bar{i} \times \bar{j}) \times \bar{j} = \bar{k} \times \bar{j} = -\bar{i}$

It is important to remember these odd properties! If you don't, you will make some very serious

errors when dealing with the cross product.

4.   <u>The triple scalar product</u> of three vectors $\bar{a}, \bar{b}$ and $\bar{c}$ is defined to be $\bar{a} \cdot (\bar{b} \times \bar{c})$.

There are four comments which we will make about this odd-looking operation:

a)   The product $\bar{a} \cdot (\bar{b} \times \bar{c})$ is very easily set up and computed as a

3 x 3 determinant:

$$\boxed{\bar{a} \cdot (\bar{b} \times \bar{c}) = \begin{vmatrix} a_1 & a_2 & a_3 \\ b_1 & b_2 & b_3 \\ c_1 & c_2 & c_3 \end{vmatrix}}$$ (Anton's (8))

b)   The algebraic rules for the cross product and the dot product can be combined to give algebraic rules for the triple scalar product. However, one important rule which is not so easily derived is

$$\overline{a} \cdot (\overline{b} \times \overline{c}) = \overline{b} \cdot (\overline{c} \times \overline{a}) = \overline{c} \cdot (\overline{a} \times \overline{b})$$ (Anton's (9))

This is verified by using non-trivial properties of the determinant.

c)   While $\|\overline{u} \times \overline{v}\|$ gives the area of the parallelogram formed by $\overline{u}$ and $\overline{v}$, the corresponding result for volume is:

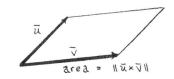

> $|\overline{a} \cdot (\overline{b} \times \overline{c})|$ is the volume
> of the parallelepiped
> formed by the three vectors
> $\overline{a}$, $\overline{b}$ and $\overline{c}$

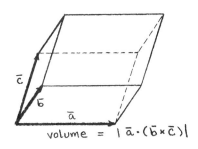

Anton proves this result right after his statement of Theorem 15.4.5. This volume result is the most important property of the triple scalar product, and is the main reason we are interested in this product. The volume result is very important in three dimensional integration theory.

d)   Two vectors $\overline{a}$ and $\overline{b}$ which have the same initial point are easily seen to be collinear (i.e., lie along the same line) if and only if they are parallel. Thus since $\overline{a}$ and $\overline{b}$ are parallel if and only if $\overline{a} \times \overline{b} = \overline{0}$ (Theorem 15.4.3) we obtain

> if $\overline{a}$ and $\overline{b}$ are vectors with the same initial point, then
> $\overline{a}$ and $\overline{b}$ are collinear if and only if $\overline{a} \times \overline{b} = \overline{0}$

Now consider whether three vectors in 3-space with the same initial point are <u>coplanar</u> (i. e. , lie on the same plane). By Theorem 15.4.6 we have

> if $\bar{a}$, $\bar{b}$ and $\bar{c}$ are vectors with the same initial point, then
>
> $\bar{a}$, $\bar{b}$ and $\bar{c}$ are coplanar if and only if $\bar{a} \cdot (\bar{b} \times \bar{c}) = 0$

If you think about it, this is just a simple corollary of the result in (c) concerning the volume of parallelepipeds.

5. <u>On real word applications.</u> There are many important scientific applications of the cross product, especially in physics. For instance, the cross product is useful in describing <u>magnetic force</u>: if a test charge q is travelling through a <u>magnetic field</u> $\bar{B}$ with velocity $\bar{v}$, then it can be shown that the <u>magnetic force</u> on the particle is

$$\bar{F} = q(\bar{v} \times \bar{B})$$

The information contained in this equation would be very cumbersome to list were it not for the cross product!

Section 15.5:   Curves in 3-Space.

1.   The more things change, the more things stay the same.   There is really little new material

to learn in this section since the calculus of curves in   3-space   differs only slightly from that

of curves in   2-space.   All of the elementary formulas differ only in one respect: the formulas

in two space involve two components,

$$\overline{r}(t) \;=\; x(t)\,\overline{i} + y(t)\,\overline{j}$$

while the formulas in   3-space   involve three components,

$$\overline{r}(t) \;=\; x(t)\,\overline{i} + y(t)\,\overline{j} + z(t)\,\overline{k}$$

For example,

differentiation in   2-space : $\overline{r}'(t) \;=\; x'(t)\,\overline{i} + y'(t)\,\overline{j}$

differentiation in   3-space : $\overline{r}'(t) \;=\; x'(t)\,\overline{i} + y'(t)\,\overline{j} + z'(t)\,\overline{k}$

arclength in   2-space : $L \;=\; \displaystyle\int_a^b \|\overline{r}'(t)\|\,dt \;=\; \int_a^b \sqrt{\left(\dfrac{dx}{dt}\right)^2 + \left(\dfrac{dy}{dt}\right)^2}\;dt$

arclength in   3-space : $L \;=\; \displaystyle\int_a^b \|\overline{r}'(t)\|\,dt \;=\; \int_a^b \sqrt{\left(\dfrac{dx}{dt}\right)^2 + \left(\dfrac{dy}{dt}\right)^2 + \left(\dfrac{dz}{dt}\right)^2}\;dt$

Almost all of the basic results and techniques of calculus in   2-space   carry over into

3-space.   In particular:

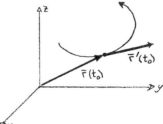

1.   The derivative vector $\overline{r}'(t_0)$ is tangent to the

graph of $\overline{r}(t)$ at the point $\overline{r}(t_0)$.

2.   All the rules of differentiation given

in Theorems 14.2.4 and 14.2.5 for

2-space   remain valid for   3-space .

3. Parametrization of a curve by arclength is done in precisely the same manner in
    3-space as in 2-space; it would be wise to reread the detailed three step process
    following Example 2 in §14.3. In particular, the crucial formula

$$ds/dt \; = \; \| \bar{r}'(t) \|$$ (A)

    remains valid in 3-space.

4. Motion problems in 3-space are handled in precisely the same way as in 2-space.
    Thus the 5-Box Diagram from §14.5 of The Companion should remain your guide
    for solving motion problems.

2. Product rules in 3-space. Let's concentrate on the vector calculus aspects which are not the
    same in 3-space as in 2-space. In §§15.3 and 15.4 we introduced two "products" in
    3-space: the dot product and the cross product. Under differentiation both of these exhibit
    the usual product rule:

$$\frac{d}{dt} \, [\, \bar{r}_1(t) \cdot \bar{r}_2(t) \,] \; = \; \bar{r}_1(t) \cdot \frac{d\bar{r}_2}{dt} + \frac{d\bar{r}_1}{dt} \cdot \bar{r}_2(t)$$ (B)

$$\frac{d}{dt} \, [\, \bar{r}_1(t) \times \bar{r}_2(t) \,] \; = \; \bar{r}_1(t) \times \frac{d\bar{r}_2}{dt} + \frac{d\bar{r}_1}{dt} \times \bar{r}_2(t)$$ (C)

Both of these formulas are verified by writing out each side in component form, and applying the
ordinary product rule to each component. As Anton remarks, the order of the terms in the
cross product formula (C) is crucial. This is because the cross product is NOT commutative;
in fact, it is anti-commutative: $\bar{u} \times \bar{w} = - \bar{w} \times \bar{u}$.

Example A. Suppose $\| \bar{r}(t) \|$ is a constant for all values of t. Prove that whenever $\bar{r}'(t)$
exists, it is perpendicular to $\bar{r}(t)$.

Solution. By Theorem 15.3.2, to prove that $\bar{r}'(t)$ and $\bar{r}(t)$ are perpendicular we have only to verify

$$\bar{r}(t) \cdot \bar{r}'(t) = 0$$

This we can do as follows: Since $\|\bar{r}(t)\|^2$ is a constant (the square of a constant remains a constant!), then its derivative must be zero:

$$0 = \frac{d}{dt} [\|\bar{r}(t)\|^2]$$

$$= \frac{d}{dt} [\bar{r}(t) \cdot \bar{r}(t)] \qquad \text{since} \quad \|\bar{v}\|^2 = \bar{v} \cdot \bar{v} \text{ by (5) §15.3.}$$

$$= \bar{r}(t) \cdot \bar{r}'(t) + \bar{r}'(t) \cdot \bar{r}(t) \qquad \text{by the product rule (B)}$$

$$= 2\,\bar{r}(t) \cdot \bar{r}'(t) \qquad \text{since the dot product is commutative.}$$

Hence $\bar{r}(t) \cdot \bar{r}'(t) = 0$ as desired. $\qquad\qquad\qquad\qquad\qquad \square$

There is a geometric interpretation of this result. If $\|\bar{r}(t)\|$ is a constant c, then the terminal point of the position vector $\bar{r}(t)$ always lies on the sphere of radius c with center at the origin of $\mathbb{R}^3$, i.e., the curve parametrized by $\bar{r}(t)$ lies on the surface of this sphere. It should be clear geometrically that the radius vector $\bar{r}(t)$ must be perpendicular to the corresponding tangent vector $\bar{r}'(t)$, and thus

$$\bar{r}(t) \cdot \bar{r}'(t) = 0$$

This is the geometric justification for the result in Example A.

Citing the product rule for the dot product (B) as a difference between calculus in 3-space and calculus in 2-space is really not accurate, for there is a perfectly valid dot product in 2-space given by

$$\overline{u} \cdot \overline{v} = u_1 v_1 + u_2 v_2$$

where $\overline{u} = \langle u_1, u_2 \rangle$ and $\overline{v} = \langle v_1, v_2 \rangle$.

It obeys all the same rules as the dot product in 3-space, including product rule (B). Anton simply did not mention the dot product back in Chapter 14, although he could have!

However, the same cannot be said for the cross product rule (C). As emphasized in §15.4, there is no analogue for the cross product in 2-space!

Example B.   Prove $\dfrac{d}{dt} [\, \overline{r}(t) \times \overline{r}'(t)] = \overline{r}(t) \times \overline{r}''(t)$.

Solution. $\dfrac{d}{dt} [\, \overline{r}(t) \times \overline{r}'(t)] = \overline{r}(t) \times \overline{r}''(t) + \overline{r}'(t) \times \overline{r}'(t)$ by product rule (C)

$\qquad\qquad\qquad\qquad = \overline{r}(t) \times \overline{r}''(t) + \overline{0}$ since $\overline{u} \times \overline{u} = \overline{0}$ by Theorem 15.4.4.   □

3.   Curvature in 3-space.   There is a major difference between curves in 2-space and curves in 3-space. A curve in 2-space divides the plane up into two pieces, and near any point on the curve we can ask which of these pieces is concave. This enables us to give a geometric meaning to the sign of the curvature κ in 2-space: κ is positive if the concave side of the curve is on the left, and κ is negative if the concave side of the curve is on the right.

This is not possible in 3-space. A curve in 3-space does not divide up space into "two pieces," and hence there is no geometric significance to give to the sign of the curvature.

For that reason the curvature κ of a curve in space is always a <u>non-negative</u> function of the parameter of the curve, i. e. ,

$$\kappa \geq 0 \quad \text{at any point on a curve in 3-space.}$$

Another problem arises in 3-space when trying to define the unit tangent and unit normal vectors $\overline{T}$ and $\overline{N}$. The definition for $\overline{T}$ is no problem: we just use the same definition as in 2-space. However, the 2-space definition of $\overline{N}$ as the <u>counterclockwise</u> rotation of $\overline{T}$ by 90° (Definition 14.3.2) makes <u>no sense</u> in 3-space. "Counterclockwise" is a meaningful concept only in 2-space. Moreover, as Anton shows in Figure 15.5.5, there are an <u>infinite</u> number of unit vectors which are perpendicular to $\overline{T}$. Which one should we pick to be $\overline{N}$?

We solve our $\overline{N}$ dilemma as follows: $\overline{N}$ is chosen to be that unit vector which is both perpendicular to $\overline{T}$ and which points in the direction in which $\overline{T}$ is turning (i. e. , in the direction of $d\overline{T}/dt$). This leads to the following equations:

$\boxed{\text{T and N} \atop \text{vectors}}$

$$\overline{T} = \frac{\overline{r}'(t)}{\|\overline{r}'(t)\|}$$

$$\overline{N} = \frac{d\overline{T}/dt}{\|d\overline{T}/dt\|} \quad \text{(Theorem 15.5.3)}$$

The fact that $\overline{N}$, when defined in this way, is perpendicular to $\overline{T}$ is proven in Theorem 15.5.2. However, this is merely a special case of Example A above: since $\|\overline{T}(t)\|$ equals the constant 1 for all values of t, then $\overline{T}'(t)$ (and hence $\overline{N}(t)$) is perpendicular to $\overline{T}$.

The curvature κ of a curve in 3-space is defined in the same way as the absolute value of curvature $|\kappa|$ was defined in 2-space:

$$\kappa = \| d\overline{T}/ds \|$$

Since curves are rarely given in the form in which they are parameterized by arc length, for computational purposes the following two formulas for κ are more useful than the defining formula:

Computation of κ

$$\kappa = \frac{\| d\overline{T}/dt \|}{\| \overline{r}'(t) \|}$$

$$\kappa = \frac{\left\| \dfrac{d\overline{r}}{dt} \times \dfrac{d^2\overline{r}}{dt^2} \right\|}{\left\| \dfrac{d\overline{r}}{dt} \right\|^3} \qquad \text{(Theorem 15.5.3)}$$

The first of these computational formulas for κ is obtained from the Chain Rule:

$$\kappa = \left\| \frac{d\overline{T}}{ds} \right\| = \left\| \frac{d\overline{T}}{dt} \cdot \frac{dt}{ds} \right\| \qquad \text{by the Chain Rule}$$

$$= \left\| \frac{d\overline{T}}{dt} \cdot \frac{1}{ds/dt} \right\| = \left\| \frac{d\overline{T}}{dt} \cdot \frac{1}{\| \overline{r}'(t) \|} \right\| = \frac{\| d\overline{T}/dt \|}{\| \overline{r}'(t) \|}$$

Anton does not mention this formula in his discussion of curvature, but it is useful, as we show in the next example.

Example C.   Compute $\overline{T}$, $\overline{N}$ and κ at $t = 0$ for the curve

$$\overline{r}(t) = \frac{1}{2} t^2 \overline{i} - \cos t \, \overline{j} + \sin t \, \overline{k}$$

Solution.   i) We first compute $\overline{r}'(t)$ and $\| \overline{r}'(t) \|$:

$$\overline{r}'(t) = \langle t, \sin t, \cos t \rangle$$

$$\|\overline{r}'(t)\| = (t^2 + \sin^2 t + \cos^2 t)^{1/2} = (t^2 + 1)^{1/2}$$

At $t = 0$ this gives $\boxed{\|\overline{r}'(0)\| = 1}$

ii)
$$\overline{T} = \frac{\overline{r}'(t)}{\|\overline{r}'(t)\|} = \left\langle \frac{t}{(t^2+1)^{1/2}}, \frac{\sin t}{(t^2+1)^{1/2}}, \frac{\cos t}{(t^2+1)^{1/2}} \right\rangle$$

At $t = 0$ this gives $\boxed{\overline{T}(0) = \langle 0, 0, 1 \rangle}$

iii)
$$\overline{N} = \frac{d\overline{T}/dt}{\|d\overline{T}/dt\|}. \quad \text{Thus we must compute } d\overline{T}/dt:$$

$$d\overline{T}/dt = \left\langle \frac{1}{(t^2+1)^{3/2}}, \frac{(t^2+1)\cos t - t \sin t}{(t^2+1)^{3/2}}, -\frac{(t^2+1)\sin t + t \cos t}{(t^2+1)^{3/2}} \right\rangle$$

At $t = 0$ this gives $d\overline{T}/dt = \langle 1, 1, 0 \rangle$.

> (Notice the order of our procedures: we first
> took the derivative of $\overline{T}$ at an arbitrary value
> of t, and then we evaluated it at $t = 0$. You
> cannot evaluate $\overline{T}$ at $t = 0$ and then take
> the derivative!)

Hence at $t = 0$ we have $\|d\overline{T}/dt\| = \|\langle 1, 1, 0 \rangle\| = \sqrt{2}$, which gives

$$\overline{N} = \frac{d\overline{T}/dt}{\|d\overline{T}/dt\|} \Bigg|_{t=0} = \frac{\langle 1, 1, 0 \rangle}{\sqrt{2}} = \boxed{\langle \frac{\sqrt{2}}{2}, \frac{\sqrt{2}}{2}, 0 \rangle = \overline{N}}$$

iv)
$$\kappa = \frac{\|d\overline{T}/dt\|}{\|\overline{r}'(t)\|} \Bigg|_{t=0} = \frac{\sqrt{2}}{1} = \boxed{\sqrt{2} = \kappa} \qquad \square$$

A use of the second computational formula for κ (Theorem 15.5.3) is illustrated in Anton's Example 3.

Helpful observation: Notice how the computations in Example C were simplified by the

use of $t = 0$ at the appropriate moments. "Appropriate" here means <u>after</u> any

desired differentiation has been carried out.

There is one last geometric reason we can give for our particular choice of the normal vector $\overline{N}$:

If their initial points are placed at the point $\overline{r}(t_0)$, then the vectors $\overline{T} = \overline{T}(t_0)$

and $\overline{N} = \overline{N}(t_0)$ will determine the plane which best "fits"

the curve parameterized by $\overline{r}(t)$ at $\overline{r}(t_0)$. This is

illustrated (as best we can!) in the diagram to the right.

The plane so determined, called the <u>osculating plane</u> of

the curve $\overline{r}(t)$ at $t = t_0$, has important applications

in physics. In fact, Anton's final remark in this section

can be reworded to state that for motion in 3-space, the

acceleration vector $\overline{a}(t_0)$ always lies in the osculating plane of $\overline{r}(t)$ at $t = t_0$.

4. <u>Graphing curves in 3-space</u>. In Exercises 1-6, Anton asks you to sketch the graphs of specific curves in 3-space. This is not an easy task for most people! You need a good sense of three-dimensional perspective, an ability which varies greatly from person to person. Nonetheless, we'll illustrate certain basic techniques by graphing the curve of Example C:

<u>Example D.</u>   Sketch the graph of the curve

$$\overline{r}(t) = \frac{1}{2} t^2 \overline{i} - \cos t\, \overline{j} + \sin t\, \overline{k}, \qquad 0 \le t \le \pi$$

<u>Solution.</u>   It's usually a good idea to start by plotting some points of the curve:

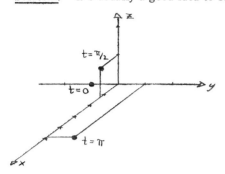

$$\bar{r}(0) \quad = -\bar{j} \qquad\qquad = \langle 0, -1, 0 \rangle$$

$$\bar{r}(\pi/2) = (\pi^2/8)\bar{i} + \bar{k} = \langle \pi^2/8, 0, 1 \rangle$$

$$\bar{r}(\pi) \quad = (\pi^2/2)\bar{i} + \bar{j} = \langle \pi^2/2, 1, 0 \rangle$$

The three points are shown to the left.   To get more of a feel of this curve, let's see what its <u>projection</u> onto the yz-plane looks like; this is obtained by disregarding the x-component of $\bar{r}(t)$.   The parametric equations so obtained are

$$y = -\cos t \qquad \text{and} \qquad z = \sin t$$

which lead to the equation $y^2 + z^2 = 1$, a circle of radius 1 and center $(0,0)$.   In fact, the restriction to $0 \le t \le \pi$ never allows

$z = \sin t$ to become negative, so we obtain

a semi-circle, parametrized as shown.   Now

reconsider the x-component,

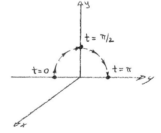

$$x = \frac{1}{2}t^2$$

As t moves from 0 to π, the (y, z) coordinates move around

the semi-circle, while the x coordinates increase

parabolically with the t variable.   Combining this

information with the previous two sketches yields the

diagram shown here.   The dark line is the graph, while

the straight lines parallel to the x-axis have been added

in to suggest a cylindrical cylinder whose border is our curve.   □

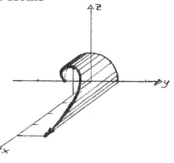

5.   <u>Tangent lines to curves in space (optional).</u>   If $\overline{r}(t) = x(t)\,\overline{i} + y(t)\,\overline{j} + z(t)\,\overline{k}$ parameterizes

a curve in space, then the derivative $\overline{r}'(t_0)$

is a <u>tangent vector</u> to the curve at the point

$\overline{r}(t_0)$. Thus the <u>tangent line</u> to the curve at

$\overline{r}(t_0)$ contains the point $P_0 = \overline{r}(t_0)$ and has

a direction vector $\overline{v} = \overline{r}'(t_0)$. From <u>The</u>

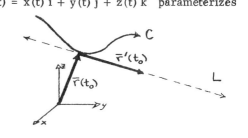

<u>Companion</u> §15. 2. 4 a parametric vector equation for this tangent line is therefore given by

$$\boxed{\;\overline{X} = \overline{r}(t_0) + u\,\overline{r}'(t_0) \qquad \text{for} \qquad -\infty < u < \infty\;}\qquad\qquad (F)$$

The tangent line is valuable as an approximation to our curve   C   near the point of

tangency. Such an approximation is useful because an "arbitrary" curve   C   might be

parameterized by some very unpleasant "non-linear" functions, while the tangent line is

parameterized in an easy "linear" fashion. In fact, many of the most important computational

techniques in applied mathematics are based on "linear" approximations for "non-linear"

problems.

We'll give two examples of the use of our tangent line formula:

<u>Example E.</u>   What is the tangent line to

$$\overline{r}(t) = t\,\overline{i} + \sin\!\left(\frac{t\pi}{2}\right)\overline{j} + t^2\,\overline{k} \quad \text{at the point}\quad X_0 = (1,1,1)\,?$$

<u>Solution.</u>   We first should determine the value $t = t_0$ for which $\overline{r}(t_0) = \overline{X}_0$, i.e.,

$$t_0\,\overline{i} + \sin\!\left(\frac{t_0\pi}{2}\right)\overline{j} + t_0^2\,\overline{k} = \langle 1,1,1\rangle$$

Clearly $t_0 = 1$. We now have only to determine $\overline{r}'(1)$:

$$\overline{r}'(t) = \overline{i} + \frac{\pi}{2}\cos\left(\frac{t\,\pi}{2}\right)\overline{j} + 2t\,\overline{k}$$

so $\overline{r}'(1) = \langle 1, 0, 2 \rangle$

From (F) we then have

$$\overline{X} = \langle 1, 1, 1 \rangle + u\langle 1, 0, 2 \rangle \quad \text{for} \quad -\infty < u < \infty$$

or $$\boxed{\overline{X} = \langle 1 + u, 1, 1 + 2u \rangle, \quad -\infty < u < \infty}$$

In scalar parametric equation form we would have

$$\begin{cases} x = 1 + u \\ y = 1 \qquad\qquad \text{for} \quad -\infty < u < \infty \\ z = 1 + 2u \end{cases}$$ □

<u>Example F.</u>   Find the angle of intersection of the two curves

$$\overline{r}_1(t) = (\sin t)\overline{i} + t^2\overline{j} + (1 - t)\overline{k}$$
$$\overline{r}_2(s) = (1 - s)\overline{i} + (s^2 - 1)\overline{j} + s\overline{k}$$

Note: The "angle of intersection" between two curves is defined to be the angle between the
the two tangent lines at the point of intersection of the curves.

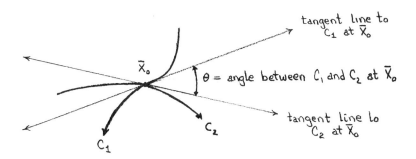

Solution.   Clearly we must first determine where the two curves intersect.   To do so we must

find values of   t   and   s   so that

$$\bar{r}_1(t) \; = \; \bar{r}_2(s) \, , \quad \text{i. e.,}$$

$$\left\{ \begin{array}{l} \sin t \; = \; 1 - s \\[4pt] t^2 \; = \; s^2 - 1 \\[4pt] 1 - t \; = \; s \end{array} \right\}$$

Using the last two equations we obtain

$$t^2 \; = \; s^2 - 1 \; = \; (1 - t)^2 - 1 \; = \; -2t + t^2$$

$$\text{or} \quad 0 \; = \; -2t \implies \boxed{t = 0} \implies \boxed{s = 1}$$

We now check these values in all three of our equations: all three equations are satisfied. Hence

the point of intersection $\bar{X}_0$ of the two curves is

$$\bar{r}_1(t = 0) \; = \; \langle 0, 0, 1 \rangle \; = \; \bar{r}_2(s = 1)$$

To compute the angle of intersection at $\bar{X}_0$ we need to compute the two tangent vectors

$\bar{u}$ and $\bar{v}$ at $\bar{X}_0$.

i. $\bar{r}_1'(t) \; = \; (\cos t)\,\bar{i} + 2t\,\bar{j} - \bar{k}$

so $\bar{r}_1'(t = 0) \; = \; \bar{i} - \bar{k} \; = \; \boxed{\langle 1, 0, -1 \rangle \; = \; \bar{u}}$

ii. $\bar{r}_2'(s) \; = \; -\bar{i} + 2s\,\bar{j} + \bar{k}$

so $\bar{r}_2'(s = 1) \; = \; -\bar{i} + 2\bar{j} + \bar{k} \; = \; \boxed{\langle -1, 2, 1 \rangle \; = \; \bar{v}}$

Thus, if θ is the desired angle, we have

15.5.13

$$\cos \theta = \frac{\overline{u} \cdot \overline{v}}{\|\overline{u}\| \, \|\overline{v}\|} = \frac{-1 + 0 - 1}{\sqrt{2} \sqrt{6}} = -\frac{2}{\sqrt{12}} \approx -.57735$$

Then $\theta \approx 125.26^{\circ}$.

Exercises.   These are problems dealing with the optional tangent line material.

1.   Does the tangent line to $\overline{r}(t) = t^2 \overline{i} + (2t + 1) \overline{j} + (1/t) \overline{k}$ at $X_0 = (1, -1, -1)$

contain the point $X_1 = (-1, 1, -2)$? Does it contain $X_2 = (0, 0, 1)$?

2.   Do the two curves $\overline{r}_1(t) = \langle (t^3 + 2)/3 , t^2 + 1 , -t \rangle$ and $\overline{r}_2(t) = \langle e^{3t} , t^2 - t + 2 , t - 1 \rangle$

intersect? Do they intersect at right angles?

3.   Find that tangent line to the curve

$\overline{r}(t) = t^2 \overline{i} + (1/t) \overline{j} - (\ln t) \overline{k}$ which

contains the point $X_1 = (3, 0, -1)$.

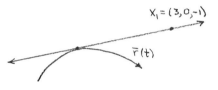

$X_1 = (3, 0, -1)$

$\overline{r}(t)$

4.   Is $\overline{X} = \langle 3, 1, -1 \rangle + w \langle -1, -1/2, 1/2 \rangle$, $-\infty < w < \infty$,   the tangent line to

$\overline{r}(t) = e^{2t} \overline{i} + \ln (t + 1) \overline{j} - t \overline{k}$ at $t = 0$?

Answers.

1. $\overline{X}_0 = \overline{r}(t_0)$ when $t_0 = -1$.

The tangent line is then $\overline{X} = \langle 1 - 2u , -1 + 2u , -1 - u \rangle$ for $-\infty < u < \infty$. When

u = 1 ,   then $\overline{X} = \overline{X}_1$,   so X_1 is on the line. However, there is no value of

u   for which $\overline{X} = \overline{X}_2$. Thus X_2 is not on the line.

2.   Yes they intersect. However, the t-values for each curve will be different at the

point of intersection, so you must change the parameter in the second curve, say to u .

Then the vector equation $\bar{r}_1(t) = \bar{r}_2(u)$ will easily yield t = 1 and u = 0 , so

the point of intersection is

$$\bar{r}_1(1) = \bar{r}_2(0) = \langle 1, 2, -1 \rangle$$

This intersection is a right angle since

$$\bar{r}_1'(1) \cdot \bar{r}_2'(0) = \langle 1, 2, -1 \rangle \cdot \langle 3, -1, 1 \rangle = 0$$

3.   We must determine a value of t for which the tangent line expression

$\bar{X} = \bar{r}(t) + u\,\bar{r}'(t)$ equals $\bar{X}_1$ for some value of u , i. e.,

$$\langle t^2 + 2tu, \ (1/t) - (u/t^2), \ -\ln t - u/t \rangle = \langle 3, 0, -1 \rangle$$

This solves to t = u = 1 , so that our answer is

$$\bar{X} = \langle 1, 1, 0 \rangle + u \langle 2, -1, -1 \rangle$$

4.   Yes! The tangent line formula derived from (F) will be

$$\bar{X} = \langle 1, 0, 0 \rangle + u \langle 2, 1, -1 \rangle, \quad -\infty < u < \infty$$

which we claim is just a reparametrization of the given line. To see this notice that

both lines contain the point $\langle 3, 1, -1 \rangle$ (u = 1) and their direction vectors are

multiples of each other.

Section 15. 6 :   Planes in 3-Space.   In $\S 15.2$ we described how to represent lines in   3-space ;

in this section we do the same for planes.   It turns out that planes are easier to deal with in

3-space   than are lines since a plane in   3-space   can be represented by a single scalar

equation.

1.   Equations for planes.   Suppose we have a plane passing through the point $P_0(x_0, y_0, z_0)$ and

perpendicular to the nonzero vector $\bar{n} = \langle a, b, c \rangle$ ($\bar{n}$ is called a normal vector for the plane).

Anton gives two forms of the equation for the plane:

(1)   The point-normal form.   Suppose $P(x, y, z)$

is any point on the plane.   Then the

vector

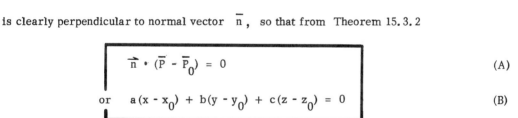

$$\overrightarrow{P_0P} = \bar{P} - \bar{P}_0 = \langle x - x_0, y - y_0, z - z_0 \rangle$$

is clearly perpendicular to normal vector $\bar{n}$,   so that from   Theorem 15. 3. 2

$$\vec{n} \cdot (\bar{P} - \bar{P}_0) = 0 \qquad\qquad (A)$$

$$\text{or} \quad a(x - x_0) + b(y - y_0) + c(z - z_0) = 0 \qquad\qquad (B)$$

Either equation is referred to as the point-normal form of the plane.

(2)   The general form.   Multiplying out   Equation (B) yields

$$ax + by + cz + d = 0 \qquad\qquad (C)$$

$$\text{where} \quad d = -(ax_0 + by_0 + cz_0)$$

This equation is called the general form of the plane.   In   Theorem 15. 6. 1   Anton shows that

any equation of form   (C)   (where at least one of the constants a, b or c is non-zero) is the equation for a plane in 3-space with normal vector $\bar{n} = \langle a, b, c \rangle$.

When attempting to solve problems involving planes in 3-space we cannot emphasize the following strongly enough:

---

*General Principle for Planes: When determining an equation for a plane in 3-space, always focus on finding a <u>point</u> P_0 on the plane and a non-zero <u>normal vector</u> $\bar{n}$ perpendicular to the plane.

Compare this with the corresponding result from §15. 2 for lines in 3-space:

---

*General Principle for Lines: When determining an equation for a line in 3-space, always focus on finding a <u>point</u> P_0 on the line and a non-zero <u>direction vector</u> $\bar{v}$ parallel to the line.

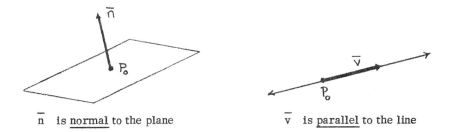

$\bar{n}$ is <u>normal</u> to the plane $\bar{v}$ is <u>parallel</u> to the line

2. <u>Line and plane problems.</u> It is very common to run across problems involving both lines and planes. They can be very tricky unless you carefully organize your thinking.

<u>Example A.</u> Determine an equation for the plane which contains the point $P_0(3, 1, -1)$ and is parallel to the two lines

$$\begin{cases} x = 1 - 3t \\ y = 2 + t \\ z = 1 \end{cases} \quad \text{and} \quad \begin{cases} x = 2 - 2t \\ y = -1 - t \\ z = t \end{cases}$$

Solution.   A quick, rough sketch such as

shown to the right can be useful in giving

geometric insights. Then turn to the General

Principle for Planes: we must find a <u>point</u> P_0

on the plane and a <u>normal vector</u> $\bar{n}$. A point

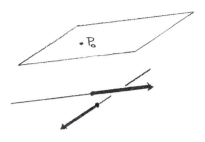

$P_0(3, 1, -1)$ is already given to us, so we must concentrate on finding a normal vector $\bar{n}$.

Clearly we must use the two given lines to find $\bar{n}$. Well, $\bar{n}$ is <u>perpendicular</u> to the plane

through P_0 which is <u>parallel</u> to the two lines; thus $\bar{n}$ must be <u>perpendicular</u> to the two lines,

which means perpendicular to the direction vectors for the lines. Thus

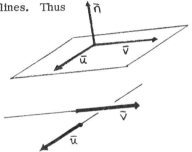

$\Vert$ find $\bar{n}$ so that $\bar{n}$ is perpendicular

$\Vert$ to both $\bar{u}$ and $\bar{v}$, where

$\Vert$ $\bar{u}$ and $\bar{v}$ are direction vectors

$\Vert$ for the two given lines.

Such a set-up cries out for one operation: the cross product! The cross product of $\bar{u}$ and $\bar{v}$

is a vector which is perpendicular to both $\bar{u}$ and $\bar{v}$, and hence is a perfect choice for $\bar{n}$.

Thus

$$\bar{n} = \bar{u} \times \bar{v}$$

$$= \langle -3, 1, 0 \rangle \times \langle -2, -1, 1 \rangle \qquad \text{from Theorem 15.2.1}$$

$$= \begin{vmatrix} \bar{i} & \bar{j} & \bar{k} \\ -3 & 1 & 0 \\ -2 & -1 & 1 \end{vmatrix}$$

$$= \begin{vmatrix} 1 & 0 \\ -1 & 1 \end{vmatrix} \bar{i} - \begin{vmatrix} -3 & 0 \\ -2 & 1 \end{vmatrix} \bar{j} + \begin{vmatrix} -3 & 1 \\ -2 & -1 \end{vmatrix} \bar{k}$$

$$= \bar{i} + 3\bar{j} + (3 + 2)\bar{k} = \langle 1, 3, 5 \rangle$$

Hence our plane is given by

$$\bar{n} \cdot (\bar{P} - \bar{P}_0) = \langle 1, 3, 5 \rangle \cdot \langle x - 3, y - 1, z + 1 \rangle = 0$$

$$(x - 3) + 3(y - 1) + 5(z + 1) = 0$$

$$\boxed{x + 3y + 5z - 1 = 0}$$ □

Example A illustrates a number of features which are common to problems of this type:

i.   The <u>cross-product</u> is oftentimes useful in computing a normal vector $\bar{n}$. In many

situations it is easy to find two non-collinear

vectors $\bar{u}$ and $\bar{v}$ which must be

perpendicular to $\bar{n}$; you can then define

$\bar{n}$ by $\bar{n} = \bar{u} \times \bar{v}$. Always look for this

situation in plane problems! (See Anton's

Examples 3 and 9, as well as Example C below.)

$\boxed{\begin{array}{c}\bar{n} \text{ from} \\ \text{cross-} \\ \text{product}\end{array}}$

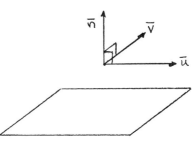

ii.   Suppose a plane has normal vector $\bar{n}$ and

a line has direction vector $\bar{v}$. Then the

plane and line are <u>parallel</u> if and only if

$\bar{n}$ and $\bar{v}$ are <u>perpendicular</u>, i.e.,

$\bar{n} \cdot \bar{v} = 0$. This sounds screwy until you

look at a picture, in which case it makes

$\boxed{\begin{array}{c}\text{parallel} \\ \text{line and} \\ \text{plane}\end{array}}$

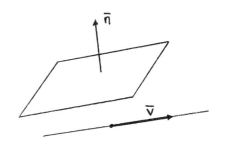

perfect sense! That many mistakes are made here should not surprise you, so be

careful! (See Anton's Example 4 as well as Example D below.)

Related to ii, although not used in Example A, is the following:

iii.   Suppose a plane has normal vector $\bar{n}$

and a line has direction vector $\bar{v}$. Then

the plane and line are <u>perpendicular</u> if and

only if $\bar{n}$ and $\bar{v}$ are <u>parallel</u>, i. e., $\bar{n}$

and $\bar{v}$ are scalar multiples of each other.

Again, another reversal of what you might

expect, although the fact is quite obvious

from the picture! (See Example B below.)

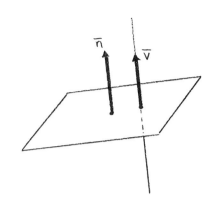

<u>Example B.</u>   Find the line through $P_0(1, 2, 2)$ which is perpendicular to the plane with general

equation $x - 2y + 3z - 2 = 0$.

<u>Solution.</u>   Since the line is perpendicular to the

plane we can take a normal vector $\bar{n}$ of the

plane to be a direction vector for the line

(iii above). Hence, with $\bar{n} = \langle 1, -2, 3 \rangle$

read off from the equation for the plane, we have

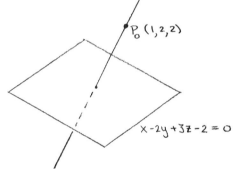

$$\bar{X} = \bar{P}_0 + t\bar{n} , \quad \text{i. e.,}$$

$$\bar{X} = \langle 1, 2, 2 \rangle + t \langle 1, -2, 3 \rangle \qquad \text{for} \qquad -\infty < t < \infty$$

or
$$\left\{ \begin{array}{l} x = 1 + t \\ y = 2 - 2t \\ z = 2 + 3t \end{array} \right\} \qquad -\infty < t < \infty$$

□

Example C.   Find a normal vector to the plane containing the three points   P(2,1,3), Q(3,3,5)

and   R(1,3,6).

Solution.   Since the vectors

$$\vec{PQ} = \langle 3-2, 3-1, 5-3 \rangle = \langle 1, 2, 2 \rangle$$

and $$\vec{PR} = \langle 1-2, 3-1, 6-3 \rangle = \langle -1, 2, 3 \rangle$$

both lie in the plane, a normal vector to the plane is their cross product (using "$\bar{n}$ from

cross product," principle (i) above):

$$\bar{n} = \vec{PQ} \times \vec{PR} = \begin{vmatrix} \bar{i} & \bar{j} & \bar{k} \\ 1 & 2 & 2 \\ -1 & 2 & 3 \end{vmatrix}$$

$$= \begin{vmatrix} 2 & 2 \\ 2 & 3 \end{vmatrix} \bar{i} - \begin{vmatrix} 1 & 2 \\ -1 & 3 \end{vmatrix} \bar{j} + \begin{vmatrix} 1 & 2 \\ -1 & 2 \end{vmatrix} \bar{k}$$

$$= 2\bar{i} - 5\bar{j} + 4\bar{k}$$

$$= \langle 2, -5, 4 \rangle \qquad \qquad \square$$

Example D.   Find a set of parametric scalar equations for the line which passes through

(3, -1, 6) and is parallel to both of the planes   x - 2y + z = 2   and   2x + y - 3z = 5.

Solution.   Using our General Principle for Lines, we

must find a point P_0 on the line and a direction

vector $\bar{v}$ parallel to the line. Since a point

$P_0(3, -1, 6)$ is already given, we only need a

direction vector $\bar{v}$.

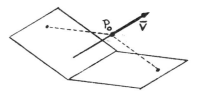

     Since $\bar{v}$ is to be parallel to each of the given planes, it must be perpendicular to the normal vector to each plane (using "parallel line and plane," principle (ii) above). We

can read off normal vectors for the two planes from their general equations using

Theorem 15.6.1:

$$\bar{n}_1 = \langle 1, -2, 1 \rangle \qquad \text{and} \qquad \bar{n}_2 = \langle 2, 1, -3 \rangle$$

Now we want $\bar{v}$ to be perpendicular to both $\bar{n}_1$ and $\bar{n}_2$ ... hmmm .... Of course!

A cross product:

Define
$$\bar{v} = \bar{n}_1 \times \bar{n}_2$$

$$= \langle 1, -2, 1 \rangle \times \langle 2, 1, -3 \rangle$$

$$= \begin{vmatrix} \bar{i} & \bar{j} & \bar{k} \\ 1 & -2 & 1 \\ 2 & 1 & -3 \end{vmatrix}$$

$$= \begin{vmatrix} -2 & 1 \\ 1 & -3 \end{vmatrix} \bar{i} - \begin{vmatrix} 1 & 1 \\ 2 & -3 \end{vmatrix} \bar{j} + \begin{vmatrix} 1 & -2 \\ 2 & 1 \end{vmatrix} \bar{k}$$

$$= 5\bar{i} + 5\bar{j} + 5\bar{k} = \langle 5, 5, 5 \rangle$$

We now have our __point__ $P_0(3, -1, 6)$ and our __direction vector__ $\bar{v} = \langle 5, 5, 5 \rangle$; the

line is therefore given by

$$\bar{X} = \bar{P}_0 + t\bar{v} \qquad -\infty < t < \infty$$

$$\langle x, y, z \rangle = \langle 3, -1, 6 \rangle + t \langle 5, 5, 5 \rangle$$

$$\begin{cases} x = 3 + 5t \\ y = -1 + 5t \\ z = 6 + 5t \end{cases} \qquad -\infty < t < \infty$$

$\square$

__Note:__ Since only the direction of $\bar{v}$ (not its length) mattered, we would have been completely justified in Example D to replace $\bar{v} = \langle 5, 5, 5 \rangle$ with $\bar{u} = \bar{v}/5 = \langle 1, 1, 1 \rangle$. (This could be viewed as a more "natural" choice for the direction vector.) Then our final set of equations for the line would have been

$$\left\{\begin{array}{l} x = 3 + t \\ y = -1 + t \\ z = 6 + t \end{array}\right\} \quad -\infty < t < \infty$$

In Theorem 15.6.2 Anton derives the formula for the distance between a point and a plane, and then gives three examples of its use. The geometric thinking which is illustrated in this discussion (especially in showing the interrelationships between the "three basic distance problems") should be carefully read and studied.

3. Planes and the calculus of curves (optional). The material in this section on planes can be combined with the calculus of curves in the previous section to handle geometric problems of a more advanced nature. We'll give a few examples:

Example E. Suppose $\bar{r}(t) = e^{2t-2}\,\bar{i} + \ln t\,\bar{j} + (1/t)\,\bar{k}$. Determine an equation for the plane which contains the point $P_0(1, 0, 1) = \bar{r}(1)$ and is perpendicular to the curve at that point.

Solution. We have a point $P_0(1, 0, 1)$ on the plane, so we only need a normal vector $\bar{n}$.

However, a plane is perpendicular to a curve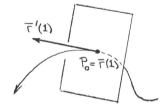

$\bar{r}(t)$ at $t = t_0$ if and only if the plane is

perpendicular to the tangent vector $\bar{r}'(t_0)$. Hence we can use $\bar{r}'(1)$ as our normal vector $\bar{n}$: since $\bar{r}'(t) = \langle 2 e^{2t-2}, 1/t, -1/t^2 \rangle$, then

$$\bar{n} = \bar{r}'(1) = \langle 2, 1, -1 \rangle$$

Thus our plane is given by

$$\bar{n} \cdot (\bar{P} - \bar{P}_0) = 0$$

$$\langle 2, 1, -1 \rangle \cdot \langle x - 1, y, z - 1 \rangle = 0$$

$$\boxed{2x + y - z - 1 = 0}$$ □

<u>Example F.</u>   Determine the angle of intersection between the curve $\bar{r}(t) = t^2 \bar{i} + 2t^2 \bar{j} - (1/t) \bar{k}$

and the plane given by $3x - 2y - z = 0$.

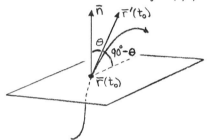

<u>Solution.</u>   From the picture it is clear that

the angle we desire is $90^\circ - \theta$, where θ

is the angle between $\bar{n}$, the normal vector

for the plane, and $\bar{r}'(t_0)$, the tangent vector

to the curve at the point of intersection $\bar{r}(t_0)$ between the plane and the curve.

i.   We can easily read off $\bar{n}$ from the equation for the plane:

$$\boxed{\bar{n} = \langle 3, -2, -1 \rangle}$$

ii.   We need the point of intersection between the plane and the curve. Any point $P(x, y, z)$

on the curve must satisfy the equations

$$x = t^2, \qquad y = 2t^2, \qquad z = -1/t$$

for some value of t. Thus find those values of t which give points on the plane by substituting

these curve equations into the plane equation and solving for t:

$$3(t^2) - 2(2t^2) - (-1/t) = 0$$

$$-t^2 + 1/t = 0$$

$$t^3 = 1$$

$$\boxed{t = 1}$$

iii.   Now find the tangent vector $\overline{r}'(t)$ at $t = 1$.

$$\overline{r}'(t) = \langle 2t, 4t, 1/t^2 \rangle$$

$$\boxed{\overline{r}'(1) = \langle 2, 4, 1 \rangle}$$

iv.   Finally, compute θ by taking the dot product of $\overline{n}$ with $\overline{r}'(1)$ (Equation (3), §15.3):

$$\cos \theta = \frac{\overline{n} \cdot \overline{r}'(1)}{\|\overline{n}\| \cdot \|\overline{r}'(1)\|} = \frac{\langle 3, -2, -1 \rangle \cdot \langle 2, 4, 1 \rangle}{\sqrt{9 + 4 + 1} \; \sqrt{4 + 16 + 1}}$$

$$= \frac{6 - 8 - 1}{\sqrt{14} \; \sqrt{21}} = -\frac{3}{\sqrt{294}} \cong -.175$$

Thus, using tables or a hand calculator, we find $\theta \cong 100.1°$, so that the desired angle is

$90° - 100.1° = \boxed{-10.1°}$. The only significance of the negative

sign is that $\overline{r}'(1)$ and $\overline{n}$

are on opposite sides of

the plane.   □

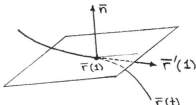

Example G.   Determine an equation for the osculating plane of the curve $\overline{r}(t) = \frac{1}{2} t^2 \overline{i} - \cos t \, \overline{j} +$

$+ \sin t \, \overline{k}$ at the point $\overline{r}(0) = \langle 0, -1, 0 \rangle$.

Solution.   As described in §15.5.3 of The Companion, the osculating plane for $\overline{r}(t)$ at

$t = t_0$ is that plane determined by the unit tangent and unit normal vectors $\overline{T}(t_0)$ and $\overline{N}(t_0)$.

    Example C of that section studied the curve under discussion here and determined that at

$t = 0$,

$$\overline{T} = \langle 0, 0, 1 \rangle \quad \text{and} \quad \overline{N} = \langle \sqrt{2}/2, \sqrt{2}/2, 0 \rangle$$

Thus our osculating plane contains the

point $P_0 = \bar{r}(0) = \langle 0, -1, 0 \rangle$ and

is parallel to the two vectors $\bar{T}$ and

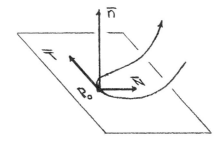

$\bar{N}$. Ahh... the cross-product rises again:

use $\bar{n} = \bar{T} \times \bar{N}$, i.e.,

$$\bar{n} = \langle 0, 0, 1 \rangle \times \langle \sqrt{2}/2, \sqrt{2}/2, 0 \rangle = \langle -\sqrt{2}/2, \sqrt{2}/2, 0 \rangle$$

Thus our plane is given by

$$\bar{n} \cdot (\bar{P} - \bar{P}_0) = 0$$

$$\langle -\sqrt{2}/2, \sqrt{2}/2, 0 \rangle \cdot \langle x, y+1, z \rangle = 0$$

$$\langle -1, 1, 0 \rangle \cdot \langle x, y+1, z \rangle = 0$$

$$\boxed{-x + y + 1 = 0}$$

□

Exercises. These are problems dealing with the optional material relating planes to the calculus

of curves.

1. Suppose $\bar{r}(t) = \dfrac{2\sqrt{3}}{3} \sin t\,\bar{i} + \cos t\,\bar{j} + \ln(t/\pi)\,\bar{k}$. Determine an equation for the plane
which contains the point $P_0(1, 1/2, -\ln 3)$ and is perpendicular to the curve at that
point.

2. Superman is flying through space with an acceleration at time t of
$\bar{a}(t) = \langle -1/t^2, -8/t^3, 2/t^2 \rangle$. If at time $t = -2$ sec, his velocity is
$\bar{v}(-2) = \langle 3/2, 2, 1 \rangle$ and his position is $\bar{r}(-2) = \langle -4 + \ln 2, 0, 1 - 2\ln 2 \rangle$, then
at what angle θ did he first hit the plane $2x + y + z = 0$? ($t < 0$ is allowed.)

3. Determine an equation for the plane which contains the point $P_1(-1, 1, 2)$ and which
is tangent to the curve $\bar{r}(t) = t\,\bar{i} + e^t\,\bar{j} + e^{-t}\,\bar{k}$ at $P_0(0, 1, 1)$.

Answers.

1. $\bar{r}(t)$ equals $P_0(1, 1/2, -\ln 3)$ when $t = \pi/3$; the problem is then just like

Example E, with an answer of

$x - (3/2)y + (3\sqrt{3}/\pi)z + 3\sqrt{3}\,(\ln 3)/\pi - 1/4 = 0.$

2. $\bar{v}(t) = \langle 2 + 1/t, 1 + 4/t^2, -2/t \rangle$

$\bar{r}(t) = \langle 2t + \ln|t|, t - 4/t, 1 - 2\ln|t| \rangle$

$t = -1$, $\cos(90^\circ - \theta) = 3/\sqrt{20}$, so that $\theta \cong 42.1^\circ$.

3. $x + z = 1$. [A plane is tangent to a curve $\bar{r}$ at $t = t_0$ if the plane contains

$\bar{r}(t_0)$ and is parallel to $\bar{r}'(t_0)$.]

Section 15. 7: Cramer's Rule (Optional).

1. Cramer's Rule: each unknown is a quotient of determinants. Here is some very useful

terminology: Given a system of linear equations

$$a_1 x_1 + b_1 x_2 + c_1 x_3 = k_1$$

$$a_2 x_1 + b_2 x_2 + c_2 x_3 = k_2$$

$$a_3 x_1 + b_3 x_2 + c_3 x_3 = k_3$$

then the coefficient matrix of this system is the three-by-three array ("3 x 3 matrix")

of the coefficients:

$$\begin{bmatrix} a_1 & b_1 & c_1 \\ a_2 & b_2 & c_2 \\ a_3 & b_3 & c_3 \end{bmatrix}$$

This array consists of three columns: the a_i's, the b_i's, and the c_i's. We also have

the column of constants formed by the k_i's:

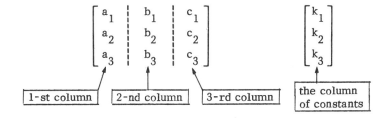

$$\begin{bmatrix} a_1 & \vdots & b_1 & \vdots & c_1 \\ a_2 & \vdots & b_2 & \vdots & c_2 \\ a_3 & \vdots & b_3 & \vdots & c_3 \end{bmatrix} \qquad \begin{bmatrix} k_1 \\ k_2 \\ k_3 \end{bmatrix}$$

| 1-st column | 2-nd column | 3-rd column | the column of constants |

With this terminology Cramer's Rule can be stated (in words) as follows:

Suppose M is the determinant of the coefficient matrix

for a system of linear equations. If $M \neq 0$, then

the system has a unique solution, and the value of

the unknown x_j ($j = 1, 2$ or 3) is given by

$$x_j = D_j / M$$

where D_j is the determinant obtained from M by

replacing the j-th column by the column of constants.

Cramer's
Rule

Thus we have

$$M = \begin{vmatrix} a_1 & b_1 & c_1 \\ a_2 & b_2 & c_2 \\ a_3 & b_3 & c_3 \end{vmatrix}$$

and

$$x_1 = \frac{1}{M} \begin{vmatrix} k_1 & b_1 & c_1 \\ k_2 & b_2 & c_2 \\ k_3 & b_3 & c_3 \end{vmatrix} \qquad x_2 = \frac{1}{M} \begin{vmatrix} a_1 & k_1 & c_1 \\ a_2 & k_2 & c_2 \\ a_3 & k_3 & c_3 \end{vmatrix} \qquad x_3 = \frac{1}{M} \begin{vmatrix} a_1 & b_1 & k_1 \\ a_2 & b_2 & k_2 \\ a_3 & b_3 & k_3 \end{vmatrix}$$

Anton illustrates the use of Cramer's Rule in Examples 4 and 5. Here is one more

example:

Example A.   Use Cramer's Rule to solve the system

$$2x + y - z = 0$$

$$x - y + z = 6$$

$$x + 2y + z = 3$$

<u>Solution.</u>   The coefficient matrix of this system is

$$\begin{bmatrix} 2 & 1 & -1 \\ 1 & -1 & 1 \\ 1 & 2 & 1 \end{bmatrix}$$

Its determinant (evaluated as in §15. 4) is found to be

$$M = -9$$

Since this is non-zero, Cramer's Rule applies and we obtain

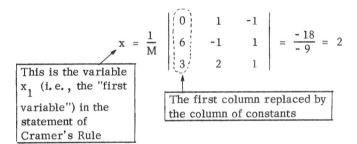

This is the variable x_1 (i. e. , the "first variable") in the statement of Cramer's Rule

The first column replaced by the column of constants

Similarly,

$$y = \frac{1}{M} \begin{vmatrix} 2 & 0 & -1 \\ 1 & 6 & 1 \\ 1 & 3 & 1 \end{vmatrix} = \frac{9}{-9} = -1$$

$$z = \frac{1}{M} \begin{vmatrix} 2 & 1 & 0 \\ 1 & -1 & 6 \\ 1 & 2 & 3 \end{vmatrix} = \frac{-27}{-9} = 3$$

Thus $(x, y, z) = (2, -1, 3)$ is the desired solution.   □

2.   <u>The importance of Cramer's Rule.</u>   As far as solving <u>specific</u> systems of linear equations, Cramer's Rule is not as important as it might seem at first.   True, given any system of n

linear equations in n unknowns (yes, there is a version of Cramer's Rule for <u>any</u> integer

n $\geq$ 2), Cramer's Rule allows you to write down and then compute the solutions, at least

in the "one unique solution" case. However, <u>there are much quicker and more efficient</u>

<u>methods for calculating these answers!</u> (One such method, called <u>Gaussian elimination</u>, is

covered in any elementary linear algebra text.) Even with a modern, high-speed computer

Cramer's Rule is a very poor technique to use when compared to Gaussian elimination.

So why study this result? Well, there are important uses for Cramer's Rule that

other techniques cannot provide. As one example, Cramer's Rule supplies a major part

of the following fact:

A system of n equations in n unknowns

has a unique solution if and only if the   (A)

determinant of its coefficient matrix

is non-zero.

(As we saw above, the determinant of the coefficient matrix of the system of equations

$$\left\{ \begin{aligned} a_1\, x + b_1\, y &= k_1 \\ a_2\, x + b_2\, y &= k_2 \end{aligned} \right\} \qquad \text{is} \qquad \begin{vmatrix} a_1 & b_1 \\ a_2 & b_2 \end{vmatrix}$$

and so on for any number of equations and unknowns. Determinants do exist for square

arrays of numbers of any number of rows and columns, although we have only treated the

2 x 2 and 3 x 3 cases.) Result A is a cornerstone of any detailed study of systems of

linear equations.

The major advantage which Cramer's Rule enjoys over other solution methods is that it gives us explicit formulas for the solution to a system. Other methods for solving systems of linear equations provide a sure-fire, efficient procedure for computing the solution, but do not give explicit formulas for the solution. Thus Cramer's Rule allows us to study properties of a solution to a system without the necessity of computing the answer past the determinant stage. Here's an example of how we use this feature to our advantage:

Example B. Verify that, for any $k \neq 0$, the solution (x, y, z) for the system of equations

$$k x + \quad y + 23 z = \sqrt{3}$$
$$- x - \quad k y + 23 z = \sqrt{3}$$
$$\sqrt{2}\, x - \sqrt{2}\, y + \quad 7 z = \quad 6$$

is such that $x = -y$, so long as the system has a unique solution.

Solution. To say that the system has a unique solution is the same as stipulating that $M \neq 0$, where M is the determinant of the coefficient matrix (Result A). Thus Cramer's Rule applies to give

$$x = \frac{1}{M} \begin{vmatrix} \sqrt{3} & 1 & 23 \\ \sqrt{3} & -k & 23 \\ 6 & -\sqrt{2} & 7 \end{vmatrix} \quad \text{and} \quad y = \frac{1}{M} \begin{vmatrix} k & \sqrt{3} & 23 \\ -1 & \sqrt{3} & 23 \\ \sqrt{2} & 6 & 7 \end{vmatrix}$$

We will need some simple determinant rules (which are easily verified from the definitions of 2×2 and 3×3 determinants):

    (1)   Interchanging two rows of a determinant,

or (2)   interchanging two columns of a determinant,

or (3)   multiplying one row or one column of a determinant by -1,

simply changes the sign of the determinant.

Thus, starting with the Cramer's Rule formula for x, we have

$$x = \frac{1}{M} \begin{vmatrix} \sqrt{3} & 1 & 23 \\ \sqrt{3} & -k & 23 \\ 6 & -\sqrt{2} & 7 \end{vmatrix} \underset{(1)}{=} -\frac{1}{M} \begin{vmatrix} 1 & \sqrt{3} & 23 \\ -k & \sqrt{3} & 23 \\ -\sqrt{2} & 6 & 7 \end{vmatrix} \underset{(2)}{=} \frac{1}{M} \begin{vmatrix} -k & \sqrt{3} & 23 \\ 1 & \sqrt{3} & 23 \\ -\sqrt{2} & 6 & 7 \end{vmatrix}$$

$$\big\downarrow +/-$$

$$\underset{(3)}{=} -\frac{1}{M} \begin{vmatrix} k & \sqrt{3} & 23 \\ -1 & \sqrt{3} & 23 \\ \sqrt{2} & 6 & 7 \end{vmatrix} = -y \, , \quad \begin{array}{l} \text{as desired, since this is} \\ \text{the Cramer's Rule formula for } y \end{array} \quad \square$$

Obviously this is a "quick" solution only if you are familiar with the stated determinant rules. Although Anton does not mention these rules, they are standard fare in linear algebra courses. Do not dismiss the techniques of Example A as being overly "abstract," and hence not "applicable to real world problems." Such is not the case; for example, many problems in economics require these exact methods for their solution!

3.   <u>The geometry of linear equations.</u>   Pay careful attention to Anton's discussion of the geometric interpretations for solutions of linear equations:

> The solutions to a system of 2 linear equations in 2 unknowns correspond to the intersection points of 2 lines in 2-space;

> The solutions to a system of 3 linear equations in 3 unknowns correspond to the intersection points of 3 planes in 3-space.

These are perfect illustrations of abstract algebra problems being converted into concrete, visual geometry problems via analytic geometry. In this way the algebraic problems should become much less confusing. You should, in particular, look over carefully the various geometric possibilities which are illustrated in Figures 15.7.1 and 15.7.2.

Section 15. 8 :   Quadric Surfaces

Visualizing and sketching surfaces in   3-space   can be very difficult for some people, while very easy for others; there seems to be a larger-than-ordinary spread of abilities in this area. Nonetheless, anyone can (with practice!) learn to visualize and sketch at least the most common surfaces in   3-space. This will prove very useful when studying double and triple integrals in Chapter 17.

The pictures which Anton gives for the quadric surfaces, many of them computer generated, are very well done. It is certainly worth the time spent to get a general feeling for each of the six surfaces described. On the other hand, more important than memorizing the details of these surfaces is the mastery of the graphing procedure which produces the sketches from the equations. Anton gives a commentary on this procedure when dealing with the general equation for an elliptic cone

$$z^2 = \frac{x^2}{a^2} + \frac{y^2}{b^2}$$

We will give a more detailed description of the general method, and illustrate it with the specific equation $2x^2 + 2y^2 - z = 2$.

1.   The graphing procedure.   Some preliminary comments are in order.   The equations which Anton studies in this section are all of the form

$$\boxed{f(x,y,z) = c}$$

where   c   is a constant and $f(x,y,z)$ is an algebraic expression using the unknowns x, y and z . (More formally, $f(x,y,z)$ is an algebraic function of the three variables x, y and z .) Examples are

$$x^2 - y^2 + z = 1 \quad \text{and} \quad 3x^2 - y/2 + z^2 = 2$$

However, the graphing procedure given below can be used to sketch the surface described by any equation $f(x,y,z) = c$, i.e., <u>even when $f(x,y,z)$ is not an algebraic expression.</u> That is, we can use it to sketch surfaces such as

$$\cos(xy) + e^z = 0 \quad \text{or} \quad x \sin y + y \ln(x + z) = 2$$

Of course such graphs can be extremely difficult to obtain if $f(x, y, z)$ is not a simple expression.

<u>General problem.</u> Sketch the surface $f(x, y, z) = c$. As a specific example, consider $2x^2 + 2y^2 - z = 2$.

<u>The Graphing Procedure.</u>

<u>Step 1.</u>   Obtain the <u>intercepts</u> of the surface with the three coordinate axes. To find the x-intercept(s) we have only to solve the equation $f(x, 0, 0) = c$, and similarly for y and z. Then place these points on the respective axes.

<u>Example:</u> $2x^2 + 2y^2 - z = 2$.

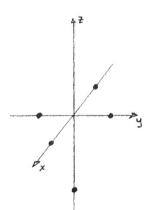

x-intercepts: $y = z = 0$ gives $2x^2 = 2$
  or $x = \pm 1$

y-intercepts: $x = z = 0$ gives $2y^2 = 2$
  or $y = \pm 1$

z-intercepts: $x = y = 0$ gives $-z = 2$
  or $z = -2$

Hence the intercepts are $(\pm 1, 0, 0)$, $(0, \pm 1, 0)$, $(0, 0, -2)$.

Obtain the coordinate traces of the surface on the three coordinate planes. To find the

xy-trace we have only to draw the curve $f(x, y, 0) = c$ on the xy-plane, and similarly

for xz and yz. At this point the drawing can get complicated, with lines running

all over the place. Many times you need to keep a lot of the graphical information "in your head"

(e.g., "this curve is in the yz plane") until you obtain the final picture. Some dotted "guide

lines" can also be useful. One helpful fact to remember: <u>your coordinate traces must include

your intercept points!</u>

Example: $2x^2 + 2y^2 - z = 2$

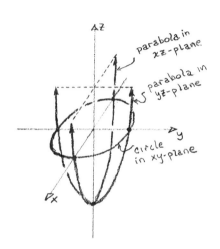

$\underline{xy\text{-trace}}$: $z = 0$ gives $2x^2 + 2y^2 = 2$

  or $x^2 + y^2 = 1$, the unit circle with

  center $(0, 0)$.

$\underline{xz\text{-trace}}$: $y = 0$ gives $2x^2 - z = 2$

  or $z = 2x^2 - 2$, an upward-turning

  parabola with vertex $(0, -2)$.

$\underline{yz\text{-trace}}$: $x = 0$ gives $2y^2 - z = 2$

  or $z = 2y^2 - 2$, an upward-turning

  parabola with vertex $(0, -2)$.

Step 3. Take some general <u>traces</u> of the surface, i.e., cuts by planes other than the coordinate

planes (which were already used in obtaining the coordinate traces). Generally planes of the

form $x = k$ or $y = k$ or $z = k$ (i.e., planes parallel to the coordinate planes) are the

best to use, although occasionally a more esoteric plane like $x = y$ might prove useful.

Simply consider the picture you have in front of you and decide where additional curves are

needed. Helpful fact: <u>your general traces, if passing through coordinate planes, must intersect

with your coordinate traces!</u>

Example: $2x^2 + 2y^2 - z = 2$.

From the picture obtained with the coordinate traces, we might expect that traces with

$z = k$ will produce circles to "tie our two parabolas together," much like metal bands around

wooden barrels. (Remember wooden barrels?)

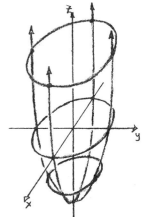

$z = k$ gives $2x^2 + 2y^2 = 2 - k$

or $x^2 + y^2 = 1 + \frac{1}{2} k$, which

is certainly a circle when

$1 + \frac{1}{2} k > 0$, or $k > -2$.

Thus our initial guess of circles for the z traces

has been verified.

Step 4.   Fill in the final picture. Generally you will need to do this on another set of xyz axes,

since your original drawing will now probably be a complete mess (although it should be

intelligible to you).   Label appropriate points and describe the surface in words as best you can.

Example. $2x^2 + 2y^2 - z = 2$.

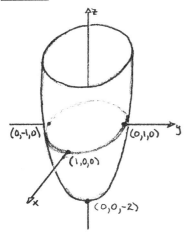

This is a <u>circular paraboloid.</u> Its

horizontal sections are circles, its vertical

sections are parabolas. Its lowest point is

the vertex at $(0, 0, -2)$, while there is

no highest point; the surface goes upward

indefinitely.

<u>Interesting fact</u>: The circular paraboloid is the reflecting surface of a flashlight! This sheds some light on the importance of quadric surfaces. We'll give one more example of our graphing procedure, this one a real hum-doozie:

<u>Example A.</u> Sketch the surface $x^2 - y^2 - z = 0$.

<u>Solution.</u> In this case $f(x,y,z) = x^2 - y^2 - z$ and $c = 0$. We will follow the four steps of our graphing procedure:

<u>Step 1.</u> Intercepts with the coordinate axes.

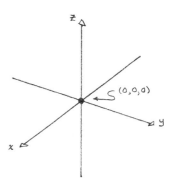

x-intercepts: $y = z = 0$ gives $x^2 = 0$

  so that $x = 0$.

y-intercepts: $x = z = 0$ gives $-y^2 = 0$

  so that $y = 0$

z-intercepts: $x = y = 0$ gives $-z = 0$

  so that $z = 0$

Hence the only intercept is $(0,0,0)$.

<u>Step 2.</u> Traces on the coordinate planes.

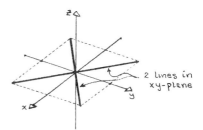

xy-trace: $z = 0$ gives $x^2 - y^2 = 0$,

  which yields $y = \pm x$, two lines

  through the origin.

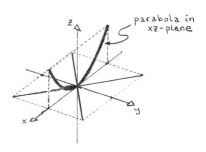

xz-trace: $y = 0$ gives $x^2 - z = 0$,

  which yields $z = x^2$, an upward-

  turning parabola with vertex at

  the origin.

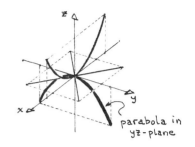

parabola in
yz-plane

<u>y z-trace</u>: $x = 0$ gives $-y^2 - z = 0$,

which yields $z = -y^2$,   a downward-

turning parabola with vertex at

the origin.

<u>Step 3.</u>   <u>General traces.</u>   Looking at the graph as it has developed to this point, it is natural

to wonder what "hangs off" the two parabolic traces, i. e., what would appear if we took cuts

parallel to the   xy-plane   by using $z = k$:

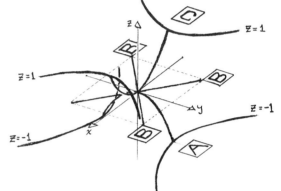

$z = k$ gives $x^2 - y^2 = k$,

which is a pair of hyperbolas

when $k \neq 0$.   Using $k = 1$

and then $k = -1$,   we get the

hyperbolas shown to the right.

At this point we need to think about what we are obtaining, or our diagram is going to over-

whelm us.   Consider the lower hyperbola $\boxed{A}$.   As it "slides up" the parabolic trace in

the   y z-plane,   it will eventually degenerate into the two straight lines $\boxed{B}$.   Similarly,

the upper hyperbola $\boxed{C}$ will "slide down" the parabolic trace in the   xz-plane   and

also degenerate into the two lines $\boxed{B}$.

"Seeing" these hyperbolas sweeping

out a surface by "sliding" on the

parabolic trace curves produces

the rough sketch shown to the right.

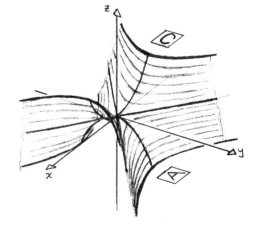

We need to "clean up the edges" of this picture; this can be done by taking cuts of the surface parallel to the yz-plane, i.e., by using x = k:

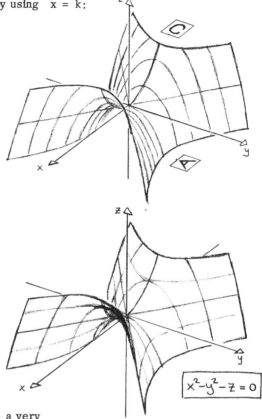

$x = k$ gives $z = -y^2 + k^2$,

which is a downward-turning parabola. These parabolas will "tie together" our $\boxed{A}$ and $\boxed{C}$ hyperbolas as shown to the right.

Step 4. The final picture. The last graph in Step 3 shows that we need to have "parabolic sides" in the next-to-last Step 3 graph. Putting this information together as carefully as we can yields the surface sketch shown to the right.

$$\boxed{x^2 - y^2 - z = 0}$$

This surface is a hyperbolic paraboloid; a very good computer-generated sketch of the surface is given in Anton's Table 15.8.1. □

2. The translation of axes. Learning how to translate coordinate axes in 3-space can save you much time and steer you away from numerous errors when sketching graphs in 3-space. Here is the general principle:

Here is the general principle:

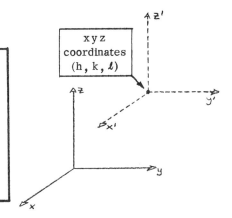

> Suppose we have two coordinate systems,
>
> x' y' z' and xyz , which are related by
>
> $x' = x - h$, $y' = y - k$, $z' = z - l$.
>
> Then the x' y' z' origin is the xyz
>
> point (h, k, l) .

Translation of Axes Principle

Here's how we can use this principle in sketching surfaces in 3-space :

Step 1.   Given an equation in x, y and z , determine constants h, k and l so that making the replacements

$$x' = x - h , \qquad y' = y - k , \qquad z' = z - l$$

yields a simpler equation in x', y' and z' . Label this new equation

$$f(x', y', z') = c$$

Example: $x^2 - 2x + y^2 - z^2 + 2x = 0$.

We get rid of the linear terms $- 2x$ and $2z$ by completing the square in both the

x and z terms:

$$(x^2 - 2x + 1) - 1 + y^2 - (z^2 - 2z + 1) + 1 = 0$$

$$(x - 1)^2 + y^2 = (z - 1)^2$$

Clearly this equation simplifies with the replacements

$$x' = x - 1 , \qquad y' = y \qquad \text{and} \qquad z' = z - 1$$

15.8.9

which yields

$$\boxed{x'^2 + y'^2 = z'^2}$$

Step 2. Sketch the graph of the new equation $f(x', y', z') = c$ in the $x'y'z'$ coordinate system. For this you can either use the sketching procedure of the previous subsection or (if applicable) use graphs of the standard quadric surfaces which you have memorized.

Example: $x'^2 + y'^2 = z'^2$.

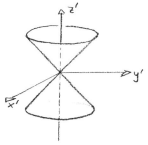

    This is a circular cone, as described in

    Anton's Table 15.8.1.

Step 3. To obtain the graph of the original equation we simply apply the Translation of Axes Principle: the $x'y'z'$ origin in our Step 2 sketch becomes the xyz point (h, k, ℓ), and from that we can accurately place the x, y and z axes into the sketch.

Example: $x^2 - 2x + y^2 - z^2 + 2z = 0$ and $x'^2 + y'^2 = z'^2$.

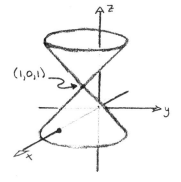

    The $x'y'z'$ origin becomes the xyz point

    $(1, 0, 1)$ as shown to the right (by setting

    $x' = y' = z' = 0$ in the "replacement"

    equations above). Thus, to place the

    xyz origin (and hence the x, y and

    z axes) correctly we should go "back" one

    x unit from the cone's vertex $(1, 0, 1)$,

    and then "down" one z unit. The result

    is as shown to the right.

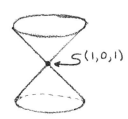

To appreciate fully the value of translation of axes, you should try to sketch the graph of the

cone above without using the axes translation.   It can be done, but it's a lot more difficult.

Anton's Examples 1 and 2 illustrate the use of the Translation of Axes Principle,

although the steps are not explicitly labeled.

Section 15.9:   Spherical and Cylindrical Coordinates.

When using polar coordinates in the plane, the elementary polar curves $r = r_0$ and

$\theta = \theta_0$ are a circle with center at the origin and a line passing through the origin.   For this

reason polar coordinates are

useful in dealing with objects

or movements in a plane which

are symmetric with respect to

a fixed point. * Similar considerations

hold true for cylindrical and spherical coordinates.

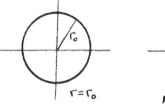

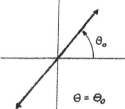

_____
*
   Even situations which lack such symmetry, but have a strong relationship with some fixed
point, can benefit by the use of polar coordinates.   For example, the derivation of Kepler's Laws
in §14.6 of The Companion used polar coordinates to handle the central gravitational force
field of the sun.

1. **Cylindrical coordinates.** Examine the elementary coordinate surfaces for the <u>cylindrical</u>

<u>coordinate system</u> in 3-space:

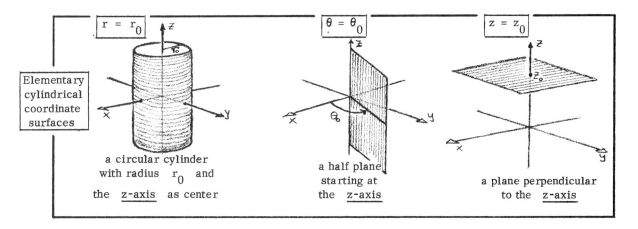

Elementary
cylindrical
coordinate
surfaces

$r = r_0$

a circular cylinder
with radius r_0 and
the <u>z-axis</u> as center

$\theta = \theta_0$

a half plane
starting at
the <u>z-axis</u>

$z = z_0$

a plane perpendicular
to the <u>z-axis</u>

Notice how all these coordinate surfaces are in some important way related to the z-axis. For

this reason <u>cylindrical coordinates are useful in dealing with objects or movements in 3-space</u>

<u>which are symmetric with respect to a fixed LINE.</u>

Since cylindrical coordinates are nothing more than polar coordinates in the xy-plane

with an added z-axis, if you understand polar coordinates, you understand cylindrical coordinates.

All the important relationships between xyz-coordinates and rθz-coordinates are nicely

remembered by the

<u>polar coordinate triangle</u>:

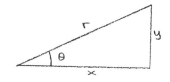

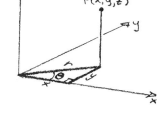

From this triangle the following formulas (for both polar and cylindrical coordinates) are easily

remembered:

$$x = r \cos \theta$$

$$y = r \sin \theta$$

$$z = z$$

$$r = \sqrt{x^2 + y^2}$$

$$\tan \theta = y/x$$

$$\text{if } x \neq 0$$

These are the formulas which we use to convert between xyz and $r\theta z$ coordinates.
Anton illustrates how they are used to change the coordinates of a point from cylindrical to
rectangular coordinates in Example 1. Here is an example showing how to change a point
from rectangular to cylindrical coordinates:

<u>Example A.</u>   Convert the point with rectangular coordinates $(3\sqrt{3}, 3, 6)$ to cylindrical
coordinates.

<u>Solution.</u>

$$r = \sqrt{x^2 + y^2} = \sqrt{(3\sqrt{3})^2 + (3)^2} = \sqrt{36} = 6$$

$$\tan \theta = \frac{y}{x} = \frac{3}{3\sqrt{3}} = \frac{1}{\sqrt{3}}$$

Since $\tan\left(\frac{\pi}{6}\right) = \frac{\sin(\pi/6)}{\cos(\pi/6)} = \frac{1/2}{\sqrt{3}/2} = \frac{1}{\sqrt{3}}$, and $(3\sqrt{3}, 3)$ is in the first quadrant,

then $\theta = \pi/6$ (see the diagram). Thus
the cylindrical coordinates are

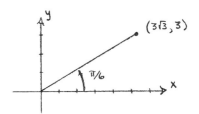

(Notice that the z-coordinate is the same in both coordinate systems.)   □

Anton's Examples 2 and 3 illustrate how the conversion formulas are used to convert equations (not just points) from one coordinate system to the other. Here are two examples:

Example B. Express $z^2 = x^2 - y^2$ in cylindrical coordinates.

Solution. $z^2 = x^2 - y^2 = r^2 \cos^2 \theta - r^2 \sin^2 \theta$

                        $= r^2 (\cos^2 \theta - \sin^2 \theta) = r^2 \cos 2\theta$

Thus $z = \pm r \sqrt{\cos 2\theta}$ □

Example C. Express $z = r \sec \theta$ in rectangular coordinates.

Solution. $z = r \sec \theta = \dfrac{r}{\cos \theta} = \dfrac{r^2}{r \cos \theta} = \dfrac{x^2 + y^2}{x}$

Thus $z = x + y^2 / x$ □

Example C and Anton's Example 3 illustrate conversion from cylindrical to rectangular coordinates. In making such a conversion, whenever possible we attempt to take the θ terms and write them as $r \cos \theta$ and $r \sin \theta$ (which become x and y, respectively).

2. Spherical coordinates. Examine the elementary coordinate surfaces for the spherical coordinate system in 3-space:

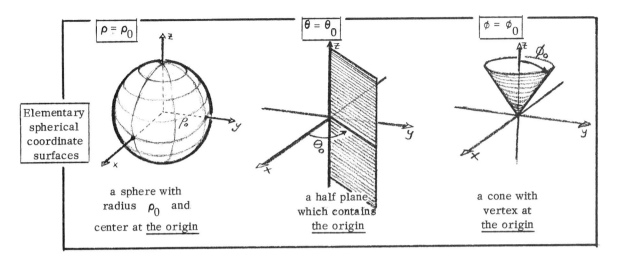

| $\rho = \rho_0$ | $\theta = \theta_0$ | $\phi = \phi_0$ |

Elementary spherical coordinate surfaces

a sphere with radius ρ_0 and center at the origin

a half plane which contains the origin

a cone with vertex at the origin

Notice how all these coordinate surfaces are in some important way related to the origin. For this reason spherical coordinates are useful in dealing with objects or movements in 3-space which are symmetric with respect to a fixed POINT.

Spherical coordinates are a bit more confusing than cylindrical coordinates because ρ and ϕ are both new to you. However, all the relationships between xyz-coordinates and $\rho\theta\phi$-coordinates are nicely remembered by coupling the polar coordinate triangle with the spherical coordinate triangle:

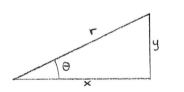

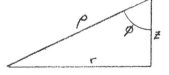

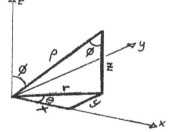

From the second (spherical coordinate) triangle we easily obtain the equations

$$r = \rho \sin \phi \quad \text{and} \quad z = \rho \cos \phi$$

Then the first (polar coordinate) triangle yields

$$x = r \cos \theta = \rho \sin \phi \cos \theta$$
$$y = r \sin \theta = \rho \sin \phi \sin \theta$$

The only other important spherical coordinate formula is easily obtained from the distance formula:

$$\rho = \sqrt{x^2 + y^2 + z^2}$$

These are the formulas which we use to convert between xyz and $\rho\theta\phi$ coordinates. In Example 4 Anton illustrates how they are used to convert from spherical to rectangular coordinates. Here is an example illustrating conversion from rectangular to spherical coordinates:

Example D.   Convert the point with rectangular coordinates $(-4\sqrt{2}, -4\sqrt{2}, 8\sqrt{3})$ into spherical coordinates.

Solution.   The easy spherical coordinate is ρ:

$$\rho = \sqrt{x^2 + y^2 + z^2} = \sqrt{(-4\sqrt{2})^2 + (-4\sqrt{2})^2 + (8\sqrt{3})^2} = \boxed{16 = \rho}$$

In order to determine θ and ψ most easily, it is convenient first to determine r:

$$r = \sqrt{x^2 + y^2} = \sqrt{(-4\sqrt{2})^2 + (-4\sqrt{2})^2} = 8$$

Then we can obtain θ and ϕ by using the easier-to-remember formulas

$$r = \rho \sin \phi \qquad \text{and} \qquad y = r \sin \theta$$

$$8 = 16 \sin \phi \qquad \text{and} \qquad -4\sqrt{2} = 8 \sin \theta$$

$$\boxed{\sin \phi = 1/2 \qquad \text{and} \qquad \sin \theta = -\sqrt{2}/2}$$

At this stage we must proceed with caution and avoid the natural urge to jump at

$$(?) \quad \phi = \pi/6 \qquad \text{and} \qquad \theta = -\pi/4 \quad (?)$$

The problem is that there are numerous angles ϕ and θ which have the given sine values. What we need to do is <u>examine where the original $P(x,y,z)$ point is located, and then pick the appropriate values for ϕ and θ.</u>

Since $z = 8\sqrt{3} > 0$, then we can take

$0 < \phi < \pi/2$ as seen to the right. Thus

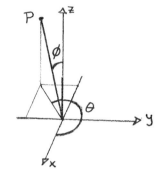

$\boxed{\phi = \pi/6}$ is the correct choice. However,

since both x and y are negative, then we

need to take θ so that $\pi < \theta < 3\pi/2$. Since

$$\sin (\psi + \pi) = -\sin \psi$$

$$\sin (\pi/4) = \sqrt{2}/2$$

then $\qquad\qquad \sin ((\pi/4) + \pi) = -\sin (\pi/4) = -\sqrt{2}/2$

Thus $\quad \boxed{\theta = 5\pi/4}$ is the choice for θ.

Our spherical coordinates are thus

$$\boxed{(\rho, \theta, \phi) = (16, 5\pi/4, \pi/6)} \qquad\qquad \square$$

Example E. Express $z^2 - x^2 - y^2 = 1$ in spherical coordinates.

Solution. $1 = z^2 - x^2 - y^2 = \rho^2 \cos^2 \phi - r^2$

⌐Introducing r as an intermediate step can be very helpful.

$$= \rho^2 \cos^2 \phi - \rho^2 \sin^2 \phi = \rho^2 (\cos^2 \phi - \sin^2 \phi)$$

$$= \rho^2 \cos 2\phi$$

Thus

$$\rho^2 = 1 / \cos 2\phi = \sec 2\phi$$

$$\rho = \pm \sqrt{\sec 2\phi}$$ □

Example F. Express $\rho = 4 \sin \phi \sin \theta$ in rectangular coordinates.

Solution.

$$\rho = 4 \sin \phi \sin \theta$$

$$\rho^2 = 4 \rho \sin \phi \sin \theta$$

$$x^2 + y^2 + z^2 = 4y$$

$$x^2 + (y^2 - 4y + 4) - 4 + z^2 = 0$$

$$x^2 + (y - 2)^2 + z^2 = 4$$ □

Converting between cylindrical and spherical coordinates is relatively easy since the θ-coordinate is the same in both. Here is an example:

Example G. Convert the point with cylindrical coordinates $(2, 2\pi/3, 6)$ to spherical coordinates.

Solution.   Since $r^2 = x^2 + y^2$,

$$\rho = \sqrt{x^2 + y^2 + z^2} = \sqrt{r^2 + z^2} = \sqrt{(2)^2 + (6)^2} = \sqrt{40} = \boxed{2\sqrt{10} = \rho}$$

Thus the equation $r \cos \theta = \rho \sin \phi \cos \theta$ becomes

$$2 \cos (2\pi/3) = 2\sqrt{10} \sin \phi \cos (2\pi/3)$$

$$\sin \phi = 1/\sqrt{10} \approx .316$$

Since $\sin (.3214) \approx .316$ (Appendix 3, Table 1) and the point $(2, 2\pi/3, 6)$ lies above the xy-plane (so that ϕ is between 0 and $\pi/2$), we have $\phi \approx .3214$ radians. Hence the spherical coordinates are

$$\boxed{(\rho, \theta, \phi) \approx (2\sqrt{10}, 2\pi/3, .3214)}$$ $\square$

Cylindrical and spherical coordinates have their most important uses with double and triple integration, to be studied in Chapter 17. Suffice it to say that we apply these coordinate systems to integration over objects which have some symmetry with respect to either a <u>line</u> or a <u>point</u> in 3-space. This will be done in §17.7, and will require you to be very comfortable with the elementary cylindrical and spherical coordinate surfaces. <u>Be sure to learn these surfaces well</u>!

Chapter 16: Partial Derivatives

Section 16.1. Functions of Two Variables

Real valued functions of two and three variables occur over and over again in the applications of calculus. In economics, the profit π of a one-product company will be a function of the unit price P , the quantity sold Q , and the costs of production C , i.e.,

$$\pi = f(P, Q, C)$$

In physics, the temperature T in a region of space will often be a function of the x, y, z position coordinates,

$$T = f(x, y, z)$$

Or, considering a more specific situation, the ideal gas law can be written as

$$P = f(n, T, V) = R \frac{nT}{V}$$

where P is the pressure of an enclosed gas, V is the volume, T is the absolute temperature (i.e., in degrees Kelvin), n is the number of moles of the gas, and R is a proportionality constant. A simple biological example comes from Poiseuille's Law for the resistance of a blood vessel. This law can be written as

$$R = f(\ell, r) = \alpha \ell r^{-4}$$

where R is the resistance of a blood vessel of length ℓ and radius r , and α is a proportionality constant.

There are also many applications of functions of two variables in the various branches of mathematics. For instance, probability and statistics depend upon the use of functions of several variables, and upon calculus techniques for differentiating and integrating them. As a specific example, any probabilistic question concerning two

varying quantities X and Y which are

independently distributed according to the

standard bell curve $\left(w = \dfrac{1}{\sqrt{2\pi}} \ e^{-t^2/2} \right)$

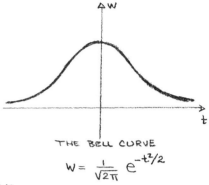

THE BELL CURVE

$w = \dfrac{1}{\sqrt{2\pi}} \ e^{-t^2/2}$

can be answered by utilizing the joint probability

density function

$$\rho_{X,Y}(x,y) \ = \ \frac{1}{2\pi} \ e^{-(x^2 + y^2)/2}$$

along with the calculus operations of "partial differentiation" and "double integration."

In the interests of simplicity, Anton considers functions of just two variables for the

first part of this chapter. Once the techniques for functions of two variables are mastered,

their generalizations to higher dimensions are quite easy to learn.

1. The domain of a function of two variables. As with functions of one variable, a function of two

variables will not necessarily accept every pair of real numbers as an input. For example,

$$f(x,y) \ = \ 1/(x-y)$$

will clearly not accept any pair of real numbers (x, y) where x = y. Such a point is not in

the domain of the function f.

In some situations the domain of a function will be stated explicitly. Such an example is

$$f(x,y) \ = \ x^3 + xy , \qquad \text{where} \qquad x^2 - y > 0$$

In this case we have been given the domain

$$D \ = \ \{ \text{all} \ (x,y) \ \text{for which} \ x^2 - y > 0 \}$$

A sketch of this domain is given in Anton's Example 2, although it arises as the domain of a different function in that example.

However, more often than not the domain of a function $z = f(x, y)$ will not be stated explicitly. In that case the following rule applies:

The Implicit Domain Rule

> If the rule for evaluating a function $z = f(x, y)$ is given by a formula, and <u>if there</u>
> <u>is no mention of the domain,</u> it is understood that the domain consists of all pairs of
> real numbers (x, y) for which the formula makes sense and yields a real value.

In particular, pairs (x, y) which yield

    (i)   negative numbers under square root signs, or

    (ii)   zeros in denominators

are <u>NOT</u> in the domain of the function. Anton's Examples 1 and 2 illustrate the use of this rule. Here is one more example:

<u>Example A.</u>   Sketch the domain of the function $f(x, y) = \sqrt{xy - 1}$.

<u>Solution.</u>   Since only non-negative numbers have square roots which are real numbers, we must restrict ourselves to (x, y) values for which $xy - 1 \geq 0$. Thus the domain of $f(x, y)$ must be

$$D = \{\text{all points } (x, y) \text{ for which } xy \geq 1\}$$

To sketch this region we notice that the pair of hyperbolic curves $xy = 1$ separate the plane into three regions, A, B and C. Take a test point in each region and determine if that point is in the domain D :

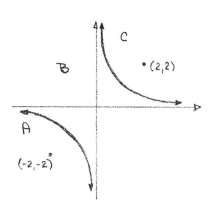

region A: The test point $(-2, -2)$ gives $xy = 4 > 1$. Thus $(-2, -2)$ is in the domain D, which shows that <u>all of A is in D</u>.

region B: The test point $(0, 0)$ gives $xy = 0 < 1$. Thus $(0, 0)$ is not in the domain D, which shows that <u>no point of B is in D</u>.

region C: The test point $(2, 2)$ gives $xy = 4 > 1$. Thus $(2, 2)$ is in the domain D, which shows that <u>all of C is in D</u>.

Since any point (x, y) <u>on</u> the hyperbolas $xy = 1$ is also clearly in the domain D, a sketch of the domain is as shown to the right. □

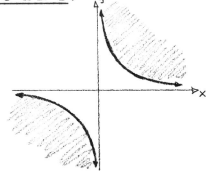

Note the use of test points in Example A. They make use of the important principle that, if an inequality is replaced by the associated equation (e.g., $xy \geq 1$ by $xy = 1$) and the graph of this equation divides the plane into regions, then <u>all points of a region will satisfy the inequality if and only if any one point of the region does.</u> *

2. <u>The graph of a function of two variables.</u> The <u>graph</u> of a function $z = f(x, y)$ is simply the collection of points (x, y, z) in 3-space which satisfy the equation $z = f(x, y)$. Thus, to every point (x, y) in the domain of f there corresponds one and only one point (x, y, z) on the

---

* A precise statement of the principle would require some assumptions about the "continuity" of the functions $f(x, y)$ involved in the inequality. However, these assumptions pose no problems with the simple inequalities arising in most problems.

graph of f, the point whose z-value is f(x, y). Thus the graph will usually be a <u>surface</u>

in 3-space, as shown in the picture on the bottom of the previous page. Note that the domain

of z = f(x, y) consists of points in the xy-plane while the points on the graph are points

above (or below) them in 3-space.

In Examples 3 through 6 Anton sketches the graphs of four different functions f(x, y).

The first is merely a plane, while the remaining three are pieces of the quadric surfaces we

studied in §15.8. In this way Anton is able to rely on the graphs already obtained in §15.8.

<u>Question</u>: What do you do when requested to sketch a graph which is <u>not</u> part of a quadric

surface? Or suppose you do have a quadric surface, but have forgotten its graph?

<u>Answer</u>: use the graphing procedure which we developed in §15.8 of <u>The Companion</u>.

Here's an example.

<u>Example B.</u>   Sketch the graph of the function $f(x, y) = 1 - xy^2$.

<u>Solution.</u>   Let $z = 1 - xy^2$. Then the equation can be written in the form $g(x, y, z) = 1 - xy^2 - z = 0$. The §15.8 graphing procedure then proceeds as follows:

<u>Step 1.</u>   Find the <u>intercepts</u> with the three coordinate axes.

x-intercepts: y = z = 0 gives 1 = 0 ,

       which is impossible. Thus the surface

       does not intersect the x-axis.

y-intercepts: x = z = 0 gives 1 = 0 , so

       there is no y-axis intersection

z-intercepts: x = y = 0 gives 1 - z = 0 ,

       or z = 1. Thus the z-axis intersection

       is (0, 0, 1).

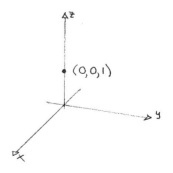

Step 2. Find the coordinate traces on the three coordinate planes.

xy-trace: $z = 0$ gives $1 - xy^2 = 0$,

   or $x = 1/y^2$. This gives a pair of

   "hyperbolic-like" curves, as shown.

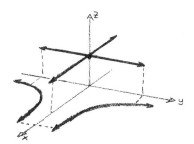

xz-trace: $y = 0$ gives $1 - z = 0$,

   or $z = 1$, a line.

yz-trace: $x = 0$ gives $1 - z = 0$, or

   $z = 1$, a line.

Step 3. Determine some general traces.

$x = k$: $1 - ky^2 - z = 0$, or

$$z = 1 - ky^2$$

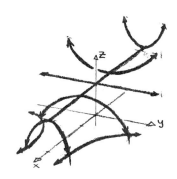

   For $k \neq 0$ these are all parabolas

   with vertices at $y = 0$, $z = 1$. If

   $k > 0$, the parabola turns down, and if

   $k < 0$, the parabola turns up.

Our traces do not make clear what is happening above the plane $z = 1$; the upward turning parabolas are not at the moment "tied together." Let's remedy that by taking the trace at

$z = 2$: $1 - xy^2 - 2 = 0$, or

$$x = -1/y^2$$

   This gives a pair of "hyperbolic-like"

   curves, as shown.

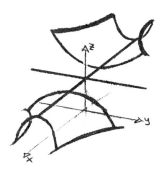

Now we tie the "right and left edges" together with traces at

y = ± 1 :   1 - x - z = 0 , or z = 1 - x

These are, of course, both lines.

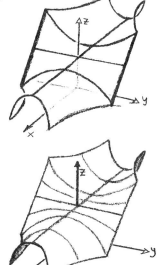

Step 4.   The <u>final sketch</u>.

The surface is one sheet,

with parabolas for the

traces which are parallel

to the yz-plane.

3.   <u>Level curves for a function of two variables.</u>   Sketching the graph in 3-space for a function

z = f(x, y) can be a nasty business, as Example B amply demonstrates. However, for many

purposes a two-dimensional graph of the <u>level curves</u> for z = f(x, y) is an adequate replacement

for the full three-dimensional graph. A <u>level curve</u> for f(x, y) is the set of points in the

xy-plane   on which the function takes on a constant value k , i. e.,

$$f(x, y) = k$$

We obtain different level curves by choosing different values of the constant k . By far

the best way to interpret level curves are as <u>contour lines on a topographical map</u>. Anton discusses

this and provides some nice pictures following Example 6 . Here is another illustration:

<u>Example C.</u>   Sketch the level curves of the function

$$f(x, y) = \frac{x^2}{2x - y^2}$$

for the following values of k :

(i) k = 0   (ii) k = 1   (iii) 0 < k < 1   (iv) 1 < k   (v) k < 0

Solution.   First observe that the domain of $f(x, y)$ consists of all points (x, y) which are not on the curve $y^2 = 2x$. Hence any of these points which, through our algebraic simplifications, end up on our level curves, must be deleted.

We simplify $\dfrac{x^2}{2x - y^2} = k$ as follows:

$$x^2 = 2kx - ky^2$$

$$x^2 - 2kx + ky^2 = 0$$

$$(x^2 - 2kx + k^2) - k^2 + ky^2 = 0$$

$$(x - k)^2 + ky^2 = k^2$$

k = 0.   Then $x^2 = 0$, or x = 0 , with $(0, 0)$ deleted.

Thus the level curve for k = 0 is the y-axis with $(0, 0)$ deleted.

k = 1.   Then $(x - 1)^2 + y^2 = 1$, which is a circle with center $(1, 0)$ and radius 1 with $(0, 0)$ deleted.

0 < k < 1.   In this case we can rewrite our level curve equation as

(∗) $$\boxed{\dfrac{(x - k)^2}{k^2} + \dfrac{y^2}{k} = 1}$$

which is an ellipse with center $(k, 0)$. (See Appendix E. 3) Its "y-diameter" is $2\sqrt{k}$ and its "x-diameter" is 2k ; hence (since 0 < k < 1 implies $k < \sqrt{k}$) we see that the ellipse is "taller than it is wide. " (Again, $(0, 0)$ is deleted.)

1 < k.   Equation (*) is again an ellipse, but (since 1 < k implies $\sqrt{k} < k$) we

see that the ellipse is "wider than it is tall. "

k < 0.   Equation (*) is now a hyperbola, with center (k, 0) , and (0, 0) deleted.   □

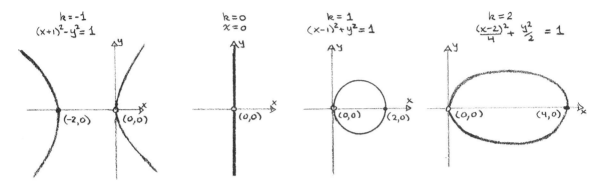

Important observation: Every point in the domain of f(x, y) must appear on one and only one

level curve for f . In particular this means that two different level curves for f cannot

intersect. Two level curves may appear to intersect at a point $P(x_0, y_0)$, but they do NOT

actually intersect. In such a case P can be an element of at most one of the two curves, or

perhaps neither. Notice how all the level curves in Example B appear to intersect at (0 , 0) ,

but (0 , 0) is not on any of them!

Section 16. 2 : Partial Derivatives

1. The computation of partial derivatives.   What you should remember is that the calculation

of $\frac{\partial f}{\partial x}$ (x, y) is just like the calculation of an ordinary derivative $\frac{df}{dx}$ (x) except that you

must treat y as a constant (and for $\frac{\partial f}{\partial y}$ (x, y) you treat x as a constant). This

difference is the only aspect of the computation of partial derivatives which causes any difficulty.

Here are some examples (note that both f_x and $\frac{\partial f}{\partial x}$ are used to denote the partial of f with respect to x):

Example A. Compute $f_x(x,y)$ for $f(x,y) = \sin x + \cos y$.

Solution. $f_x(x,y) = \dfrac{\partial}{\partial x}(\sin x + \cos y)$

$$= \cos x - \sin y \quad ??$$

FALSE

It is tempting to say "the derivative of the cosine is the minus sine," and if your derivative was with respect to y, you'd be correct. However, the derivative is with respect to x, and hence <u>cos y</u> <u>is a constant</u> which has a zero derivative.

Thus

$$f_x(x,y) = \frac{\partial}{\partial x}(\sin x + \cos y)$$

$$= \cos x + 0$$

$$= \cos x \qquad\qquad \square$$

Example B. Compute $\dfrac{\partial w}{\partial s}$ for $w = w(r,s) = r^2 e^{s \tan r}$

Solution. $\dfrac{\partial w}{\partial s} = \dfrac{\partial}{\partial s}(r^2 e^{s \tan r})$

$$= 2re^{s \tan r} + r^2 e^{s \tan r}\frac{\partial}{\partial s}(s \tan r)$$

FALSE

Ahh, it is so tempting to use the product rule, along with "the derivative of r^2 is $2r$." WRONG! The derivative is taken with respect to s, not r, and thus r^2 is simply <u>a multiplicative constant</u> which pulls out of the partial derivative.

Thus

$$\frac{\partial w}{\partial s} = \frac{\partial}{\partial s} (r^2 e^{s \tan r}) = r^2 \frac{\partial}{\partial s} (e^{s \tan r})$$

$$= r^2 e^{s \tan r} \frac{\partial}{\partial s} (s \tan r)$$

Now be careful! The $\tan r$ term is simply a multiplicative constant when differentiating by s. Thus

$$= r^2 e^{s \tan r} \tan r \frac{\partial}{\partial s} (s)$$

$$= r^2 e^{s \tan r} \tan r \qquad\qquad \square$$

Just as with single variable functions, a multivariate function need not have partial derivatives defined at every point in its domain. Put in another way, the domain of a partial derivative function might be "smaller" than the domain of the original function, as the following example shows:

Example C.   Compute the first-order partial derivatives for

$$f(x, y) = (xy)^{1/3} = x^{1/3} y^{1/3}$$

and determine the domain for each function so obtained.

Solution.   The domain D of $f(x, y)$ consists of all pairs of real numbers (x, y). However, notice that

$$f_x(x, y) = \frac{\partial f}{\partial x}(x, y) = \frac{1}{3 x^{2/3}} y^{1/3} = \frac{1}{3} \left[\frac{y}{x^2} \right]^{1/3}$$

has domain $D_x = \{(x, y) | x \neq 0\}$,   and

$$f_y(x, y) = \frac{\partial f}{\partial y}(x, y) = \frac{1}{3 y^{2/3}} x^{1/3} = \frac{1}{3} \left[\frac{x}{y^2} \right]^{1/3}$$

has domain $D_y = \{(x, y) | y \neq 0\}$

Thus D_x is "smaller" than D in the sense that every pair (x, y) in D_x is also in D, but every pair in D is not in D_x. Similarly, D_y is also "smaller" than D. □

2. Implicit differentiation is carried out in the same way with partial derivatives as with ordinary derivatives. The only new aspect is the need to identify those variables which are to be held constant.

Example D. Suppose $r^2 \ln(st - r^2) + \sqrt{s + r} = \sqrt{2}$. Determine $\partial s/\partial t$ at the point $(r, s, t) = (1, 1, 2)$.

Solution. Since solving the equation for s in terms of r and t seems hopeless, we are forced to use implicit differentiation, in this case treating r as a constant. Thus

$$0 = \frac{\partial}{\partial t} (\sqrt{2}) = \frac{\partial}{\partial t} [r^2 \ln(st - r^2) + \sqrt{s + r} \,]$$

$$= r^2 \frac{\partial}{\partial t} \ln(st - r^2) + \frac{\partial}{\partial t} (s + r)^{1/2}$$

$$= r^2 \left(\frac{1}{st - r^2} \right) \frac{\partial}{\partial t} (st - r^2) + \frac{1}{2} (s + r)^{-1/2} \frac{\partial}{\partial t} (s + r)$$

But $\frac{\partial}{\partial t} (st) = t \frac{\partial s}{\partial t} + s$ by the product rule,

and $\frac{\partial}{\partial t} (r^2) = 0$ since r is being held constant.

Thus

$$0 = \frac{r^2}{st - r^2} (t \frac{\partial s}{\partial t} + s) + \frac{1}{2} (s + r)^{-1/2} \frac{\partial s}{\partial t}$$

Therefore, at the point $(r, s, t) = (1, 1, 2)$ we have

$$0 = (\frac{1}{1}) (2 \frac{\partial s}{\partial t} + 1) + \frac{1}{2} (2)^{-1/2} \frac{\partial s}{\partial t}$$

which solves to $\qquad \dfrac{\partial s}{\partial t} = - \dfrac{2\sqrt{2}}{4\sqrt{2} + 1} \approx -.4249$ □

3. <u>Higher order partial derivatives</u> are simply partial derivatives of partial derivatives, and as
    such pose no additional difficulties <u>except in their notation.</u> Notice, however, that the variables
    which are held constant can vary from step-to-step in a higher order partial, depending upon
    which variable you are differentiating with respect to.

<u>Example E.</u>   Compute f_{uuv} for $f(u, v) = v e^{uv}$

<u>Solution.</u> $$f_{uuv} = \frac{\partial^3 f}{\partial v \, \partial u^2} = \frac{\partial^2}{\partial v \, \partial u} \left(\frac{\partial f}{\partial u} \right) = \frac{\partial^2}{\partial v \, \partial u} (v^2 e^{uv})$$

$$\boxed{v \quad \text{held constant}}$$

$$= \frac{\partial}{\partial v} (v^3 e^{uv}) \qquad = 3v^2 e^{uv} + v^3 u e^{uv}$$

$$\boxed{v \quad \text{held constant}} \qquad \boxed{u \quad \text{held constant}} \qquad \qquad \square$$

<u>Notation.</u>   Just before Example 7, Anton makes an important remark concerning a difference
in the $\dfrac{\partial^2 f}{\partial x \, \partial y}$ and f_{xy} notations for higher order partial derivaties:

> In ∂ notation, the order of differentiation is read right to left.
>
> In subscript notation, the order of differentiation is read left to right.

That is, $\dfrac{\partial^2 f}{\partial x \, \partial y}$ means "first differentiate with respect to y and then differentiate with
respect to x " while f_{xy} means "first differentiate with respect to x and the differentiate
with respect to y. " If you remember the equation

$$\boxed{\dfrac{\partial^2 f}{\partial y \, \partial x} = f_{xy}}$$

you should be able to avoid confusion. Notational inconsistencies such as this exist because different mathematicians have grown accustomed to different notations over the years and changing them now would be even more confusing than living with the differences. Your best bet as a newcomer to partial derivatives is not to waste your energy fighting this inconsistency, but, instead, to accept it as an unpleasant fact of mathematical life.

4. <u>Geometric interpretation of partial derivatives.</u> As Anton carefully points out in Figure 16.2.1, the two partial derivatives

$$f_x(x_0, y_0) = \frac{\partial f}{\partial x}(x_0, y_0) \quad \text{and} \quad f_y(x_0, y_0) = \frac{\partial f}{\partial y}(x_0, y_0)$$

are the <u>slopes of two tangent lines</u> to the surface

$$z = f(x, y) \quad \text{at the point} \quad (x_0, y_0, f(x_0, y_0))$$

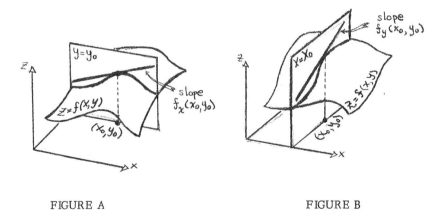

FIGURE A   FIGURE B

This geometric observation is important because the two tangent lines in question will be used in §16.6 to determine an equation for the <u>tangent plane</u> to the surface $z = f(x, y)$ at the point $(x_0, y_0, f(x_0, y_0))$. (See diagram on following page.)

Example F.   Find the slope of the tangent line

at $(0, 1, 2)$ to the curve of intersection

of the surface $xy^2 + yz^2 + zx^2 = 4$

and the plane $y = 1$.

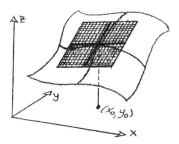

Solution.   As indicated by Figure A, the slope we desire is given by $\partial z/\partial x$ evaluated at the

point $(0, 1, 2)$. This is most easily found by implicit differentiation:

$$0 = \frac{\partial}{\partial x}[4] = \frac{\partial}{\partial x}[xy^2 + yz^2 + zx^2]$$

$$= y^2 + y\left(2z\,\frac{\partial z}{\partial x}\right) + \left(\frac{\partial z}{\partial x}\right)x^2 + z(2x)$$

$$= 2xz + y^2 + (2yz + x^2)\frac{\partial z}{\partial x}$$

Thus, at $(x, y, z) = (0, 1, 2)$ we obtain

$$0 = 1 + 4\,\frac{\partial z}{\partial x}, \quad \text{or} \quad \frac{\partial z}{\partial x} = -1/4$$

□

5. Rate of change interpretation of partial derivatives.   From one-variable calculus we know that

$f'(x_0)$ is the rate of change of the function $f(x)$ at the point x_0. However, for a function

$f(x, y)$, the partial derivative $f_x(x_0, y_0)$ is simply an "ordinary" derivative for which y

is held at the constant value $y = y_0$. Thus we see that

> $f_x(x_0, y_0)$ is the rate of change of $f(x, y)$ at (x_0, y_0)
>
> with respect to the quantity x (i.e., with y held at
>
> the constant value $y = y_0$).

A similar statement holds true for $f_y(x_0, y_0)$. Here is a good example of how the rate of change interpretation of partial differentiation arises in applications, in this case in economic theory:

A <u>production function</u> for a one-product firm expresses the output Q as a function of the capital input K and the labor input L, i.e.,

$$Q = f(K, L)$$

Then Q_K, the rate of change of output with respect to capital input, is called the marginal physical product of capital (abbreviated MPP_K), while

Q_L, the rate of change of output with respect to labor input, is called the marginal physical product of labor (abbreviated MPP_L).

The graph in 3-space of $Q = f(K, L)$ is called the production surface, while the level curves in 2-space are termed the production isoquants.

<u>Example G.</u> Suppose the production function for a one-product firm is given by $Q = K^{1/2} \ln L$. If the current daily capital and labor input are $K = 5000$ and $L = 3000$, then which of the two inputs should be increased to achieve the largest increase in output? *

<u>Solution.</u> We need to compare $\partial Q / \partial K$ and $\partial Q / \partial L$, i.e., MPP_K and MPP_L.

$$MPP_K = \frac{\partial Q}{\partial K} = \frac{1}{2} K^{-1/2} \ln L = \frac{1}{2} (5000)^{-1/2} \ln 3000 \cong .0566$$

$$MPP_L = \frac{\partial Q}{\partial L} = K^{1/2} / L = (5000)^{1/2} / 3000 \cong .0236$$

Since $MPP_K > MPP_L$, increasing the capital input will yield the largest increase in output.

<u>Example H.</u> The oxygen consumption X of a fur-bearing animal can be approximated by the

---

* A better question to ask would be "In what <u>proportion</u> should <u>both</u> K and L be increased to achieve the largest increase in Q?" The answer to this question requires <u>gradient</u> techniques, and we will return to it in Examples E and H of §16.5.

formula

$$X = k(T_b - T_f) w^{-2/3}$$

where

T_b = internal body temperature

T_f = outside temperature of the fur

w = weight of the animal (in kg)

k = a constant depending on the units used for temperature and oxygen consumption

Determine formulas for the rate of change of X with respect to T_b, T_f and w when the remaining two variables are held constant.

<u>Solution.</u> We wish to calculate $\partial X/\partial T_b$, $\partial X/\partial T_f$ and $\partial X/\partial w$.

$$\frac{\partial X}{\partial T_b} = kw^{-2/3} \qquad\qquad \frac{\partial X}{\partial T_f} = -kw^{-2/3}$$

$$\frac{\partial X}{\partial w} = -\frac{2k}{3}(T_b - T_f) w^{-5/3} \qquad\qquad \Box$$

6. <u>Limit Definitions of Partial Derivatives.</u> (Optional) Using the definition of the derivative of a function of one variable

$$f'(x) = \lim_{h \to 0} \frac{f(x+h) - f(x)}{h} \qquad\qquad \text{(Definition 3.1.2)}$$

we can write a "limit definition" of the partial derivative $f_x(x, y)$ as follows: When y is held constant and x is allowed to vary, then z = f(x, y) becomes a function of the <u>one</u> variable x. Therefore Definition 3.1.2 can be applied to express its derivative, which is the partial derivative $f_x(x, y)$, as a limit:

$$f_x(x,y) = \lim_{h \to 0} \frac{f(x+h,y) - f(x,y)}{h}$$

Similarly

$$f_y(x,y) = \lim_{h \to 0} \frac{f(x,y+h) - f(x,y)}{h}$$

These are the expressions Anton is asking for in Exercise 48 of Exercise Set 16.2.

Section 16.3: <u>Limits, Continuity and Differentiability</u>

To state precisely most of the important calculus results for functions of two variables, we need a stronger definition of differentiability than merely the existence of partial derivatives. This stronger concept is given in Anton's Definition 16.3.4, and it can be difficult to understand when first encountered. We will summarize Anton's discussion, highlighting the most important aspects of the section. We hope this will make the concepts and their applications easier to understand.

1. <u>Limits.</u> Suppose $z = f(x,y)$ is defined for all points (x,y) in a circular region centered on (x_0, y_0) except possibly at the point (x_0, y_0) itself. Then

$f(x, y)$ has <u>limit</u> L as (x, y) approaches (x_0, y_0)

written $\qquad \lim_{(x,y) \to (x_0, y_0)} f(x, y) = L$

if the numbers $f(x, y)$ become close to L as the

points (x, y) become close to (x_0, y_0).

A precise "epsilon-delta" definition is given

in Anton's 16.3.1 for the limit concept,

but the intuitive statement is all that is

needed for our level of work (more advanced

work <u>must</u> use the $\epsilon - \delta$ formulation).

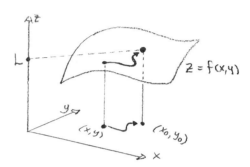

Example A.

(i) $\lim_{(x,y) \to (0,0)} x + y = 0$, for it is clear that as x and y become small, their

sum becomes small.

(ii) $\lim_{(x,y) \to (1,2)} x^2/y = 1/2$, for it is clear that if x approaches 1 and

y approaches 2, then x^2/y approaches $1^2/2 = 1/2$.

(iii) $\lim_{(x,y) \to (0,0)} \dfrac{1}{x^2 + y^2} = \ldots$ undefined, since $\dfrac{1}{x^2 + y^2}$ can assume arbitrarily large

values as x and y approach 0.

(iv) $\lim_{(x,y) \to (0,0)} \dfrac{xy}{x^2 + y^2} = \ldots$ hmm... This limit looks more complicated since both

numerator and denominator approach 0 as x and y approach 0.

We will show that the limit does <u>not exist</u> because

$$f(x,y) = \frac{xy}{x^2 + y^2}$$

approaches different values along different paths leading to $(0,0)$ (see Anton's Example 1, and the remarks previous to it). For suppose (x,y) approaches $(0,0)$ along the line $y = mx$. Then

$$f(x,y) = \frac{x(mx)}{x^2 + (mx)^2} = \frac{mx^2}{x^2 + m^2 x^2} = \frac{m}{1 + m^2}$$

Thus $f(x,y)$ is constant along each line through the origin, and we obtain

$$\lim_{\substack{(x,y) \to (0,0) \\ \text{along } y = mx}} \frac{xy}{x^2 + y^2} = \frac{m}{1 + m^2}$$

In particular, $f(x,y)$ appraoches $1/2$ along the line $y = x$ $(m = 1)$, while it approaches 0 along the line $y = 0$ $(m = 0)$. Since $1/2 \neq 0$, then $\displaystyle\lim_{(x,y) \to (0,0)} \frac{xy}{x^2 + y^2}$ does not exist.

(v) $\displaystyle\lim_{(x,y) \to (0,0)} \frac{e^{xy} - 1}{xy} = \ldots$ hmm... Another "zero over zero." In this case, however, since x and y appear only in the form of the product xy, we can change our two-variable limit into a one-variable limit by using $h = xy$ and observing that $h \to 0$ as $(x,y) \to (0,0)$. Thus

$$\lim_{(x,y) \to (0,0)} \frac{e^{xy} - 1}{xy} = \lim_{h \to 0} \frac{e^h - 1}{h}$$

$$= \lim_{h \to 0} \frac{e^h}{1} \qquad \text{by L'Hôpital's Rule (§10.2)}$$

$$= e^0 = 1$$

Note: If you wish to avoid using L'Hôpital's Rule, you can evaluate

the h-limit as follows:

$$\lim_{h \to 0} \frac{e^h - 1}{h} = \lim_{h \to 0} \frac{e^{0+h} - e^0}{h}$$

$$= \frac{d}{dx}[e^x]\Big|_{x = 0} \qquad \text{by the definition of the derivative (\S3.1)}$$

$$= e^x\Big|_{x = 0} = e^0 = 1 \qquad \qquad \square$$

The computation of more complicated limits will be discussed in a moment using the concept

of continuity.

2. Continuity. Suppose $z = f(x,y)$ is defined for all points (x,y) in a circular region

centered on (x_0, y_0). Then

$$\boxed{\begin{array}{l} f(x,y) \text{ is continuous at } (x_0, y_0) \\[2mm] \text{if} \qquad \lim_{(x,y) \to (x_0, y_0)} f(x,y) = f(x_0, y_0) \end{array}}$$

i. e., if the numbers $f(x,y)$ become close to $f(x_0, y_0)$ as the points (x,y) become close

to (x_0, y_0). Loosely stated, $f(x,y)$ is continuous at (x_0, y_0) if there is no "break" or

"gap" in the graph of f over (x_0, y_0).

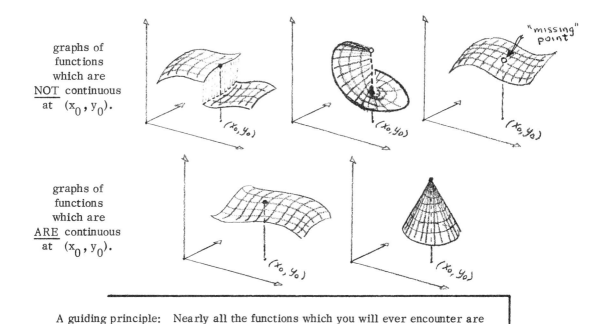

graphs of
functions
which are
NOT continuous
at (x_0, y_0).

"missing" point

(x_0, y_0)

(x_0, y_0)

(x_0, y_0)

graphs of
functions
which are
ARE continuous
at (x_0, y_0).

(x_0, y_0)

(x_0, y_0)

A guiding principle: Nearly all the functions which you will ever encounter are

continuous at the interior points of their domains.

When requested to determine where a specific function $z = f(x, y)$ is continuous, you can use

Theorem 16.3.3 and the remarks following Example 4 concerning the sum, difference, product

and quotient of continuous functions (i. e. , the analogue for functions of two variables of Theorem

3.7.3). Anton's Examples 2 through 6 illustrate the use of these results in verifying continuity.

Here is another example:

Example C. Sketch the region where $f(x, y) = \sqrt{1 - x^{-2} \ln y}$ is continuous.

Solution. We first must determine the domain of f (using the Implicit Domain Rule of

§16.1.1 of course). The function $1 - x^{-2} \ln y$ is defined only when $\boxed{y > 0 \quad \text{and} \quad x \ne 0}$;

moreover, the square root of $1 - x^{-2} \ln y$ is a real number only when $1 - x^{-2} \ln y \ge 0$, i. e.,

$$1 \ge x^{-2} \ln y$$

$$x^2 \ge \ln y \qquad\qquad \text{since} \quad x^2 > 0$$

$$e^{(x^2)} \ge e^{\ln y} = y \qquad\qquad \text{since} \quad e^a \ge e^b \quad \text{whenever} \quad a \ge b$$

Thus the domain of f consists of all

(x, y) such that $\boxed{e^{(x^2)} \geq y > 0}$

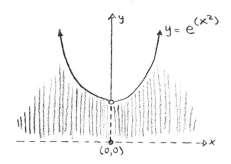

and $\boxed{x \neq 0}$. This is the shaded region

shown to the right.

To determine where f is continuous we apply the results cited above.

Since $g(x) = x^{-2}$ is continuous for $x \neq 0$, and $h(y) = \ln y$ is continuous for $y > 0$,

then by Theorem 16.3.3 (a) we find $g(x)\,h(y) = x^{-2}\ln y$ is continuous for $x \neq 0$ and $y > 0$.

Moreover, since the constant function $c(x, y) = 1$ is clearly continuous everywhere, then

$1 + x^{-2}\ln y$ is continuous on its full domain since it is a <u>sum</u> of continuous functions. Finally,

$f(x, y)$ is the composition of $1 + x^{-2}\ln y$ with the square root function $k(u) = \sqrt{u}$, and

$k(u) = \sqrt{u}$ is continuous for all $u > 0$. * Thus, by Theorem 16.3.3 (b) we see that

$f(x, y) = \sqrt{1 + x^{-2}\ln y}$ is continuous when

$1 + x^{-2}\ln y > 0$, i.e.,

$$\boxed{e^{(x^2)} > y > 0 \quad \text{and} \quad x \neq 0}$$

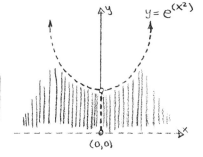

This region is shown to the right.   □

Limits of many common functions now become easy: if you know that $f(x, y)$ is

continuous at (x_0, y_0) , then its limit as $(x, y) \to (x_0, y_0)$ is just $f(x_0, y_0)$ (i.e., you

just "plug in") by the definition of continuity at (x_0, y_0) ! Example 7 illustrates a typical

application of this fact.

---

*
   Technical point: In this section we do not allow a function to be called continuous at a <u>boundary</u>
   <u>point</u> of its domain. Thus $k(u) = \sqrt{u}$ is <u>not</u> continuous at $u = 0$.

3. <u>Differentiability.</u> Suppose $f(x, y)$ is a function defined for all points (x, y) in a circular region centered on (x, y). Anton defines the concept of differentiability for f at (x_0, y_0) in Definition 16.3.4. This is certainly not an easy concept to understand, even with the careful motivation given prior to the definition. The intuitive geometric meaning of the statement "$f(x, y)$ is differentiable at (x_0, y_0)" is that

the surface $z = f(x, y)$ has a non-vertical tangent plane over the point (x_0, y_0).

As you will see in subsequent work, most of the theorems concerning functions of two variables will include the hypothesis that the functions involved be differentiable. This could be a nasty problem when we wish to apply our theorems to particular situations, since verifying a function to be differentiable via Definition 16.3.4 can be <u>very unpleasant</u>! Fortunately Theorem 16.3.6 comes to the rescue:

Criteria for Differen- tiability

Suppose $f(x, y)$ has <u>continuous</u> first order partial derivatives $f_x(x, y)$ and $f_y(x, y)$ for all points (x, y) in a circular region centered on (x_0, y_0).

Then $f(x, y)$ is differentiable at (x_0, y_0).

Ahh..., we now have a condition for differentiability which is relatively easy to verify in specific situations. Here's an example:

Example C.   Show that $f(x, y) = x e^{xy}$ is a differentiable function.

Solution.   The partial derivatives $f_x(x, y) = e^{xy} + xy e^{xy}$ and $f_y(x, y) = x^2 e^{xy}$ are seen by our earlier methods to be defined and continuous everywhere in the xy-plane.   Thus the hypotheses of Theorem 16.3.6 apply to every point (x_0, y_0),   so that $f(x, y) = x e^{xy}$ is everywhere differentiable.   □

Warning.   The mere existence of partial derivatives is NOT enough to ensure that a function is differentiable. As indicated in the hypothesis of Theorem 16.3.6, the partials need to be continuous as well.

Do not become nervous about differentiability. You need an intuitive understanding of the concept because it will appear among the hypotheses of many future results. However, unless told otherwise, or unless dealing with a very strange function, it is almost always safe to assume a given function is differentiable. You will not be asked (at this level of calculus) actually to verify that a given function is differetiable.

However, it should be made clear that the theoretical material just discussed is of immeasurable importance in more advanced courses.

Section 16.4 :   The Chain Rule for Functions of Two Variables.

Given the importance of the chain rule for one variable functions, it is not surprising that there is an equally important chain rule for functions of two variables. Actually there are several "chain rules," depending on the initial function set-up.

1.   Computations with the basic chain rule.   Suppose $z = f(x, y)$ is a differentiable function of x and y,   while x and y are in turn differentiable functions of t :

$$z = f(x, y), \quad \text{with} \quad x = x(t), \quad y = y(t)$$

Then

$$\frac{dz}{dt} = \frac{\partial z}{\partial x}\frac{dx}{dt} + \frac{\partial z}{\partial y}\frac{dy}{dt}$$

(A)

This is really a pretty easy formula to use in specific cases, ... assuming you have mastered the computation of partial derivatives! Moreover, in many circumstances, the use of this formula can yield quite a savings in time over more direct methods. Anton's Examples 1 and 2 illustrate simple uses of the chain rule; here is a more complex example:

<u>Example A.</u>   Compute $\frac{dz}{dt}$ at $t = 1$ if

$$z = x^{3/2}\, y^2 (y - x^2)^{4/3}, \qquad x = \ln(2t - 1) + 1, \qquad y = e^{t^2 - 1}.$$

<u>Absurd Solution.</u>   You might have the thought "Express z as a function of t and differentiate directly. " Well, you will discard that idea quickly once you stare at

$$z = [\ln(2t - 1) + 1]^{3/2}\, e^{2t^2 - 2}\, [e^{t^2 - 1} - (\ln(2t - 1) + 1)^2]^{4/3}$$

<u>Smart Solution.</u>   Use the chain rule of course.

(i) $\dfrac{dx}{dt} = \dfrac{2}{2t - 1}$;   if $t = 1$,   then $\boxed{\dfrac{dx}{dt} = 2}$

$\dfrac{dy}{dt} = 6t\,e^{t^2 - 1}$;   if $t = 1$,   then $\boxed{\dfrac{dy}{dt} = 2}$

Now use $z = x^{3/2}\, y^2 (y - x^2)^{4/3}$.

(ii) $\dfrac{\partial z}{\partial x} = \dfrac{3}{2} x^{1/2}\, y^2 (y - x^2)^{4/3} + x^{3/2}\, y^2 \left(\dfrac{4}{3}\right)(y - x^2)^{1/3}(-2x)$

At $t = 1$ we find $x = 2$ and $y = 3$.   Thus

evaluating $\dfrac{\partial z}{\partial x}$ at $x = 2$ and $y = 3$ yields $\boxed{\dfrac{\partial z}{\partial x} = \dfrac{219}{2}\sqrt{2}}$

(iii) $\dfrac{\partial z}{\partial y} = x^{3/2} (2y) (y - x^2)^{4/3} + x^{3/2} y^2 (\tfrac{4}{3}) (y - x^2)^{1/3}$

At $t = 1$ we find $x = 2$ and $y = 3$. Thus

evaluating $\dfrac{\partial z}{\partial y}$ at $x = 2$ and $y = 3$ yields $\boxed{\dfrac{\partial z}{\partial y} = -12 \sqrt{2}}$

Hence, at $t = 1$ we obtain

$$\frac{dz}{dt} = \frac{\partial z}{\partial x} \frac{dx}{dt} + \frac{\partial z}{\partial y} \frac{dy}{dt} = \left(\frac{219}{2} \sqrt{2} \right) 2 + (-12 \sqrt{2})(6)$$

or $\boxed{\dfrac{dz}{dt} = 147 \sqrt{2}}$ □

2.   <u>The chain rule for partial derivatives.</u>   Suppose $z = f(x, y)$ is a differentiable function of x and y, while x and y are in turn differentiable functions of u and v:

$$z = f(x, y), \qquad \text{with} \qquad x = x(u, v), \qquad y = y(u, v)$$

Then

$$\boxed{\begin{aligned} \frac{\partial z}{\partial u} &= \frac{\partial z}{\partial x} \frac{\partial x}{\partial u} + \frac{\partial z}{\partial y} \frac{\partial y}{\partial u} \\[2mm] \frac{\partial z}{\partial v} &= \frac{\partial z}{\partial x} \frac{\partial x}{\partial v} + \frac{\partial z}{\partial y} \frac{\partial y}{\partial v} \end{aligned}} \qquad \text{(B)}$$

As Anton carefully points out in the proof of Theorem 16.4.2, these formulas are obtained directly from the more basic chain rule (A). Their use is the same as in the formula (A), and is illustrated in Example 4. We'll give a more exotic example, similar to Anton's Exercises 31 through 36 of Exercise Set 16.4.

Example B. Let $z = f(cx - y, x - c^{-1}y)$. Show that if f is differentiable, then

$$\frac{\partial z}{\partial x} + c \frac{\partial z}{\partial y} = 0 .$$

Solution. Let $u = cx - y$ and $v = x - c^{-1}y$. Then $z = f(u, v)$. Using the chain rule (B) (with a reversal of roles for x, y and u, v) we obtain

$$\frac{\partial z}{\partial x} = \frac{\partial z}{\partial u} \frac{\partial u}{\partial x} + \frac{\partial z}{\partial v} \frac{\partial v}{\partial x} = f_u(u, v)c + f_v(u, v)1$$

$$\frac{\partial z}{\partial y} = \frac{\partial z}{\partial u} \frac{\partial u}{\partial y} + \frac{\partial z}{\partial v} \frac{\partial v}{\partial y} = f_u(u, v)(-1) + f_v(u, v)(-c^{-1})$$

Thus $\frac{\partial z}{\partial x} + c \frac{\partial z}{\partial y} = cf_u(u, v) + f_v(u, v) - cf_u(u, v) - f_v(u, v) = 0$, as desired. □

3. Implicit differentiation and related rates. The chain rules introduced in this section are probably more useful in "theoretical" settings than they are in specific computations. In subsequent material the chain rule will provide the key step in establishing important results (e. g. , it is used to justify Definition 16. 5. 1 for the directional derivative in the next section). However, in many specific computations its use might be helpful but not necessarily crucial.

For example, suppose $F(x, y) = 0$ defines y implicitly as a differentiable function of x . Then, using the chain rule, Anton shows

$$\boxed{\frac{dy}{dx} = - \frac{\partial F / \partial x}{\partial F / \partial y}} \tag{C}$$

Although this is certainly a time saving formula, our standard method of implicit differentiation will yield the same result in specific situations, and does not require the chain rule. Here's an illustration:

Example C.   Find dy/dx given $x^3 y + x e^{xy} = 0$.

Solution 1.   Using (C) with $F(x, y) = x^3 y + x e^{xy}$ yields

$$\frac{dy}{dx} = -\frac{3x^2 y + e^{xy} + xy e^{xy}}{x^3 + x^2 e^{xy}}$$

Solution 2.   Using standard implicit differentiation yields

$$\frac{d}{dx}(x^3 y + x e^{xy}) = \frac{d}{dx} 0 = 0$$

$$3x^2 y + x^3 \frac{dy}{dx} + e^{xy} + x\left(y + x\frac{dy}{dx}\right)e^{xy} = 0$$

$$3x^2 y + e^{xy} + xy e^{xy} + \left(x^3 + x^2 e^{xy}\right)\frac{dy}{dx} = 0$$

$$\frac{dy}{dx} = -\frac{3x^2 y + e^{xy} + xy e^{xy}}{x^3 + x^2 e^{xy}} \qquad \square$$

Anton cites the use of the new chain rules with related rates problems when the rates of three or more quantities are involved. However, in most of the specific cases which you will see, using the chain rule is engaging in overkill. For instance, in Example 5, Anton uses the chain rule on $A = xy$ to obtain

$$\frac{dA}{dt} = y\frac{dx}{dt} + x\frac{dy}{dt}$$

However, this formula is also easily obtained by differentiating $A = xy$ by d/dt and using the product rule. Here's another example:

Example D.   Two straight roads intersect at right angles. Car A, moving on one of the roads, approaches the intersection at 40 mph, and car B, moving at 55 mph, moves away from the intersection on the other road. At what rate is the distance between the cars changing when A is 4 miles from the intersection and B is 3 miles from the intersection?

Solution 1. A diagram for the situation described, with appropriately labelled quantities, is shown to the right. Translating our problem into equations gives:

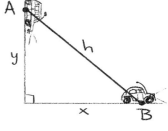

$$\frac{dy}{dt} = -40 \qquad \frac{dx}{dt} = 55$$

What is $\frac{dh}{dt}$ when $y = 4$ and $x = 3$?

We need an equation relating our three quantities x, y and h. This is easily obtained:

$$h = \left(x^2 + y^2\right)^{1/2} \tag{D}$$

We now differentiate this equation by d/dt:

$$\frac{dh}{dt} = \frac{1}{2}\left(x^2 + y^2\right)^{-1/2}\left(2x\,\frac{dx}{dt} + 2y\,\frac{dy}{dt}\right) \tag{E}$$

Finally, plug in our given values:

$$\frac{dh}{dt} = \frac{1}{2}(9 + 16)^{-1/2}(6\,(55) + 8\,(-40))$$

$$= \frac{1}{10}(330 - 320) = \boxed{1 \text{ mph}}$$

Solution 2. If we wish to use the chain rule, we return to equation (D) and compute its derivative as follows:

$$\frac{dh}{dt} = \frac{\partial h}{\partial x}\frac{dx}{dt} + \frac{\partial h}{\partial y}\frac{dy}{dt}$$

$$= \frac{1}{2}\left(x^2 + y^2\right)^{-1/2}(2x)\frac{dx}{dt} + \frac{1}{2}\left(x^2 + y^2\right)^{-1/2}(2y)\frac{dy}{dt}$$

$$= \frac{1}{2}\left(x^2 + y^2\right)^{-1/2}\left(2x\frac{dx}{dt} + 2y\frac{dy}{dt}\right)$$

As you can see, we have just obtained equation (E) again, and the rest of the solution proceeds

as above.   □

4. <u>Geometric justification of the chain rule (optional).</u> Anton proves the chain rule (A) in

Theorem 16.4.1; the proof given uses the rigorous definition of differentiability in a very precise

analytic argument. We'll present here a less rigorous but more geometric argument for (A)

based on the tangent plane concept. This should help you to understand the chain rule better,

and will reinforce in your mind the following tangent plane properties which are important for

later work.

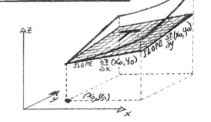

Suppose $z = f(x, y)$ is differentiable

at (x_0, y_0) . Then there is a tangent plane T to

the surface $z = f(x, y)$ over the point (x_0, y_0) .

Moreover

(1)   the slope of T in the x direction is $\frac{\partial z}{\partial x}(x_0, y_0)$, and   (F)

(2)   the slope of T in the y direction is $\frac{\partial z}{\partial y}(x_0, y_0)$.   (G)

If $x = x(t)$ and $y = y(t)$ are differentiable functions of t , then we wish to prove

$$\frac{dz}{dt} = \frac{\partial z}{\partial x}\frac{dx}{dt} + \frac{\partial z}{\partial y}\frac{dy}{dt}$$ (A)

From the definition of the one-variable derivative we know

$$\frac{dz}{dt} = \lim_{\Delta t \to 0} \frac{\Delta z}{\Delta t} \tag{H}$$

We must convert equation (H) into equation (A).

Let's work at a fixed t value $t = t_0$, with $x_0 = x(t_0)$ and $y = y(t_0)$. A change Δt in t_0 will cause changes Δx in x_0 and Δy in y_0.

Can we express Δz in terms of Δx and Δy, at least approximately? Yes! First observe

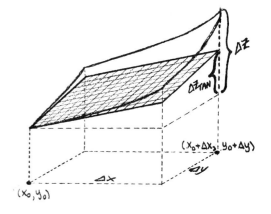

$\Delta z = f(x_0 + \Delta x, y_0 + \Delta y) - f(x_0, y_0)$

= the change in z-values on

the surface $z = f(x, y)$ as

(x, y) changes from (x_0, y_0)

to $(x_0 + \Delta x, y_0 + \Delta y)$.

However, for small Δt the quantity Δz
is approximately equal to the quantity

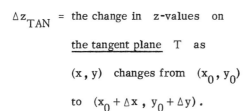

Δz_{TAN} = the change in z-values on

the tangent plane T as

(x, y) changes from (x_0, y_0)

to $(x_0 + \Delta x, y_0 + \Delta y)$.

Thus $\Delta z \cong \Delta z_{TAN}$. See the diagram above. (I)

However, the following equality can be "seen" from the tangent plane diagram (this is the key relationship in our discussion):

$$\Delta z \cong \Delta z_{TAN} = \frac{\partial z}{\partial x} \Delta z + \frac{\partial z}{\partial y} \Delta y \qquad *$$

(J)

To justify this, consider the tangent

plane diagram to the right.

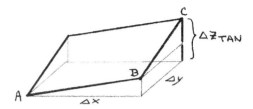

    (i) The line segment $\overline{AB}$ has

slope $\partial z / \partial x$ from (F). Thus

$$\frac{\partial z}{\partial x} = \text{slope } (\overline{AB}) = \frac{\text{rise in z-value along tangent plane from A to B}}{\Delta x}$$

from which we see that the rise in z-value along the tangent plane from A to B is

$\left(\dfrac{\partial z}{\partial x}\right) \Delta x$.

    (ii) The line segment $\overline{BC}$ has slope $\dfrac{\partial z}{\partial y}$ from (G) ; thus the rise in the z-value

along the tangent plane from B to C is $\left(\dfrac{\partial z}{\partial y}\right) \Delta y$.

    (iii) In total, Δz_{TAN} , the change in the z-value along the tangent plane from A

to C , is the sum of the changes in (i) and (ii). This gives equation (J), as desired.

    Dividing by Δt , equations (I) and (J) yield

$$\frac{\Delta z}{\Delta t} \cong \frac{\partial z}{\partial x} \cdot \frac{\Delta x}{\Delta t} + \frac{\partial z}{\partial y} \cdot \frac{\Delta y}{\Delta t}$$

As Δt approaches zero, this approximation becomes better and better; in the limit we have

equality. ** Thus

---

*   Compare this equation with Anton's equation (4) in his proof of the Chain Rule.

** To prove this claim rigorously you must use the precise argument which Anton gives for the
Chain Rule, in particular, his equation (4).

$$\frac{dz}{dt} = \lim_{\Delta t \to 0} \frac{\Delta z}{\Delta t} = \lim_{\Delta t \to 0} \left(\frac{\partial z}{\partial x} \cdot \frac{\Delta x}{\Delta t} + \frac{\partial z}{\partial y} \cdot \frac{\Delta y}{\Delta t} \right)$$

$$= \frac{\partial z}{\partial x} \lim_{\Delta t \to 0} \frac{\Delta x}{\Delta t} + \frac{\partial z}{\partial y} \lim_{\Delta t \to 0} \frac{\Delta y}{\Delta t}$$

$$= \frac{\partial z}{\partial x} \cdot \frac{dx}{dt} + \frac{\partial z}{\partial y} \cdot \frac{dy}{dt}$$

Ahh, we have succeeded in converting equation (H) into equation (A), which justifies the chain rule.

Section 16. 5: Directional Derivatives; Gradient

Our major concern in this section is the <u>gradient.</u> The gradient $\nabla f(x, y)$ plays much the same role for a function $z = f(x, y)$ as the ordinary derivative $h'(x)$ plays for a function $y = h(x)$. With this principle in mind, most of what follows in the next few sections will not be so surprising, and will be easier to understand and remember.

1. <u>The gradient: definition and basic results.</u> The definition of the gradient of a function couldn't be easier:

Suppose $z = f(x, y)$ is differentiable at $P(x, y)$.

Then the <u>gradient</u> of f at $P(x, y)$ is defined by

$$\nabla f(x, y) = f_x(x, y) \, \overline{i} + f_y(x, y) \, \overline{j}$$

Gradient
∇f

The symbol ∇f is read as "del f."

Example A.   Compute the gradient for $f(x, y) = x^2 \ln(1 + xy^2)$ at $P(e - 1, 1)$.

Solution.   The partial derivatives of f are

$$f_x(x, y) = 2x \ln(1 + xy^2) + x^2(y^2/(1 + xy^2))$$

$$f_y(x, y) = x^2(2xy/(1 + xy^2))$$

so that

$$f_x(e - 1, 1) = 2(e - 1) \ln e + (e - 1)^2(1/e)$$

$$= 3e - 4 + e^{-1} \cong 4.5227$$

$$f_y(e - 1, 1) = (e - 1)^2(2(e - 1)/e)$$

$$= 2(e - 1)^3/e \cong 3.7327$$

Thus $\nabla f(e - 1, 1) \cong 4.5227\,\overline{i} + 3.7327\,\overline{j}$ □

That's all there is to it! If you have mastered the computation of partial derivatives in §16.2, then the computation of a gradient is nothing new or difficult. Moreover, because the gradient is defined in terms of the partial derivatives (which are themselves not much more than "ordinary" derivatives), all the usual rules of differentiation carry over:

Proposition.   Suppose $f(x, y)$ and $g(x, y)$ are differentiable. Then

i. $\nabla(f + g) = \nabla f + \nabla g$

ii. $\nabla(cf) = c\nabla f$ (c any constant)

iii. $\nabla(fg) = f\nabla g + g\nabla f$

iv. $\nabla\left(\dfrac{f}{g}\right) = \dfrac{g\nabla f - f\nabla g}{g^2}$ wherever $g \neq 0$

16.5.3

(This is Anton's Problem 48, Exercise Set 16.5.)

Proof. The proofs of all four results are done in the same way: convert to partial derivatives and observe that the property under consideration is valid for partials. We illustrate this by proving i:

$$\nabla (f + g) (x, y) = \frac{\partial}{\partial x} (f + g) (x, y) \bar{i} + \frac{\partial}{\partial y} (f + g) (x, y) \bar{j}$$

by definition of $\nabla (f + g)$

$$= \left(\frac{\partial f}{\partial x} (x, y) + \frac{\partial g}{\partial x} (x, y) \right) \bar{i} + \left(\frac{\partial f}{\partial y} (x, y) + \frac{\partial g}{\partial y} (x, y) \right) \bar{j}$$

since partial derivatives obey the usual addition rule

$$= \left(\frac{\partial f}{\partial x} (x, y) \bar{i} + \frac{\partial f}{\partial y} (x, y) \bar{j} \right) + \left(\frac{\partial g}{\partial x} (x, y) \bar{i} + \frac{\partial g}{\partial y} (x, y) \bar{j} \right)$$

by regrouping terms

$$= \nabla f (x, y) + \nabla g (x, y)$$

by definition of ∇f and ∇g

Since these computations are valid for any point (x, y) where f and g are differentiable, then we have established

$$\nabla (f + g) = \nabla f + \nabla g$$

as desired. Although tedious and a bit messy, you should observe that all the steps in this verification are elementary. □

We have seen that the gradient of a function is easy enough to calculate. However, why do we <u>want</u> to calculate it? What good is it? This is the question that Anton devotes the rest of §16.5 (and much of Chapter 16), to answering. The gradient for a function f is essentially the derivative of f . In the preceeding proposition we showed that ∇f obeys the same

algebraic rules as does an "ordinary derivative;" in the work which follows we will show that

∇f solves the same type of problems for functions $f(x, y)$ that $h'(x)$ solves for "ordinary"

functions $h(x)$. In fact, ∇f will do even more for us ... simply because there are more

questions that can be asked for functions of two variables than for functions of one variable.

2. <u>Gradient Property #1: The Directional Derivative Theorem.</u>   The gradient makes the computa-

tion of directional derivatives an easy matter. In fact, Anton uses the gradient to <u>define</u> them

(Definition 16.5.1). However, since directional derivatives are so important, we will give a

slightly different, but more natural, definition. (Then Anton's "definition" will be a "theorem"

for us: the Directional Derivative Theorem.)

Suppose $z = f(x, y)$. One interpretation of $\dfrac{\partial f}{\partial x}(x_0, y_0)$ is the rate of change of

$f(x, y)$ at $P(x_0, y_0)$ in the x direction. Similarly,

$\dfrac{\partial f}{\partial y}(x_0, y_0)$ is the rate of change in the y

direction. But why restrict ourselves to just

these two particular directions? Surely other

directions can also be important. Hence, let us

take any vector $\bar{u} = u_1 \bar{i} + u_2 \bar{j}$ of unit length

(i.e., $\|\bar{u}\| = 1$). Then

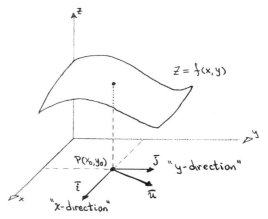

what is the rate of change of

$f(x, y)$ at $P(x_0, y_0)$ <u>in the</u>

direction of $\bar{u}$?

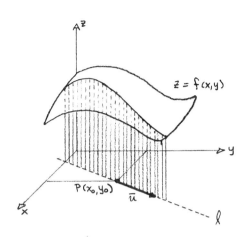

Well... let ℓ be the line in the
xy plane through $P(x_0, y_0)$ and in
the direction of $\bar{u}$. From Theorem 15.2.1
applied to 2-space we see that ℓ
has parametric equations

$$x = x_0 + s u_1$$
$$y = y_0 + s u_2$$

Then consider the values of f
restricted to ℓ, i.e.,

$$f(x_0 + s u_1, y_0 + s u_2)$$

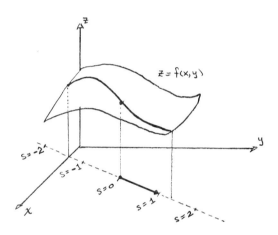

We are thus looking only at the points
of the graph of f which lie directly
above the line ℓ. Now the rate of
change of $f(x, y)$ at $P(x_0, y_0)$ in
the direction of $\bar{u}$ is simply the rate
of change of

$$f(x_0 + s u_1, y_0 + s u_2)$$

with respect to the line parameter s
at $s = 0$, i.e.,

$$\left. \frac{d}{ds} f(x_0 + s u_1, y_0 + s u_2) \right|_{s = 0}$$

We have therefore motivated the following definition:

> The <u>directional derivative of</u> $f(x, y)$ at the point $P(x_0, y_0)$
>
> and in the direction of the <u>unit</u> vector $\bar{u} = u_1 \bar{i} + u_2 \bar{j}$ is defined by
>
> $$D_{\bar{u}} f(x_0, y_0) \;=\; \frac{d}{ds}\, f(x_0 + su_1,\, y_0 + su_2)\Big|_{s = 0}$$
>
> whenever the indicated derivative with respect to s exists.

Directional Derivative

Here are some basic facts about the directional derivative:

<u>Rate of change interpretation</u>: as shown above, $D_{\bar{u}} f(x_0, y_0)$ is the rate of change of $f(x, y)$

at $P(x_0, y_0)$ in the direction of $\bar{u}$.

<u>Geometric interpretation</u>: as done with partial derivatives, it is not hard to show that

$D_{\bar{u}} f(x_0, y_0)$ is the slope of the tangent line to the graph of $z = f(x, y)$ at $P(x_0, y_0)$

in the direction of $\bar{u}$. See Anton's Figure 16. 5. 3 .

<u>Relation to partial derivatives</u>: we claim that

$$D_{\bar{i}} f(x_0, y_0) \;=\; \frac{\partial f}{\partial x}\,(x_0, y_0)$$

$$D_{\bar{j}} f(x_0, y_0) \;=\; \frac{\partial f}{\partial y}\,(x_0, y_0)$$

This is easy to see from

$$D_{\bar{i}} f(x_0, y_0) \;=\; \frac{d}{ds}\, f(x_0 + s \cdot 1,\, y_0 + s \cdot 0)\Big|_{s = 0}$$

$$=\; \frac{d}{ds}\, f(x_0 + s,\, y_0)\Big|_{s = 0}$$

$$=\; \frac{d}{ds}\, f(x, y_0)\Big|_{x = x_0}$$

$$=\; \frac{\partial f}{\partial x}\,(x_0, y_0)$$

A similar computation works for $D_{\bar{j}} f(x_0, y_0)$. Thus

> partial derivatives are merely directional
>
> derivatives in the directions $\bar{u} = \bar{i}$ and $\bar{u} = \bar{j}$

So how do we compute a directional derivative? Such a computation can often be made directly from the definition for $D_{\bar{u}} f(x, y)$ given above. Here is an example:

Example B. Suppose $f(x, y) = x^2 + y^2$. Using the definition, compute the directional derivative of f at $P(2, 1)$ in the direction of $\bar{z} = 3\bar{i} - 4\bar{j}$.

Solution. First note that $\bar{z}$ is <u>not a unit vector</u>. We must instead convert to $\bar{u} = \bar{z} / \|\bar{z}\| = (3/5)\bar{i} - (4/5)\bar{j}$. Then

$$f(x_0 + su_1, y_0 + su_2) = f(2 + \tfrac{3}{5} s, 1 - \tfrac{4}{5} s)$$

$$= (2 + \tfrac{3}{5} s)^2 + (1 - \tfrac{4}{5} s)^2$$

$$= 5 + \tfrac{4}{5} s + s^2$$

Hence, taking the derivative with respect to s at $s = 0$ gives

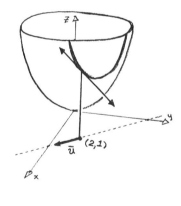

$$D_{\bar{u}} f(2, 1) = \frac{d}{ds} (5 + \tfrac{4}{5} s + s^2)\Big|_{s = 0}$$

$$= (\tfrac{4}{5} + 2s)\Big|_{s = 0} = \tfrac{4}{5}$$

Thus $4/5$ is the rate of change of $f(x, y)$ at $P(2, 1)$ in the direction of $\bar{z} = 3\bar{i} - 4\bar{j}$ (i. e., in the direction of the unit vector $\bar{u} = (3/5)\bar{i} - (4/5)\bar{j}$).

Computing $D_{\bar{u}} f(x_0, y_0)$ directly from its definition can often be messy and unpleasant.

Fortunately we have a much better method at our disposal! If f is differentiable at $P(x, y)$

(which, as previously noted, will be true 99.9% of the time), then the gradient of f comes to

the rescue via the chain rule:

GP #1 :
The Directional
Derivative Theorem *

Suppose $f(x, y)$ is differentiable at $P(x, y)$.

Then the directional derivative of f at $P(x, y)$ in

the direction of any unit vector $\bar{u}$ is given by

$$D_{\bar{u}} f(x, y) = \nabla f(x, y) \cdot \bar{u}$$

The proof of the theorem is very easy:

$$D_{\bar{u}} f(x, y) = \frac{d}{ds} f(x + su_1, y + su_2)\Big|_{s=0} \quad \text{by definition of } D_{\bar{u}} f$$

$$= \frac{\partial f}{\partial x} (x, y) u_1 + \frac{\partial f}{\partial y} (x, y) u_2 \quad \text{by the Chain Rule}$$

$$\text{(Theorem 16.4.1) since} \quad \frac{d}{ds} (x + su_1) = u_1$$

$$\frac{d}{ds} (y + su_2) = u_2$$

$$= \nabla f(x, y) \cdot \bar{u} \quad \text{by definition of } \nabla f \text{ and the dot product.}$$

Hence, for a differentiable function, we need only compute the gradient to obtain all of the

directional derivatives. Quite a computational shortcut!

Example C. Redo Example B using the Directional Derivative Theorem.

Solution. We first compute the gradient of $f(x, y) = x^2 + y^2$:

---

*
Anton uses the Directional Derivative Theorem as his definition for the directional derivative
(Definition 16.5.1).

$$\nabla f(x, y) = f_x(x, y)\overline{i} + f_y(x, y)\overline{j} = 2x\,\overline{i} + 2y\,\overline{j}$$

Thus, with $\overline{u} = (3/5)\overline{i} - (4/5)\overline{j}$ and $(x, y) = (2, 1)$, we obtain

$$D_{\overline{u}} f(2, 1) = \nabla f(2, 1) \cdot ((3/5)\overline{i} - (4/5)\overline{j}) \quad \text{by} \quad GP\,\#1$$

$$= (4\overline{i} + 2\overline{j}) \cdot ((3/5)\overline{i} - (4/5)\overline{j})$$

$$= (12/5) - (8/5) = 4/5, \quad\quad \text{as obtained earlier.} \quad \square$$

If you think about it, the Directional Derivative Theorem has an amazing consequence: if $f(x, y)$ is <u>differentiable</u> at a point $P(x, y)$, then knowing the directional derivatives for $f(x, y)$ in the $\overline{i}$ and $\overline{j}$ directions (i.e., the partial derivatives) determines the directional derivatives of $f(x, y)$ in <u>all</u> directions! This is not true for a function $f(x, y)$ which is not differentiable; in fact, there are examples of functions where the partials are both zero at a specific point, but all the other directional derivatives are non-zero.

3.   <u>Gradient Property #2: The Maximum Increase Theorem.</u>   We'll motivate this result with an important application:

Suppose the temperature at any point on a metal plate is given by

$$T(x, y) = 1 - x^2 - y^2/4 .$$

a.   Find the direction in which heat will flow from the point $P(1, 2)$.

b.   Find an equation for the path of heat flow starting at $P(1, 2)$.

We need a simple fact from physics to do this problem: heat flows in the direction of the maximum rate of <u>decrease</u> in the temperature (which, as will be seen, is the direction directly opposite to the direction of maximum <u>increase</u> in the temperature). This example (which we

will solve momentarily) is a special case of the following general question:

Suppose $f(x, y)$ is a differentiable function at $P(x_0, y_0)$.

In what direction $\bar{u}_{max}$ from $P(x_0, y_0)$ does the maximum

rate of increase of f occur? And what <u>is</u> this maximum

rate of increase?

Worded in terms of directional derivatives, we wish to find that unit vector $\bar{u} = \bar{u}_{max}$ which

maximizes the directional derivative $D_{\bar{u}} f(x_0, y_0)$. This is easily done by the Directional

Derivative Theorem:

$$D_{\bar{u}} f(x_0, y_0) = \nabla f(x_0, y_0) \cdot \bar{u} \qquad\qquad \text{by the DDT}$$

$$= \|\nabla f(x_0, y_0)\| \, \|\bar{u}\| \cos \theta$$

where θ is the angle between $\nabla f(x_0, y_0)$ and $\bar{u}$.

Thus $D_{\bar{u}} f(x_0, y_0) = \|\nabla f(x_0, y_0)\| \cos \theta$, since $\|\bar{u}\| = 1$.

This shows that the maximum value of $D_{\bar{u}} f(x_0, y_0)$ for a fixed

point $P(x_0, y_0)$ occurs when $\cos \theta = 1$, i.e., when $\theta = 0$. But this means that

$\bar{u}_{max}$ and $\nabla f(x_0, y_0)$ must point in

the same direction!

Since $\bar{u}_{max}$ is a unit vector, then we see that dividing $\nabla f(x_0, y_0)$ by its own length

(assuming $\nabla f(x_0, y_0) \neq \bar{0}$) must yield $\bar{u}_{max}$, i.e.,

$$\bar{u}_{max} = \nabla f(x_0, y_0) / \|\nabla f(x_0, y_0)\|$$

Moreover, what is the value of $D_{\overline{u}_{max}} f(x_0, y_0)$? Well, for $\overline{u} = \overline{u}_{max}$ we have $\|u\| = 1$

and $\cos \theta = 1$. Thus the equation

$$D_{\overline{u}} f(x_0, y_0) = \|\nabla f(x_0, y_0)\| \, \|\overline{u}\| \cos \theta$$

becomes

$$\boxed{D_{\overline{u}_{max}} f(x_0, y_0) = \|\nabla f(x_0, y_0)\|}$$

We have therefore established the second major gradient result:

<table>
<tr><td>GP #2
The Maximum
Increase Theorem</td><td>Suppose $f(x, y)$ is differentiable at $P(x_0, y_0)$, and

$\nabla f(x_0, y_0)$ is not zero. Then:

   $\nabla f(x_0, y_0)$ gives the <u>direction</u> of the maximum

      rate of <u>increase</u> of $f(x, y)$ at $P(x_0, y_0)$, and

$\|\nabla f(x_0, y_0)\|$ is the <u>value</u> of the maximum rate

of increase.</td></tr>
</table>

It is easy to modify the proof of GP #2 to show that $- \nabla f(x_0, y_0)$ gives the direction of the maximal rate of <u>decrease</u> of $f(x, y)$ at $P(x_0, y_0)$, and that $\|\nabla f(x_0, y_0)\|$ is still the value of this maximum rate of decrease (use $\cos \theta = -1$ when $\theta = \pi$) . We now return to the example from the beginning of this subsection.

Example D. Suppose the temperature at any point on a metal plate is given by $T(x, y) = 1 - x^2 - y^2/4$.

   a.   Find the direction in which heat will flow from the point $P(1, 2)$.

b.   Find an equation for the path of heat flow starting at   P(1, 2).

Solution.   (a)   As observed earlier, heat flows in the direction of the maximum rate of decrease

in the temperature.   Thus we need to find $-\nabla T(1, 2)$.

$$-\nabla T(1, 2) = -\langle -2x, -y/2 \rangle \big|_{x=1, y=2}$$
$$= \langle 2, 1 \rangle$$

Thus heat will flow (instantaneously) from   P(1, 2)   in the direction of the vector $\langle 2, 1 \rangle$.

        (b)   To determine the actual path of heat flow starting from   P(1, 2)   is a much more

sophisticated problem.   The techniques needed for its solution are treated in our (optional)

Subsection 5.   □

        We give one additional example of the Maximum Increase Theorem, this one containing

an economic application started in   Example G   of §16.2.

Example E.   Suppose the production function for a one-product firm is given by $Q = K^{1/2} \ln L$,

        where   K   is the capital input,   L   is the labor input, and   Q   is the output.   If the current

        daily capital and labor input are   K = 5000   and   L = 3000,   in what proportion should both

        K   and   L   be increased to achieve the largest increase in   Q ?

        Solution.   We first determine the direction of the maximum rate of increase of   Q   at the

        point   P(5000, 3000)   in the   KL-plane.   This is given by $\nabla Q(5000, 3000)$.   So we compute:

$$\nabla Q(K, L) = \frac{\partial Q}{\partial K}\,\overline{i} + \frac{\partial Q}{\partial L}\,\overline{j}$$

$$= \frac{1}{2}\, K^{-1/2} \ln L\,\overline{i} + (K^{1/2}/L)\,\overline{j}$$

$$\nabla Q(5000, 3000) \cong .0566\,\overline{i} + .0236\,\overline{j}$$

Thus, if we increase K by $\nabla K \cong .0566$,

then we should increase L by $\nabla L \cong .0236$.

The desired proportion is therefore

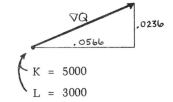

K = 5000

L = 3000

$$\nabla K/\nabla L \cong .0566/.0236 \cong 2.40 \qquad \square$$

The proportion obtained in Example E will be useful only for relatively small changes in K and L. How to determine the optimal proportional increases in K and L without this size restriction will be considered in our (optional) Subsection 5 on flow lines.

4. <u>Gradient Property #3: The Level Curve Theorem.</u> The approximation of "curved" objects by "flat" objects (or of "non-linear" objects by "linear" objects) is of great importance in applications. For example, approximating the (curved) graph of $y = f(x)$ by a (flat) tangent line leads in §3.4 to the method of approximation by differentials, concisely summed up by Equation (12) of that section: $\Delta y \approx dy$ (see Figure 3.4.4.).

Computing tangent lines to <u>graphs</u> of functions $y = f(x)$ by use of the <u>derivative</u> $f'(x_0)$ was discussed in §2.3 (in particular, see Example 1) and §3.1 (see Example 2).

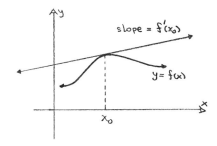

Computing tangent lines to parametrized

curves $\bar{r}(t)$ by the use of the vector

derivative $\bar{r}'(t_0)$ was discussed in

§ 15. 5. 5 of The Companion (see

Equation (F)) .

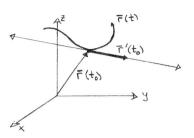

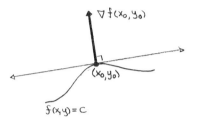

We now compute the tangent line

to level curves $f(x, y) = c$ by use of

the gradient $\nabla f(x_0, y_0)$. This is done

via our third major gradient property:

<table>
<tr><td>GP #3
The Level
Curve Theorem</td><td>Suppose $f(x, y)$ is differentiable at (x_0, y_0), and

$\nabla f(x_0, y_0)$ is not zero. Then $\nabla f(x_0, y_0)$ is

perpendicular to the level curve of $f(x, y)$

which passes through (x_0, y_0).</td></tr>
</table>

Remember that to say a vector is "perpendicular

to a curve" at a point P_0 means that the vector

is perpendicular to the tangent line of the curve

at the point P_0.

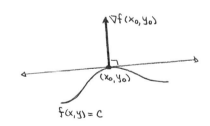

Example F.   Suppose $f(x, y) = x^2/(x + y)$.

    i.   Determine a vector which is perpendicular at $P_0(2, 2)$ to the level curve of

        $f(x, y)$ passing through $P_0(2, 2)$.

    ii. Determine the tangent line to the level curve of $f(x, y)$ at the point $P_0(2, 2)$.

Solution. i. The level curve of f passing through $P_0(2, 2)$ is given by

$$x^2/(x + y) = f(2, 2) = 4/(2 + 2) = 1$$

and, by the Level Curve Theorem, a vector perpendicular to this curve at $P_0(2, 2)$ is

$$\nabla f(2, 2) = \frac{x^2 + 2xy}{(x + y)^2}\, \overline{i} - \frac{x^2}{(x + y)^2}\, \overline{j}\, \Bigg|_{x = 2, y = 2}$$

$$= (3/4)\,\overline{i} - (1/4)\,\overline{j}$$

ii. The tangent line we desire consists of precisely those points

$P(x, y)$ for which the vector $\overrightarrow{P_0 P} = \langle x - 2, y - 2 \rangle$
is perpendicular to $\nabla f(2, 2)$, i.e.,

$$\nabla f(2, 2) \cdot \overrightarrow{P_0 P} = 0$$

Thus $\langle 3/4, -1/4 \rangle \cdot \langle x - 2, y - 2 \rangle = 0$

$$(3/4)(x - 2) - (1/4)(y - 2) = 0$$

or $y = 3x - 4$

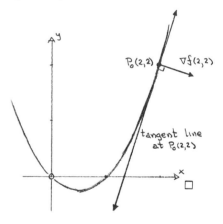

5. Flow lines (optional). In many applications we need to determine the path $\overline{r}(t)$ taken by an object in a plane which is following the maximum increase or decrease of a quantity $z = f(x, y)$. The path $\overline{r}(t)$ so obtained will be called a flow line for $z = f(x, y)$. Here is an illustration:

Example G. Howard Ant is placed at a point $(2, 1)$ on a heated plate, where each point (x, y) has a temperature of

$$T(x, y) = x^2 + 4y^2$$

If Howard attempts to walk in the direction of maximal increase of temperature, what is the path he traces?

Solution. Let Howard's path be given by

$$\bar{r}(t) = x(t)\,\bar{i} + y(t)\,\bar{j}, \quad \text{where} \quad x(0) = 2 \quad \text{and} \quad y(0) = 1$$

To walk in the direction of maximal increase of temperature

means that the tangent vector $\bar{r}'(t)$ to his path must be

a positive multiple of the gradient vector $\nabla T(x, y)$, i.e.,

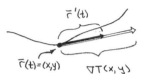

$$\bar{r}'(t) = \lambda \nabla T \qquad \text{for some} \qquad \lambda > 0$$

Let's suppose $\lambda = 1$ (the choice of this multiplicative factor will affect the parameter t, but it will not affect the actual path $\bar{r}(t)$). Summarizing, we must

$$\boxed{\begin{array}{l} \text{find} \quad \bar{r}(t) = x(t)\,\bar{i} + y(t)\,\bar{j} \quad \text{so that} \\[2mm] \bar{r}'(t) = \nabla T \quad \text{for all} \quad t, \quad \text{and} \\[2mm] x(0) = 2, \; y(0) = 1 \end{array}}$$

Be sure you understand these equations, for the rest of the problem is just plugging in and solving these equations. We easily obtain

$$\nabla T(x, y) = \langle 2x, 8y \rangle \quad,$$

$$\bar{r}'(t) = \langle x'(t), y'(t) \rangle, \quad \text{where} \quad x(t) \quad \text{and} \quad y(t) \quad \text{are the}$$

functions we wish to determine.

Thus the equations

$$\left\{\begin{array}{l} \bar{r}'(t) = \nabla T \\[2mm] x(0) = 2, \; y(0) = 1 \end{array}\right\}$$

become $\left\{ \begin{array}{l} x'(t) = 2x \\ x(0) = 2 \end{array} \right\}$ and $\left\{ \begin{array}{l} y'(t) = 8y \\ y(0) = 1 \end{array} \right\}$

From the discussion in §7.7 (in particular, Equation 24) we see that these differential equations have as their solutions the functions $x(t) = 2e^{2t}$ and $y(t) = e^{8t}$. Thus the path we are looking for is given by

$$\overline{r}(t) = \langle 2e^{2t}, e^{8t} \rangle, \quad -\infty < t < \infty$$

If we wish to have an equation for this curve in terms of x and y, i.e., with the parameter t removed, we can continue as follows:

$$x^4 = (2e^{2t})^4 = 16 e^{8t} = 16 y$$

Hence $\boxed{y = x^4/16, \text{ with } x > 0}$ is the desired equation. $\qquad \square$

Let's use the flow lines technique to complete the analysis of our economic Example E :

Example H. Suppose the production function for a one-product firm is given by $Q = K^{1/2} \ln L$, where K is the capital input, L is the labor input, and Q is the output. If the current daily capital and labor input are $K = 5000$ and $L = 3000$, then in what way should both K and L be increased to achieve the largest increase in Q ?

Solution. Let $\overline{r}(t) = \langle K(t), L(t) \rangle$. We wish to determine the path $\overline{r}(t)$ so that $K(0) = 5000$, $L(0) = 3000$, and the tangent vector $\overline{r}'(t)$ to the path must be a positive multiple of the gradient vector $\nabla Q(K, L)$, i.e.,

$$\overline{r}'(t) = \lambda \nabla Q \qquad \text{for some} \qquad \lambda > 0$$

Computing ∇Q and equating components we obtain

$$K'(t) = \frac{\lambda}{2} K^{-1/2} \ln L \quad \text{and} \quad L'(t) = \lambda K^{1/2}/L$$

These do not look easy to solve! They are coupled differential equations in that the two unknown functions K and L appear in both equations. There is a way out, however: we can choose the multiple λ to be any convenient positive number, and it can vary with t. In this case choosing $\lambda = K^{-1/2}$ (K is a function of t) is very convenient, for we obtain

$$K'(t) = (2K)^{-1} \ln L \quad \text{and} \quad L'(t) = 1/L$$

The solution of the second equation is easy (it's a separable differential equation):

$$L = (2t + c_1)^{1/2} \qquad \text{for some constant} \qquad c_1$$

Plugging this into the first equation and separating variables we will find

$$4K^2 = (2t + c_1) \ln (2t + c_1) - 2t + c_2$$

where c_1 and c_2 are constants. Eliminating the variable t from the two equations for L and K yields

$$K = \frac{1}{2} (2L^2 \ln L - L^2 + c)^{1/2}$$

for some constant c. The initial conditions $K = 5000$ and $L = 3000$ show $c \cong -3.511 \times 10^7$. We thus have learned how to increase Q most efficiently by increasing the variables K and L from their starting values at $K = 5000$ and $L = 3000$. $\square$

Exercises. These problems deal with the optional material on flow lines.

1. Suppose each point (x, y) on a metal plate has a potential energy of

$$V(x, y) = x^2 - 4y$$

If a particle starts at the point $P(2, 1)$, determine the path it will take if it moves in the direction of maximal decrease of V.

2. Suppose a mountain climber is at the point $(1, 2)$ of a mountain whose height at point (x, y) is given by

$$T(x, y) = 2x^2 + y^2$$

(a mighty tall mountain!). If the climber wishes to take the most challenging route up the mountain, what path in the xy-plane will she choose?

3. The rate of reaction S of a certain chemical process decreases with the addition of either chemical A or chemical B according to the following formula:

$$S = e^{-x}/y$$

where x = amount of chemical A in solution

y = amount of chemical B in solution

If initially we have $x = 3$ and $y = 4$, determine how to increase jointly both x and y to achieve the most efficient slowdown in the chemical process.

4. Solve part (b) of Example D.

Answers.

1. $y = 1 - 2 \ln (x/2)$

2. $y = 2 x^{1/2}$

3. $y = (2x + 10)^{1/2}$

      (You can use either $\lambda = e^x$ or $\lambda = y$).

4. $y = (x/2)^{1/4}$

Section 16.6:   Tangent Planes

1.   Tangent planes to graphs of $z = f(x, y)$.   Recall from §15.6 that to compute the equation of

a plane in space we only need a point $P_0(x_0, y_0, z_0)$ on the plane and a non-zero normal vector

$\bar{n} = \langle a, b, c \rangle$. Then any point $P(x, y, z)$ on the plane is determined by the equation

$$\bar{n} \cdot \overrightarrow{P_0 P} = 0$$

   i.e., $\langle a, b, c \rangle \cdot \langle x - x_0, y - y_0, z - z_0 \rangle = 0$

$$\boxed{a(x - x_0) + b(y - y_0) + c(z - z_0) = 0}$$

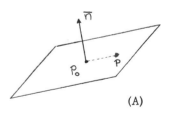

(A)

In particular, to find the equation of the tangent plane to the graph of $z = f(x, y)$ at a specified

point $P_0(x_0, y_0, z_0)$ (where $z_0 = f(x_0, y_0)$) we need only determine a normal vector $\bar{n}$

to the plane. This is done by Theorem 16.6.2:

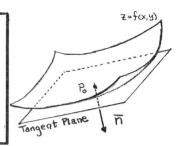

If $z = f(x, y)$ is differentiable at (x_0, y_0), then

a normal vector $\bar{n}$ to the tangent plane at

$P(x_0, y_0, z_0)$ is given by

$$\bar{n} = \langle f_x(x_0, y_0), f_y(x_0, y_0), -1 \rangle$$

normal
to a
tangent
plane

If you simply remember this formula for the normal vector, then all the other parts of Theorem 16.6.2 follow directly from previous work! For example, from (A), the equation for the tangent plane to $z = f(x, y)$ at $P_0(x_0, y_0, z_0)$ is given by

$$f_x(x_0, y_0)(x - x_0) + f_y(x_0, y_0)(y - y_0) - (z - z_0) = 0 \qquad \text{(B)}$$

Example A. Find equations for the tangent plane and normal line to the graph $z = \ln(1 + xy^2)$ over the point $(e - 1, 1)$.

Solution. Let $f(x, y) = \ln(1 + xy^2)$. For the tangent plane we need to find a <u>point</u> P_0 and a <u>normal</u> $\bar{n}$. The point is given by

$$x_0 = e - 1$$
$$y_0 = 1$$
$$z_0 = f(x_0, y_0) = \ln(1 + (e - 1)1^2) = \ln e = 1$$

For the normal vector we need the partial derivatives of $f(x, y)$:

$$f_x(x, y) = \frac{y^2}{1 + xy^2} \quad \text{and} \quad f_y(x, y) = \frac{2xy}{1 + xy^2}$$

so $f_x(e - 1, 1) = 1/e$ and $f_y(e - 1, 1) = 2(e - 1)/e$. Thus a normal to the surface at $P_0(e - 1, 1, 1)$ is given by

$$\bar{n} = \langle f_x(e - 1, 1), f_y(e - 1, 1), -1 \rangle = \langle 1/e, 2(e - 1)/e, -1 \rangle$$

The equation for the tangent plane now follows easily from Equation (B) :

$$(1/e)(x - (e - 1)) + (2(e - 1)/e)(y - 1) - (z - 1) = 0$$

$$\boxed{x + 2(e - 1)y - ez = 2e - 3}$$

For the normal line to the graph at P_0 we
have only to observe that the direction vector for
this line is our normal vector $\bar{n}$. Thus by
Theorem 15. 2. 1 , using

$$\bar{P}_0 = \langle e - 1, 1, 1 \rangle$$

and
$$\bar{n} = \langle 1/e, 2(e - 1)/e, -1 \rangle$$

we see that the normal line has the parametric scalar equations

$$\boxed{\begin{cases} x = e - 1 + t/e \\ y = 1 + 2t(e - 1)/e \\ z = 1 - t \end{cases}}$$ □

2. <u>The total differential and its use in approximations.</u>   The tangent plane to the graph $z = f(x, y)$

at $P_0(x_0, y_0, z_0)$ is intuitively supposed to be the plane passing through P_0 which <u>best</u>

<u>approximates the surface $z = f(x, y)$ near</u> P_0. It should not be surprising therefore that

values of the function $z = f(x, y)$ for (x, y) near (x_0, y_0) could be approximated by the

corresponding z values <u>on the tangent plane.</u>

Suppose (x, y) is a point which

is close to (x_0, y_0), i. e.,

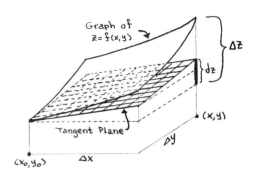

Graph of $z = f(x, y)$

Tangent Plane

(x, y)

(x_0, y_0)

$$x = x_0 + \Delta x$$

$$y = y_0 + \Delta y$$

where Δx and Δy are small. We'll denote

the difference in the function values at (x, y)

and (x_0, y_0) by Δz (see the figure above), i. e.,

$$\Delta z = f(x, y) - f(x_0, y_0)$$

$$\boxed{\Delta z = f(x_0 + \Delta x, y_0 + \Delta y) - f(x_0, y_0)}$$ (C)

On the other hand, let

$$dz = z - z_0$$

where $z - z_0$ is the difference of the z-values on the <u>tangent plane</u> at (x, y) and

(x_0, y_0) (see the figure above). From (B) the equation for our tangent plane is

$$z - z_0 = f_x(x_0, y_0)(x - x_0) + f_y(x_0, y_0)(y - y_0)$$

Using

$$dz = z - z_0$$

$$dx = \Delta x = x - x_0$$

$$dy = \Delta y = y - y_0$$

changes this equation into

$$dz = f_x(x_0, y_0)\, dx + f_y(x_0, y_0)\, dy \qquad\qquad (D)$$

This quantity is called the <u>total differential</u> of z at (x_0, y_0); its importance stems from the

following:

when $\Delta x = dx$ and $\Delta y = dy$ are both small quantities,

then $\Delta z \cong dz$.

Computing total differentials is quite easy.

<u>Example B.</u> Compute dz at an arbitrary point (x, y) for the function

$$z = (1 - x/y)^{1/2}$$

Then compute dz at the point $(3, 4)$.

<u>Solution.</u> The total differential dz at (x, y) is given by

$$dz = f_x(x, y)\, dx + f_y(x, y)\, dy$$

$$= (1/2)(1 - x/y)^{-1/2}\,(-1/y)\, dx$$

$$+ (1/2)(1 - x/y)^{-1/2}\,(x/y^2)\, dy$$

or

$$dz = -\frac{dx}{2y(1 - x/y)^{1/2}} + \frac{x\, dy}{2y^2(1 - x/y)^{1/2}}$$

At the point $(x, y) = (3, 4)$ we obtain

$$dz = -\frac{dx}{4} + \frac{3\, dy}{16} \qquad\qquad \square$$

The approximation equation $\Delta z \cong dz$ can be expanded in the following useful way using (C) and (D).

The Method of Differentials

Suppose $z = f(x, y)$ is differentiable at (x_0, y_0). Then

$$f(x_0 + \Delta x, y_0 + \Delta y) \cong f(x_0, y_0) + dz$$

where $dz = f_x(x_0, y_0)\, \Delta x + f_y(x_0, y_0)\, \Delta y$

and Δx and Δy are small.

Here's how the method of differentials is used: We wish to approximate a quantity $f(x, y)$, where x and y may be values for which $f(x, y)$ is difficult to compute. So we look for near-by values x_0 and y_0 for which $f(x_0, y_0)$, $f_x(x_0, y_0)$ and $f_y(x_0, y_0)$ are easy to compute. Then, determining Δx and Δy by $\Delta x = x - x_0$ and $\Delta y = y - y_0$, we approximate $f(x, y)$ by the method of differentials formula. Here's a typical example:

<u>Example C.</u> Approximate the quantity $7.84^{-1/3}\, 4.04^{1/2}$.

<u>Solution.</u> The quantity under consideration is $f(7.84, 4.04)$, where $f(x, y) = x^{-1/3} y^{1/2}$.
Notice that $(8, 4)$ is near $(7.84, 4.04)$, and $f(8, 4)$ is easy to compute:

$$f(8, 4) = 8^{-1/3}\, 4^{1/2} = (1/2)2 = 1$$

So we use the method of differentials formula

$$f(x_0 + \Delta x, y_0 + \Delta y) \cong f(x_0, y_0) + dz \qquad (E)$$

with

$$x_0 + \Delta x = 7.84, \qquad x_0 = 8$$
$$y_0 + \Delta y = 4.04, \qquad y_0 = 4$$

Thus

$$\Delta x = 7.84 - 8 = -.16$$
$$\Delta y = 4.04 - 4 = .04 \qquad (F)$$

To compute dz we have only to compute the two partial derivatives

$$f_x(x, y) = (-1/3) x^{-4/3} y^{1/2}$$

$$f_y(x, y) = (1/2) x^{-1/3} y^{-1/2}$$

Thus, at $(8, 4)$ we obtain

$$f_x(8, 4) = (-1/3) 8^{-4/3} 4^{1/2} = -1/24$$

$$f_y(8, 4) = (1/2) 8^{-1/3} 4^{-1/2} = 1/8$$

(G)

and hence

$$dz = f_x(8, 4) \Delta x + f_y(8, 4) \Delta y$$

$$= +.16/24 + .04/8 \qquad \text{from (F) and (G)}$$

$$= .011\overline{6}$$

Therefore, from (E) we obtain

$$7.84^{-1/3} 4.04^{1/2} \cong 1 + .011\overline{6} = \boxed{1.011\overline{6}}$$

For comparison, a pocket calculator will yield the value 1.01178. $\square$

You should compare this two-variable method of differentials with the one variable version presented in §3.4 .

A common use of the method of differentials is with <u>percentage error estimators</u>. In this type of problem, we are given maximum percentage measurement errors in two variables x and y and we need to determine the maximum percentage error possible in a third quantity z which depends on x and y, i.e., $z = f(x, y)$.

We let x, y and z denote the true values of the given quantities, and let Δx, Δy and Δz denote the errors in these quantities. We are thus given bounds on the percentage errors

$$\left| \Delta x/x \right| \quad \text{and} \quad \left| \Delta y/y \right|$$

and we need to determine a bound on the percentage error

$$\left| \Delta z/z \right|$$

The method of solution is to use the approximation $\Delta z \cong dz$ to find an upper bound for $\left| \Delta z/z \right|$ in terms of $\left| \Delta x/x \right|$ and $\left| \Delta y/y \right|$. Anton's Example 5 illustrates this procedure. Here is another:

Example D. According to the ideal gas law, the pressure, temperature and volume of a confined gas are related by $P = kT/V$ where k is a constant. Suppose T can be measured with an error of at most 2%, and V can be measured with an error of at most 3%. Then what is the maximum possible percentage error in the calculated pressure P ?

Solution. Let T, V and P denote the true values of the given quantities, and let ΔT, ΔV and ΔP be the errors in these quantities. We are given

$$\left| \Delta T/T \right| \leq .02 \quad \text{and} \quad \left| \Delta V/V \right| \leq .03$$

We want to find the maximum possible value of $\left| \Delta P/P \right|$. We thus use $\Delta P \cong dP$, where

$$dP = \frac{\partial P}{\partial T} \, \Delta T + \frac{\partial P}{\partial V} \, \Delta V$$

or

$$dP = (k/V) \, \Delta T + (- kT/V^2) \, \Delta V$$

Dividing by $P = kT/V$ yields

$$dP/P = \Delta T/T - \Delta V/V$$

so that

$$\left|\frac{\Delta P}{P}\right| \cong \left|\frac{dP}{P}\right| \leq \left|\frac{\Delta T}{T}\right| + \left|\frac{\Delta V}{V}\right|$$

(by the triangle inequality)

Hence $\left|\Delta P/P\right| \leq .02 + .03 = .05$. Thus the maximum percentage error in P is

approximately 5%.   □

This type of percentage error computation is used over and over again in scientific

calculations.

Section 16.7:   Functions of Three Variables

Reading this section should give you a severe case of déja vu: it does nothing but restate

for functions of three variables $w = f(x, y, z)$ all of the results given in the previous 6 sections

for functions of two variables $z = f(x, y)$. All the definitions, formulas and computations are

the obvious generalizations from the two variable case; there is very little new for you to learn!

Accordingly we will re-examine (briefly!) each of the six previous sections, highlighting any

aspects which change significantly in passing from two variable functions to three.

1.   Functions of three variables... as in §16.1.   Most people visualize a function of two variables

$z = f(x, y)$ by visualizing the <u>graph</u> of f,

i. e., the collection of points (x, y, z) in

3-space for which $z = f(x, y)$. However,

for functions of three variables $w = f(x, y, z)$

no such easy visualization of the "graph" of

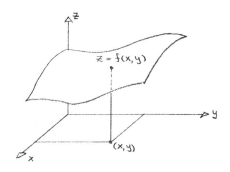

$w = f(x, y, z)$ is possible ... it would be

a collection of points (x, y, z, w) in "4-space," hardly something that can be easily pictured!

One good way to develop a feeling for a function of three variables is to think of the example of

<u>temperature</u> of a region D in 3-space. To each point (x, y, z) in our region D we

assign a number $T = f(x, y, z)$ called

the temperature of the region at that point.

You should be able to "feel" this example if

not exactly "see" it.

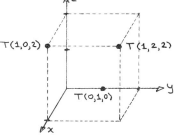

      Describing a function of two variables $z = f(x, y)$ by its level curves $f(x, y) = k$

was an important technique developed in §16. 1. Similarly, describing a function of three

variables $w = f(x, y, z)$ by its <u>level surfaces</u> $f(x, y, z) = k$ is an important technique

developed in this section. A level surface will in general be a curved surface in 3-space, a

much harder object to draw than a level curve in 2-space. Techniques for obtaining such

sketches were carefully laid out in §15. 8 and §16. 1 of <u>The Companion</u>, and should be reviewed.

<u>Example A.</u> Sketch the level surface $2x^2 + 2y^2 - z = 2$.

<u>Solution.</u> This sketch is carefully worked out in §15. 8 of <u>The Companion</u>.   □

Level curves and level surfaces are not used solely to draw pictures of functions of two or three variables; these curves and surfaces occur naturally in many important applications, such as "equipotential curves and surfaces " in physics and "indifference curves and surfaces " in economics. Level curves and surfaces are also central to the study of Lagrange multipliers, a topic with important applications in economics which will be discussed in §16.10.

2. <u>Parial derivatives... as in §16.2.</u> A function of three variables $w = f(x, y, z)$ has three partial derivatives,

$$f_x(x, y, z) = \frac{\partial f}{\partial x}(x, y, z)$$

$$f_y(x, y, z) = \frac{\partial f}{\partial y}(x, y, z)$$

$$f_z(x, y, z) = \frac{\partial f}{\partial z}(x, y, z)$$

In each case the calculation is performed as an ordinary differentiation where <u>all variables except the "differentiating variable" are treated as constants.</u>

<u>Example B.</u>   Compute $f_z(2, 3, 2)$ for $f(x, y, z) = z \ln(1 + xz) + \ln y$.

<u>Solution.</u> $f_z(x, y, z) = \frac{\partial}{\partial z}(z \ln(1 + xz) + \ln y)$

$$= \frac{\partial}{\partial z}(z \ln(1 + xz)) + \frac{\partial}{\partial z} \ln y$$

Careful here! You may be tempted to say "the derivative of $\ln y$ is $1/y$." <u>DON'T</u> ... because we are differentiating with respect to z , <u>not y</u> . Thus, with respect to z , $\ln y$ is a <u>constant</u>, and hence has a zero derivative!

Thus, using the product rule on the first term yields

$$f_z(x, y, z) = \ln(1 + xz) + z \frac{\partial}{\partial z} \ln(1 + xz)$$

$$= \ln(1 + xz) + z\left(\frac{1}{1 + xz}\right) x$$

This is from the Chain Rule, since

$$\frac{\partial}{\partial z}(1 + xz) = x$$

Finally,

$$f_z(2, 3, 2) = \ln 5 + 4/5 \qquad \square$$

3. <u>Limits, continuity and differentiability ... as in §16.3.</u> The notions of <u>limit</u> and <u>continuity</u> for a function of three variables are essentially the same as for a function of two variables. Nearly all functions of three variables which you will ever encounter are continuous at the interior points of their domain. The techniques for carefully verifying where a function is continuous are exactly the same for functions of three variables as for functions of two variables.

The definition of <u>differentiability</u> for $w = f(x, y, z)$ at (x_0, y_0, z_0) is the same as in the 2 variable case, and hence remains a difficult concept to understand. Fortunately, the <u>verification</u> of differentiability for a function $w = f(x, y, z)$ is made easier by the fundamental Theorem 16.7.5:

<table>
<tr><td>Criteria
for
Differen-
tiability</td><td>Suppose $f(x, y, z)$ has first order partial derivatives

$f_x(x, y, z)$, $f_y(x, y, z)$ and $f_z(x, y, z)$ for all points

(x, y, z) in a spherical region centered on (x_0, y_0, z_0), and

these partial derivatives are continuous at (x_0, y_0, z_0). Then

$f(x, y, z)$ is differentiable at (x_0, y_0, z_0).</td></tr>
</table>

You may be reassured to know that you will not be asked (at this level of calculus) to verify differentiability by using this criteria: unless dealing with a very strange function it is generally safe to assume differentiability. Most of our subsequent results will include as a hypothesis that the functions under consideration are differentiable.

4. **The chain rule... as in §16.4.** In the present section we only generalize the most basic chain rule of §16.4 (the more exotic variations to appear in the next section). Suppose $w = f(x, y, z)$ is a differentiable function of x, y and z, while x, y and z are in turn differentiable functions of t:

$$w = f(x, y, z), \text{ with } x = x(t), y = y(t) \text{ and } z = z(t)$$

Then

$$\boxed{\frac{dw}{dt} = \frac{\partial w}{\partial x}\frac{dx}{dt} + \frac{\partial w}{\partial y}\frac{dy}{dt} + \frac{\partial w}{\partial z}\frac{dz}{dt}}$$

<u>Example C.</u> Compute $\dfrac{dw}{dt}$ at $t = 1$ if

$$w = x \ln(x + yz^2), \quad x = e^{t^2} - 1, \quad y = t + \ln t, \quad z = t^2$$

Solution.

(i) $\dfrac{dx}{dt} = 2te^{t^2}$; if $t = 1$, then $\boxed{\dfrac{dx}{dt} = 2e}$

$\dfrac{dy}{dt} = 1 + 1/t$; if $t = 1$, then $\boxed{\dfrac{dy}{dt} = 2}$

$\dfrac{dz}{dt} = 2t$; if $t = 1$, then $\boxed{\dfrac{dz}{dt} = 2}$

(ii) At $t = 1$ we find $x = e - 1$, $y = 1$, $z = 1$.

$$\frac{\partial w}{\partial x} = \ln(x + yz^2) + \frac{x}{x + yz^2}$$

$$\frac{\partial w}{\partial y} = \frac{xz^2}{x + yz^2} \qquad\qquad \frac{\partial w}{\partial z} = \frac{2xyz}{x + yz^2}$$

Thus, at $t = 1$ we have

$$\boxed{\frac{\partial w}{\partial x} = 2 - \frac{1}{e}, \quad \frac{\partial w}{\partial y} = 1 - \frac{1}{e}, \quad \frac{\partial w}{\partial z} = 2 - \frac{2}{e}}$$

(iii) Hence, at $t = 1$, the Chain Rule yields

$$\frac{dw}{dt} = (2 - \frac{1}{e})2e + (1 - \frac{1}{e})2 + (2 - \frac{2}{e})2 = \boxed{4e + 4 - 6/e} \qquad \square$$

5. <u>Directional derivatives; gradient... as in §16.5.</u> The definition of the gradient of $w = f(x, y, z)$ is just what you would expect:

Gradient
∇f

Suppose $w = f(x, y, z)$ is differentiable at $P(x, y, z)$.

Then the <u>gradient</u> of f at $P(x, y, z)$ is defined by

$$\nabla f(x, y, z) = f_x(x, y, z)\,\overline{i} + f_y(x, y, z)\,\overline{j} + f_z(x, y, z)\,\overline{k}$$

All the basic algebraic properties of ∇f are the same here as in §16.5. Moreover, <u>the three gradient properties which we highlighted in §16.5</u> of <u>The Companion</u> <u>remain valid when generalized to functions of three variables!</u> We'll elaborate on this:

We first need to define directional derivatives.

<table>
<tr><td>

Directional
Derivative

</td><td>

The directional derivative of $f(x, y, z)$ at the point (x_0, y_0, z_0)

and in the direction of the <u>unit</u> vector $\bar{u} = u_1 \bar{i} + u_2 \bar{j} + u_3 \bar{k}$ is

defined by

$$D_{\bar{u}} f(x_0, y_0, z_0) = \frac{d}{ds} f(x_0 + su_1, y_0 + su_2, z_0 + su_3)\Big|_{s=0}$$

whenever the indicated derivative with respect to s exists.

</td></tr>
</table>

<u>Rate of change interpretation:</u> $D_{\bar{u}} f(x_0, y_0, z_0)$ is the rate of change of $f(x, y, z)$ at

(x_0, y_0, z_0) in the direction of $\bar{u}$.

<u>Relation to partial derivatives:</u> $D_{\bar{u}} f(x_0, y_0, z_0)$ equals the partial derivatives f_x, f_y, and

f_z when $\bar{u} = \bar{i}, \bar{j}$ and $\bar{k}$ respectively. Thus

<table>
<tr><td>

partial derivatives are merely directional

derivatives in the directions $\bar{u} = \bar{i}, \bar{j}$ and $\bar{k}$.

</td></tr>
</table>

Now we can state the generalizations of the three Gradient Properties of §16.5:

<table>
<tr><td>

GP #1 :
The Directional
Derivative Theorem*

</td><td>

Suppose $f(x, y, z)$ is differentiable at (x, y, z).

Then the directional derivative of f at (x, y, z) in

the direction of any unit vector $\bar{u}$ is given by

$$D_{\bar{u}} f(x, y, z) = \nabla f(x, y, z) \cdot \bar{u}$$

</td></tr>
</table>

---

* This is Anton's <u>definition</u> for $D_{\bar{u}} f(x, y, z)$. See §16.5.2 of <u>The Companion</u>.

Suppose $f(x, y, z)$ is differentiable at (x_0, y_0, z_0), and $\nabla f(x_0, y_0, z_0)$ is not zero. Then:

$\nabla f(x_0, y_0, z_0)$ gives the __direction__ of the maximum

rate of __increase__ of $f(x, y, z)$ at (x_0, y_0, z_0), and

$\| \nabla f(x_0, y_0, z_0) \|$ is the __value__ of the maximum

rate of increase.

GP #2 :
The Maximum
Increase Theorem

Suppose $f(x, y, z)$ is differentiable at (x_0, y_0, z_0), and $\nabla f(x_0, y_0, z_0)$ is not zero. Then $\nabla f(x_0, y_0, z_0)$ is

perpendicular to the level surface of $f(x, y, z)$

which passes through (x_0, y_0, z_0).

GP #3 :
The Level
Surface Theorem

There is really nothing new to be said about GP #1 and GP #2 : their meaning and use are exactly the same when applied to functions of three variables as to functions of two variables. See Anton's Example 6.

Some comments are in order, however, concerning GP #3, the Level Surface Theorem. As a strict generalization of the earlier Level Curve Theorem, the result sounds plausible: $\nabla f(x_0, y_0, z_0)$ is perpendicular to the level surface $f(x, y, z) = c$ which contains (x_0, y_0, z_0). This means that $\nabla f(x_0, y_0, z_0)$ is perpendicular __to the__ tangent plane to the level surface $f(x, y, z) = c$ at (x_0, y_0, z_0).

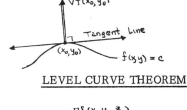

LEVEL CURVE THEOREM

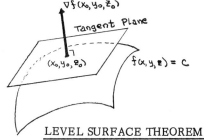

LEVEL SURFACE THEOREM

Ahh... this gives us a method for computing the <u>tangent plane</u> to a level surface since

$\overline{n} = \nabla f(x_0, y_0, z_0)$ will be a normal vector:

> The equation for the tangent plane to $f(x, y, z) = c$
>
> at $P_0(x_0, y_0, z_0)$ is given by
>
> $$\nabla f(x_0, y_0, z_0) \cdot \overrightarrow{P_0 P} = 0$$
>
> i. e. ,
>
> $$\boxed{f_x(x_0, y_0, z_0)(x - x_0) + f_y(x_0, y_0, z_0)(y - y_0) + f_z(x_0, y_0, z_0)(z - z_0) = 0}$$

<u>Example D.</u> Determine an equation for the tangent plane to the level surface $x^2 \ln(y + xz) = 1$

at $P_0(1, e, 0)$.

<u>Solution.</u> We first determine $\nabla f(1, e, 0)$:

$$\nabla f(x, y, z) = \langle 2x \ln(y + xz) + \frac{x^2 z}{y + xz}, \frac{x^2}{y + xz}, \frac{x^3}{y + xz} \rangle$$

Thus

$$\nabla f(1, e, 0) = \langle 2, 1/e, 1/e \rangle$$

The tangent plane is therefore given by

$$0 = \nabla f(1, e, 0) \cdot \overrightarrow{P_0 P}$$

$$= \langle 2, 1/e, 1/e \rangle \cdot \langle x - 1, y - e, z \rangle$$

$$= 2x - 2 + y/e - 1 + z/e$$

Thus $\boxed{2ex + y + z = 3e}$ is the desired tangent plane.   □

In many instances you can alternately compute the tangent plane to a level surface $f(x, y, z) = c$ by the methods of §16. 6:

Example E.   Redo Example D using the methods of §16. 6.

Solution.   The equation $x^2 \ln (y + xz) = 1$ can be solved for z to yield

$$z = g(x, y) = (\frac{1}{x})\left(e^{1/x^2} - y\right)$$

The partial derivatives of g are computed as follows:

$$g_x(x, y) = \left(-\frac{1}{x^2}\right)\left(e^{1/x^2} - y\right) + (\frac{1}{x})\left(-\frac{2}{x^3} e^{1/x^2}\right)$$

$$g_y(x, y) = -1/x$$

Thus, at $(x_0, y_0) = (1, e)$ we obtain

$$\boxed{g_x(1, e) = -2e} \qquad \boxed{g_y(1, e) = -1}$$

and our normal vector to the tangent plane at $(1, e, 0)$ is

$$\bar{n} = \langle g_x(1, e), g_y(1, e), -1 \rangle = \langle -2e, -1, -1 \rangle$$

The plane is therefore given by

$$0 = \bar{n} \cdot \overrightarrow{P_0 P}$$

$$= \langle -2e, -1, -1 \rangle \cdot \langle x - 1, y - e, z \rangle$$

$$= -2ex + 2e - y + e - z$$

or $\boxed{2ex + y + z = 3e}$,   as obtained in Example D.

Note that in Example D a normal vector to the tangent plane was found by computing

$\nabla f(1, e, 0)$ and in Example E such a vector was found by computing $\langle g_x(1, e), g_y(1, e), -1 \rangle$.

Of course these two vectors are scalar multiples of each other, i. e.,

$\langle g_x(1, e), g_y(1, e), -1 \rangle = \langle -2e, -1, -1 \rangle = -e \langle 2, 1/e, 1/e \rangle = -e \nabla f(1, e, 0)$.

6. <u>Tangent (hyper?) planes...?</u>   In §16. 6 we devised a method for computing the tangent plane

to a graph of a function $z = f(x, y)$. The analogue in our present situation would be devising

a method for computing the "tangent hyperplane to the graph of $w = f(x, y, z)$ in 4-space"!

Well... we won't touch that one (although it actually can be done)!

However, §16. 6 also introduced the concept of the total differential for a function

$z = f(x, y)$. This does carry over easily to functions of three variables:

The <u>total differential</u> of $w = f(x, y, z)$ at (x_0, y_0, z_0) is

$$dw = f_x(x_0, y_0, z_0)dx + f_y(x_0, y_0, z_0)dy + f_z(x_0, y_0, z_0)dz$$

The total differential can now be used in approximation problems.

The Method of Differentials

Suppose $w = f(x, y, z)$ is differentiable at (x_0, y_0, z_0). Then

$f(x_0 + \Delta x, y_0 + \Delta y, z_0 + \Delta z) \cong f(x_0, y_0, z_0) + dw$

where $dw = f_x(x_0, y_0, z_0)\Delta x + f_y(x_0, y_0, z_0)\Delta y + f_z(x_0, y_0, z_0)\Delta z$

and $\Delta x, \Delta y$ and Δz are small.

This three-variable Method of Differentials is used in exactly the same way as the two-variable

version of §16. 6. You should carefully study Anton's Example 8 and refer to Examples C

and D in §16. 6. 2 of <u>The Companion</u>.

Section 16. 8 :   Functions of n Variables; More on the Chain Rule

1.   Basic tree diagrams.   The use of tree diagrams to obtain chain rules is really a pretty easy

procedure as long as you are careful in setting up your tree.   In §16. 4 you learned the following

chain rule:

Suppose $z = f(x, y)$, with $x = x(u, v)$ and $y = y(u, v)$.   Then

$$\frac{\partial z}{\partial u} = \frac{\partial z}{\partial x} \frac{\partial x}{\partial u} + \frac{\partial z}{\partial y} \frac{\partial y}{\partial u}$$

$$\frac{\partial z}{\partial v} = \frac{\partial z}{\partial x} \frac{\partial x}{\partial v} + \frac{\partial z}{\partial y} \frac{\partial y}{\partial v}$$

Here's a detailed description of how you would obtain these rules from a tree diagram

1.   Start with the "final variable" - in this

case   z - at the top of the diagram.

On the line under   z   list all the variables

upon which z   directly depends - in this

case   x   and y - and connect   z   by lines

to each of these variables.

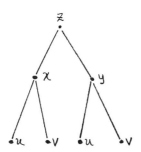

2.   On the line under   x   list all the variables

upon which x   directly depends - in

this case   u   and   v - and conncect x

by lines to each of these variables.   Do the

same for   y .

3. Label each line in the diagram with

the appropriate partial derivative. For

example, since x depends on v ,

the line connecting x to v should

be labelled with $\partial x/\partial v$.

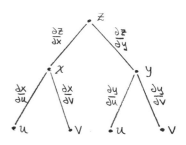

Chain rules can now be read off of the diagram. For example, to find the chain rule for

$\partial z/\partial u$, simply find all paths in

the tree diagram which lead from

z   to   u - in this case there are

two. Each such path yields a term

in the chain rule for $\partial z/\partial u$ by

multiplying together the partial

derivatives along the path. The full

chain rule is then just the sum of the

terms for each path. We thus obtain

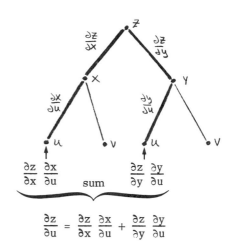

$$\frac{\partial z}{\partial u} = \begin{pmatrix} \text{term from} \\ \text{path 1} \end{pmatrix} + \begin{pmatrix} \text{term from} \\ \text{path 2} \end{pmatrix}$$

or

$$\frac{\partial z}{\partial u} = \frac{\partial z}{\partial x}\frac{\partial x}{\partial u} + \frac{\partial z}{\partial y}\frac{\partial y}{\partial u}$$

which is the chain rule of Theorem 16. 4. 2.

Although this tree diagram construction procedure is centered on a particular example,

it should be clear how the procedure generalizes to other situations.

16. 8. 3

Example A. Suppose $u = r \ln (s + t^2)$, $r = x + y^2$, $s = e^{xy}$ and $t = (x - 1)/y$. Find $\partial u/\partial x$ and $\partial u/\partial y$ at $x = 1$ and $y = 1$.

Solution. The "final variable" in this scheme is u. This variable depends directly on r, s and t, each of which in turn depends on

x and y. Our tree diagram is thus as

appears to the right, and we read off the

chain rules

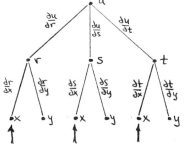

$$\frac{\partial u}{\partial x} = \frac{\partial u}{\partial r} \frac{\partial r}{\partial x} + \frac{\partial u}{\partial s} \frac{\partial s}{\partial x} + \frac{\partial u}{\partial t} \frac{\partial t}{\partial x}$$

$$\frac{\partial u}{\partial y} = \frac{\partial u}{\partial r} \frac{\partial r}{\partial y} + \frac{\partial u}{\partial s} \frac{\partial s}{\partial y} + \frac{\partial u}{\partial t} \frac{\partial t}{\partial y}$$

These paths used to compute $\partial u/\partial x$.
Remaining paths used for $\partial u/\partial y$.

From these we obtain

$$\frac{\partial u}{\partial x} = \ln (s + t^2) (1) + \frac{r}{s + t^2} (y\,e^{xy}) + \frac{2\,r\,t}{s + t^2} (\frac{1}{y})$$

and

$$\frac{\partial u}{\partial y} = \ln (s + t^2) (2\,y) + \frac{r}{s + t^2} (x\,e^{xy}) + \frac{2\,r\,t}{s + t^2} \left(- \frac{x - 1}{y^2}\right)$$

However, at $x = 1$ and $y = 1$ we obtain $r = 2$, $s = e$ and $t = 0$. Thus

$$\boxed{\frac{\partial u}{\partial x} = 3}$$ and $$\boxed{\frac{\partial u}{\partial y} = 4}$$ □

2. Mixed level tree diagrams. Tree diagrams are especially useful when there is a "mixing up" of the levels of certain variables, i. e., a variable on two different horizontal lines in the tree

diagram. This happens when some of the variables in a function are themselves functions of

the remaining variables. Anton's Example 3 is a straightforward illustration of this situation.

Here is a more unusual example:

Example B. Suppose $z = x^2 \ln (s + xy)$, where x and y are both differentiable functions

of s such that

$$\text{when} \quad s = 2, \quad \text{then} \quad \left\{ \begin{array}{l} x = 1, \ y = e - 2 \\[2mm] \dfrac{dx}{ds} = 2, \ \text{and} \ \dfrac{dz}{ds} = 6 \end{array} \right\}$$

Determine dy/ds when s = 2 .

Solution. We first set up the appropriate tree diagram

as shown to the right. Note that s appears at

two different levels.

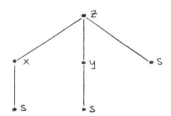

We must exercise caution in writing the partial derivatives because <u>some of the partial</u>

<u>derivatives are really ordinary derivatives</u>! For instance, both x and y are functions of

s alone, and hence there are no "partial" derivatives of x and y , only ordinary derivatives

dx/ds and dy/ds .

Handling the dependence of z on s is trickier! The variable z depends on s

in two different ways. First, from the equation

$$z = x^2 \ln (s + xy)$$

z is seen to be a function of the three variables x, y and s . Thus we have three partial

derivatives

$$\partial z/\partial x \ , \quad \partial z/\partial y \ , \quad \partial z/\partial s$$

and these symbols can be
placed on the first level
of branches in our tree
diagram, as shown to the
right.

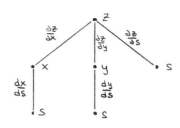

Notice, however, that z can be considered simply a function of s alone! To see this, take any value of s (say s = 2). Then values of x and y will be specified since they are functions of s (when s = 2 we are given that x = 1 and y = e - 2). But three specified values of x, y and s will determine a value for z (when x = 1, y = e - 2 and s = 2, we obtain z = 1). In short, specifying a value of s will determine a value of z, and that means z is a function of s. Thus we can consider the (ordinary) derivative of z with respect to s, dz/ds.

But dz/ds does not appear in our tree diagram. We have to use the tree diagram to express dz/ds through a chain rule. We count three paths leading from z to s. Thus

$$\frac{dz}{ds} \ = \ \left(\begin{array}{c}\text{term from}\\ \text{path 1}\end{array}\right) \ + \ \left(\begin{array}{c}\text{term from}\\ \text{path 2}\end{array}\right) \ + \ \left(\begin{array}{c}\text{term from}\\ \text{path 3}\end{array}\right)$$

or

$$\boxed{\frac{dz}{ds} \ = \ \frac{\partial z}{\partial x}\frac{dx}{ds} \ + \ \frac{\partial z}{\partial y}\frac{dy}{ds} \ + \ \frac{\partial z}{\partial s}}$$

Notice that dz/ds and $\partial z/\partial s$ are different, and they both appear in our chain rule! It is absolutely crucial to distinguish between the two terms, and to choose properly between them

when solving problems.

We now can plug into our chain rule and solve the given problem. Since $z = x^2 \ln (s + xy)$

we obtain

$$\frac{dz}{ds} = \frac{\partial z}{\partial x} \frac{dx}{ds} + \frac{\partial z}{\partial y} \frac{dy}{ds} + \frac{\partial z}{\partial s}$$

or

$$\frac{dz}{ds} = \left(2x \ln (s + xy) + \frac{x^2 y}{s + xy}\right) \frac{dx}{ds} + \left(\frac{x^3}{s + xy}\right) \frac{dy}{ds} + \frac{x^2}{s + xy}$$

Thus, using the given values when $s = 2$ yields

$$6 = \left(2 \ln e + \frac{e - 2}{e}\right) 2 + \left(\frac{1}{e}\right) \frac{dy}{ds} + \frac{1}{e}$$

which can be solved for dy/ds to yield

$$\boxed{\left.\frac{dy}{ds}\right|_{s = 2} = 3}$$
□

Note that in Anton's Example 3 the same crucial distinction between partial and

ordinary derivatives arises as we have in Example B.

3. **Variables occurring in fixed combinations.** In many partial differential equations we need to

use functions whose variables occur in fixed combinations. For instance, in the function

$$z = g(x, y) = \ln(xy) + e^{-xy}$$

the x and y appear only in the fixed combination xy. Such situations almost always call for use of a chain rule, where the "fixed combination" of variables is denoted collectively by another variable. In our example we could use $t = xy$, so that

$$z = f(t) = \ln t + e^{-t}$$

Anton gives a typical application of this technique in Example 4. Here is another:

Example C.   Assuming that the derivatives exist, show that a function of the form $z = f(x^2 - y^2)$ satisfies the equation

$$y \frac{\partial z}{\partial x} + x \frac{\partial z}{\partial y} = 0$$

Solution.   Let $t = x^2 - y^2$, so that

$$z = f(t), \quad \text{where} \quad t = x^2 - y^2$$

The tree diagram for this scheme is shown to the right. The two chain rules which we read off of the diagram are

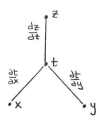

$$\frac{\partial z}{\partial x} = \frac{dz}{dt} \frac{\partial t}{\partial x} = 2x \frac{dz}{dt}$$

$$\frac{\partial z}{\partial y} = \frac{dz}{dt} \frac{\partial t}{\partial y} = -2y \frac{dz}{dt}$$

Thus

$$y \frac{\partial z}{\partial x} + x \frac{\partial z}{\partial y} = 2xy \frac{dz}{dt} - 2xy \frac{dz}{dt} = 0$$

as desired.   □

Section 16.9 :   Maxima and Minima of Functions of Two Variables

Finding relative extremum for a function $z = f(x, y)$ depends on Theorem 16.9.4, which we can reword as our fourth major gradient property:

GP #4 : The Critical Point Theorem	Suppose the partial derivatives for $f(x, y)$ exist at (x_0, y_0). Then f can have a relative extremum at (x_0, y_0) only if (x_0, y_0) is a <u>critical point</u> of f, i.e., $$\nabla f(x_0, y_0) = \overline{0}$$

Thus the first step in attempting to find extreme values is to <u>find the critical points.</u> This comes down to solving two simultaneous (in general, nonlinear) equations

$$f_x(x_0, y_0) = 0 \quad \text{and} \quad f_y(x_0, y_0) = 0$$

In fact, Anton's Theorem 16.9.4 is written in terms of these two equations, and does not even mention the gradient.

Computing the solutions to such a set of simultaneous equations might or might not be difficult; the examples in the text are all relatively straightforward, but many examples in the real world can only be solved "by approximation", generally with aid from a computer. The techniques for such approximation are studied in the branch of mathematics known as numerical analysis.

Once the critical points for a function have been found we then are faced with the <u>classification problem:</u>

16.9.2

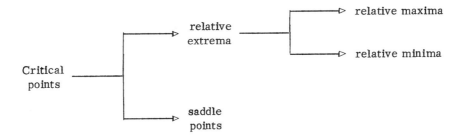

Into which category does a given critical point go? There are essentially two methods which can be used to answer this question:

1. <u>Classification by inspection.</u> Suppose (x_0, y_0) is a critical point for $z = f(x, y)$. Then consider the expression $\epsilon(h, k)$ defined by

$$\epsilon(h, k) = f(x_0 + h, y_0 + k) - f(x_0, y_0)$$

where h and k are two arbitrary variables.

Suppose that $\epsilon(h, k)$ is less than or equal to zero for all small values of h and k. Then $f(x_0 + h, y_0 + k)$ is less than or equal to $f(x_0, y_0)$ for all small h and k. Ah ha... then $f(x_0, y_0)$ must be a relative maximum value for f! We have shown

$$\left. \begin{array}{c} \epsilon(h, k) \leq 0 \text{ for all} \\[6pt] \text{small values of } h \text{ and } k \end{array} \right\} \text{ implies } \left\{ \begin{array}{l} \text{a relative maximum} \\[6pt] \text{occurs at } (x_0, y_0) \end{array} \right.$$

Similarly we can show

$$\left. \begin{array}{c} \epsilon(h, k) \geq 0 \text{ for all} \\[6pt] \text{small values of } h \text{ and } k \end{array} \right\} \text{ implies } \left\{ \begin{array}{l} \text{a relative minimum} \\[6pt] \text{occurs at } (x_0, y_0) \end{array} \right.$$

$$\left. \begin{array}{l} \epsilon(h, k) \text{ can be either positive} \\[6pt] \text{or negative for arbitrarily} \\[6pt] \text{small values of } h \text{ and } k \end{array} \right\} \text{ implies } \left\{ \begin{array}{l} \text{a saddle point} \\[6pt] \text{occurs at } (x_0, y_0) \end{array} \right.$$

"Classification by inspection" just means playing around with the expression $\varepsilon(h, k)$ until it can be determined into which of the above three categories it belongs. Anton's Example 1 is essentially a "classification by inspection," although he does not explicitly write out $\varepsilon(h, k)$. He considers the function $f(x, y) = 9 - x^2 - y^2$, and shows that it has only one critical point $(0, 0)$. For this point $(0, 0)$ the function $\varepsilon(h, k)$ would be

$$\varepsilon(h, k) = f(h, k) - f(0, 0) = -h^2 - k^2$$

Clearly $\varepsilon(h, k)$ is always less than or equal to zero, and thus $(0, 0)$ is a relative maximum (in fact, an absolute maximum).

Let's consider a more complicated example:

Example A.   Locate all the relative maxima, relative minima and saddle points of $f(x, y) = 2x^3 + 3y^2 - 6xy$.

Solution.   We first compute the partial derivatives for $f(x, y)$:

$$f_x(x, y) = 6x^2 - 6y \; , \quad f_y(x, y) = 6y - 6x$$

To find the critical points of f we set these partial derivatives equal to zero

$$6x^2 - 6y = 0$$

$$6y - 6x = 0$$

The second of these equations shows $y = x$; the first equation then becomes $x = x^2$, so that $x = 0$ or $x = 1$. We therefore have two critical points, $(0, 0)$ and $(1, 1)$, and we will classify each one "by inspection."

For $(0, 0)$ the function $\varepsilon(h, k)$ becomes

$$\varepsilon(h, k) = f(h, k) - f(0, 0) = 2h^3 + 3k^2 - 6hk$$

As written it is not clear what the sign of $\varepsilon(h, k)$ might be for small values of h and k. However, completing the square in the k terms will make things clearer:

$$\varepsilon(h, k) = 2h^3 + 3(k^2 - 2hk + h^2) - 3h^2$$
$$= h^2(2h - 3) + 3(k - h)^2$$

When $k = h$ and $0 < h < 3/2$ we obtain

$$\varepsilon(h, k) = h^2(2h - 3) < 0$$

However, when $h = 0$ and $k \neq 0$ we obtain

$$\varepsilon(h, k) = 3k^2 > 0$$

Thus $\varepsilon(h, k)$ can be either positive or negative for small values of h and k, proving that $(0, 0)$ is a saddle point for f.

Now consider $(1, 1)$. For this critical point the function $\varepsilon(h, k)$ is given by

$$\varepsilon(h, k) = f(h + 1, k + 1) - f(1, 1)$$
$$= 2(h + 1)^3 + 3(k + 1)^2 - 6(h + 1)(k + 1) - 2 - 3 + 6$$
$$= 2h^3 + 6h^2 + 3k^2 - 6hk$$

Again, it is not obvious what the sign of $\varepsilon(h, k)$ might be for small values of h and k, but completing the square in the k terms clears up the mystery:

$$\varepsilon(h, k) = 2h^3 + 6h^2 + 3(k^2 - 2hk + h^2) - 3h^2$$
$$= 2h^3 + 3h^2 + 3(k - h)^2$$
$$= h^2(2h + 3) + 3(k - h)^2$$

The second term is clearly always non-negative. The first term is also non-negative when

$h \geq -3/2$. Since this includes all the small values of h and k, we see that (1, 1)

must give a <u>relative minimum</u> value for f.   □

It should be clear that the success of "classification by inspection" in a given problem

depends on the simplicity of the expression $\epsilon(h, k)$, which in turn depends on the simplicity

of the original function f. Even a slightly complicated function f can prove too difficult to

handle "by inspection." Such functions are generally more successfully treated by using our

second classification method.

2.   <u>Classification by the Second-Partials Test.</u>

The Second-Partials test (Theorem 16.9.5) allows us to classify most critical points

for functions of two variables simply by examining the second partial derivatives evaluated at

the critical point in question. The test is admittedly weird looking, and it is hard to "justify"

the result without giving the full proof. That we won't do ... it's too difficult at this level. (The

proof depends on generalizing Taylor series expansions to functions of two variables.) We simply

point out that the strange expression

$$D = f_{xx}(x_0, y_0) f_{yy}(x_0, y_0) - f_{xy}^2(x_0, y_0)$$

comes from a determinant:

$$D = \det \begin{bmatrix} f_{xx}(x_0, y_0) & f_{xy}(x_0, y_0) \\ f_{xy}(x_0, y_0) & f_{yy}(x_0, y_0) \end{bmatrix}$$

This determinant does not come about as an accident; it plays an important role in the proof of

the result.

Using the Second-Partials test is quite straightforward; three very good examples are given in Anton's Examples 2, 3 and 4. We give two more:

<u>Example B.</u>   Redo Example A using the Second-Partials test.

<u>Solution.</u>   We have $f(x, y) = 2x^3 + 3y^2 - 6xy$, and the two critical points are $(0, 0)$ and $(1, 1)$. Since

$$f_x(x, y) = 6x^2 - 6y \; , \quad f_y(x, y) = 6y - 6x$$

then

$$f_{xx}(x, y) = 12x \; , \quad f_{xy}(x, y) = -6 \; , \quad f_{yy}(x, y) = 6$$

Hence

$$D = \det \begin{bmatrix} 12x & -6 \\ -6 & 6 \end{bmatrix} = 72x - 36$$

We can now classify each critical point:

$\boxed{(0, 0)}$ $D = 72(0) - 36 = -36 < 0$

Thus $(0, 0)$ is a <u>saddle point.</u>

$\boxed{(1, 1)}$ $D = 72(1) - 36 = 36 > 0$

$f_{xx}(1, 1) = 12(1) = 12 > 0$

Thus $(1, 1)$ gives a <u>relative minimum.</u>

As you can see this is a quicker method than "classification by inspection. "   □

We next consider an important application in economics. Consider a two product firm, making wigits and whackits, under circumstances of pure competition (i. e. , the prices are

"exogenous" -- determined by outside forces, in this case the marketplace). Suppose P_1

and P_2 are the prices, and Q_1 and Q_2 the daily output levels, of wigits and whackits

respectively. The daily <u>revenue function</u> of the firm will be

$$R = P_1 Q_1 + P_2 Q_2$$

and the cost will depend just on Q_1 and Q_2 , i. e. , there will be some <u>cost function</u>

$$C = f(Q_1, Q_2)$$

The <u>profit function</u> will therefore be

$$\pi = R - C = P_1 Q_1 + P_2 Q_2 - f(Q_1, Q_2)$$

Assuming P_1 and P_2 are non-varying, how do we choose output levels Q_1 and Q_2 so

as to maximize π ?

<u>Example C.</u>   Maximize the profit function $\pi(x, y)$ if

$$P_1 = 6 \quad, \quad P_2 = 10$$

and

$$C(Q_1, Q_2) = Q_1^3 - 6Q_1 Q_2 + Q_2^2 + 20Q_2$$

<u>Solution.</u>   Letting $x = Q_1$ and $y = Q_2$, we wish to maximize the function

$$\pi(x, y) = 6x + 10y - x^3 + 6xy - y^2 - 20y$$

or

$$\boxed{\pi(x, y) = -x^3 + 6x + 6xy - 10y - y^2}$$

We first determine the critical points for π. We have

$$\pi_x(x, y) = -3x^2 + 6 + 6y = 0$$

$$\pi_y(x, y) = 6x - 10 - 2y = 0$$

which simplifies to

$$2y = x^2 - 2$$

$$2y = 6x - 10$$

Thus $x^2 - 2 = 6x - 10$, which solves to $x = 2$ or $x = 4$. Our second equation then shows that

$$x = 2 \quad \text{gives} \quad y = 1, \quad \text{and} \quad x = 4 \quad \text{gives} \quad y = 7$$

Our two critical points are therefore $(2, 1)$ and $(4, 7)$.

We classify our points by the Second-Partials test: Since

$$\pi_{xx}(x, y) = -6x, \quad \pi_{xy}(x, y) = 6, \quad \pi_{yy}(x, y) = -2$$

then

$$D = \det \begin{bmatrix} -6x & 6 \\ 6 & -2 \end{bmatrix} = 12x - 36$$

We can now classify each critical point:

$\boxed{(2, 1)}$ $D = 12(2) - 36 = -12 < 0$

Thus $(2, 1)$ is a <u>saddle point.</u>

$\boxed{(4, 7)}$ $D = 12(4) - 36 = 12 > 0$

$\pi_{xx}(4, 7) = -6(4) = -24 < 0$

Thus $(4, 7)$ gives a <u>relative maximum.</u>

Thus $Q_1 = 4$ and $Q_2 = 7$ are the daily output levels which maximize the profit.   □

3.   A brief warning.   There is one limitation in the Critical Point Theorem which you should be

aware of:

> The Critical Point Theorem does not give any information about extreme
>
> points of $f(x, y)$ at which the partial derivatives fail to exist, nor
>
> about extreme points of   f   which occur on the boundary of its domain
>
> of definition.

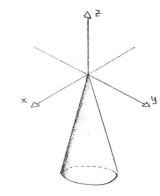

Here are two examples:

1.   Consider $f(x, y) = -\sqrt{x^2 + y^2}$.   The graph

of this function clearly shows a maximum

value at $(0, 0)$,   but $(0, 0)$ is not a critical

point for   f   since $f_x(x, y)$ and $f_y(x, y)$

are not even defined at $(0, 0)$.

2.   Consider $f(x, y) = x^2 + y^2$ with the domain

restricted to the square

$$\{(x, y)   \text{where}   -1 \leq x \leq 1 ,   -1 \leq y \leq 1\}$$

The graph of this function shows four maximum

values at the corner points of the square. However,

these points are clearly not critical points for   f .

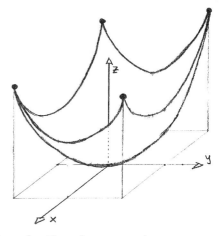

Moral: If you need to find all the possible extreme values for a function   f ,   you need to examine all

  (i) critical points,

  (ii) points where the partials f_x and f_y do not exist, and

  (iii) points on the boundary of the domain of f.

In this text, however, you will only be required to examine the critical points, utilizing the methods previously described.

4. <u>Functions of three variables.</u> It is possible to generalize all of the methods of this section to functions of three (or more!) variables, $w = f(x, y, z)$. The Critical Point Theorem holds true without change, except that to find critical points we must simultaneously solve three equations

$$f_x(x, y, z) = 0, \quad f_y(x, y, z) = 0 \quad \text{and} \quad f_z(x, y, z) = 0$$

The method of "classification by inspection" carries over in the obvious way by considering the expression

$$\epsilon(h, k, \ell) = f(x_0 + h, y_0 + k, z_0 + \ell) - f(x_0, y_0, z_0)$$

at each critical point (x_0, y_0, z_0). This expression can, of course, be pretty messy to deal with.

Even the Second-Partials test carries over to functions of three variables. However, we need to consider a third term in addition to $f_{xx}(x_0, y_0)$ and D. The new term turns out to be a 3×3 determinant (again, it is no accident). The interested student is encouraged to explore these mysteries in an advanced vector calculus course.

Section 16. 10 :   Lagrange Multipliers

Optimization problems <u>with constraints</u> occur quite often in applications, perhaps most commonly in Economics. We'll develop such an example in detail, both for its own sake and to highlight certain key aspects of the subject.

Let   x = daily output of margarine in America (in tons)

   y = daily output of MX missiles (in pounds) .

(This will be a modern version of the "guns and butter" story...) We assume there is some "relationship" between   x   and   y , i.e., produce more margarine and you must reduce the production of missiles. Let

$$g(x, y) \;=\; 0$$

be the equation expressing this relationship. For a specific example we'll use

$$g(x, y) \;=\; -y - x^2 + 1 \;=\; 0$$

Thus the points (x, y) which satisfy $g(x, y) = 0$ represent all the possible "maximum joint output levels" of margarine and missiles.

However, society will have preferences among the various possible joint output levels (x, y) , which we may attempt to quantify as a function,

> $z \;=\; f(x, y) \;=\;$ the "satisfaction" of society given
>
>   x   output of margarine and   y   output
>
>   of missiles.
>
> (The <u>utility</u> function of x   and   y .)

For a specific example we'll use

$$f(x, y) = x^2 y$$

Our goal is now easy to state: find the output levels of margarine and missiles which most please society and which is actually attainable, i.e.,

$$\begin{array}{l} \text{maximize} \quad z = f(x, y) \quad \text{relative to} \\[2mm] \text{the constraint} \quad g(x, y) = 0 \end{array}$$

In our specific example we wish to

$$\begin{array}{l} \text{maximize} \quad f(x, y) = x^2 y \quad \text{relative to} \\[2mm] \text{the constraint} \quad -y - x^2 + 1 = 0 \end{array}$$

We have a choice of two ways to proceed:

1. Absorb the constraint to yield an unconstrained problem. In this method we take $g(x, y) = 0$ and solve for one of the two variables x, y in terms of the other (assuming this is possible). We then plug the result into $z = f(x, y)$ to obtain an unconstrained optimization problem in just one variable.

Example A. Maximize the utility function $f(x, y) = x^2 y$ relative to the constraint $g(x, y) = -y - x^2 + 1 = 0$ by absorbing the constraint.

Solution. The expression $g(x, y) = -y - x^2 + 1 = 0$ gives

$$y = -x^2 + 1$$

which, when plugged into $z = f(x, y) = x^2 y$, yields

$$\boxed{z \;=\; -x^4 + x^2}$$

Thus we have an ordinary optimization problem which we solve by setting dz/dx equal to zero:

$$dz/dx \;=\; -4x^3 + 2x \;=\; 0$$

which gives $x = -\sqrt{1/2}$, 0 or $\sqrt{1/2}$. However, $x = -\sqrt{1/2}$ is not possible since output levels cannot be negative, and $x = 0$ certainly does not maximize $-x^4 + x^2$. Thus $x = \sqrt{1/2}$ is the only possible maximizing value. We can use the second derivative test to check this:

$$d^2z/dx^2 \;=\; -12x^2 + 2 \;=\; -4 \quad \text{at} \quad x = \sqrt{1/2}$$

Hence $x = \sqrt{1/2}$ does give a local maximum for $z = -x^4 + x^2$. Using $y = -x^2 + 1$ we see that $(\sqrt{1/2}, 1/2)$ is our desired answer. $\square$

Unfortunately our "absorption method" suffers from two serious problems:

(1) Many times it is extremely difficult (or impossible) to solve $g(x, y) = 0$ for one variable in terms of the other. For example, how would you handle

$$g(x, y) \;=\; \frac{e^{xy} \sin x}{x^2 y + y^2} \;=\; 0 \qquad ?$$

(2) Even if we can solve $g(x, y) = 0$ for, say, y as a function of x, there is no guarantee that substitution into $z = f(x, y)$ will yield a particularly manageable expression.

These problems make us seek out a "sneakier" method of optimization relative to a constraint.

2. **The Lagrange Multiplier Rule.**   Again the gradient comes to our aid, with its fifth major

property:

<table>
<tr>
<td>

GP #5 :
The Constrained
Extremum
Principle

</td>
<td>

Suppose $z = f(x, y)$ and $z = g(x, y)$ both have continuous first partial

derivatives on an open set containing the curve $g(x, y) = 0$, and that

$\nabla g(x, y) \neq \overline{0}$ at any point along this curve.

If (x_0, y_0) is a constrained relative extrema point for $f(x, y)$ on the

constraint curve $g(x, y) = 0$, then

$\nabla f(x_0, y_0) = \lambda \nabla g(x_0, y_0)$ for some number λ .

</td>
</tr>
</table>

This result is also known as the <u>Lagrange Multiplier Rule</u>. The number λ is called a Lagrange

multiplier.

Anton gives the basics of the proof of the Lagrange Multiplier

Rule in §16. 10. The **heart** of the proof lies in

showing that both $\nabla f(x_0, y_0)$ and $\nabla g(x_0, y_0)$

are perpendicular to the level curve $g(x, y) = 0$

at the maximizing (or minimizing) point (x_0, y_0) .

Then $\nabla f(x_0, y_0)$ and $\nabla g(x_0, y_0)$ must be

parallel to each other, which means one is a

scalar multiple of the other.

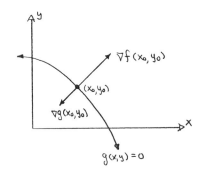

To use the Lagrange Multiplier Rule in actual computations can be somewhat confusing unless you are very organized in your work! To maximize or minimize $z = f(x, y)$ subject

to the constraint $g(x, y) = 0$ we must solve the simultaneous equations

$$\begin{cases} \nabla f(x, y) = \lambda \nabla g(x, y) \\ g(x, y) = 0 \end{cases}$$

for x, y and λ. Each resulting point (x, y) is a possible constraint extreme point. Also

note that our first equation is a vector equation which is equivalent to two scalar equations.

Hence our total system really consists of three scalar equations:

$$\begin{aligned} f_x(x, y) &= \lambda g_x(x, y) \\ f_y(x, y) &= \lambda g_y(x, y) \\ g(x, y) &= 0 \end{aligned}$$

Warning! The most common **error** in using the Lagrange Multiplier rule is to forget the constraint

equation $g(x, y) = 0$! If you do this you will have only two equations in three unknowns. This

will not work!

We illustrate the Lagrange Multiplier Rule by redoing the "margarine and missiles"

example:

Example B. Maximize the utility function $f(x, y) = x^2 y$ relative to the constraint $g(x, y) =$

$-y - x^2 + 1 = 0$ by the Lagrange Multiplier Rule.

Solution. At a constrained relative extremum we must have

$$\begin{cases} \nabla f(x, y) & = & \lambda \, \nabla g(x, y) \\ g(x, y) & = & 0 \end{cases}$$

or

$$\begin{cases} \langle 2xy, x^2 \rangle & = & \lambda \langle -2x, -1 \rangle \\ -y - x^2 + 1 & = & 0 \end{cases}$$

which is equivalent to the three equations

$$\begin{cases} 2xy & = & -2x\lambda \\ x^2 & = & -\lambda \\ y & = & -x^2 + 1 \end{cases}$$

A very common procedure in solving "Lagrange Multiplier equations" such as these is to solve the first two equations for λ and then to equate the resulting expressions. In this case we obtain

$$\begin{cases} \lambda & = & -y \qquad * \\ \lambda & = & -x^2 \\ y & = & -x^2 + 1 \end{cases}$$

which gives

$$\begin{cases} y & = & x^2 \\ y & = & -x^2 + 1 \end{cases}$$

Solving this pair of equations we obtain $x^2 = -x^2 + 1$, so $2x^2 = 1$, giving $x = \pm\sqrt{1/2}$. However, since $x < 0$ is not possible for an output level, we see that $x = \sqrt{1/2}$ is the only

---

* Division by x in this case is acceptable since $x = 0$ certainly does not give a constrained maximum value for $f(x, y) = x^2 y$.

possible answer. Since $y = x^2$, we see that $(\sqrt{1/2}, 1/2)$ is the final answer, as obtained

earlier.   □

There are some important comments to be made concerning the Lagrange Multiplier Rule:

(i)   The points (x_0, y_0) obtained by this procedure (sometimes called constrained critical

points) need not always be actual constrained extreme points. Just as with ordinary critical

points we can have "constrained saddle points. "

(ii)   There is a "Second-Partials test" for classifying the constrained critical points, but

because of its complicated nature, it is of more theoretical than computational value, at least

at our level. For this reason classifying the constrained critical points obtained from the

Lagrange Multiplier Rule is generally done in a more ad hoc fashion. Rarely will this cause

serious difficulties.

3. Constrained optimization in three variables. All our previous work with functions of two

variables carries over to functions of three variables. The basic problem we now want to solve

is

> maximize or minimize $w = f(x, y, z)$ relative
>
> to the constraint $g(x, y, z) = 0$.

We again have two ways to proceed. The first way is to "absorb the constraint," i.e.,

take $g(x, y, z) = 0$ and solve for one of the three variables in terms of the other two (if

possible). Then plug the result into $w = f(x, y, z)$ to obtain an unconstrained optimization problem is just <u>two variables</u>. Then we can use the techniques of §16.9, i.e., the Critical

Point Theorem and the Second-Partial Test. A good illustration of this technique is given in

Anton's Example 4 of §16.9. In that example we have to minimize

$$S = f(x, y, z) = xy + 2xz + 2yz$$

subject to the constraint

$$g(x, y, z) = xyz - 32 = 0$$

Anton solves the constraint for z to give

$$z = 32/xy$$

which he then substitutes into $S = f(x, y, z)$ to obtain

$$S = xy + 64/y + 64/x$$

The Critical Point Theorem and the Second-Partials Test now can be applied to S as a function

of x and y. This will yield x = 4, y = 4 and z = 2.

The second way to solve the problem is to use the Constrained Extremum Principle (i.e.,

the Lagrange Multiplier Rule), which carries over to functions of three variables without change.

To maximize or minimize $w = f(x, y, z)$ subject to the constraint $g(x, y, z) = 0$ we must

solve the simultaneous equations

$$\begin{cases} \nabla f(x, y, z) = \lambda \nabla g(x, y, z) \\ g(x, y, z) = 0 \end{cases}$$

for x, y, z and λ. Since the first equation is a vector equation in 3-space, then it is

equivalent to three scalar equations. Hence our total system really consists of four scalar

equations:

$$f_x(x,y,z) = \lambda\, g_x(x,y,z)$$

$$f_y(x,y,z) = \lambda\, g_y(x,y,z)$$

$$f_z(x,y,z) = \lambda\, g_z(x,y,z)$$

$$g(x,y,z) = 0$$

Anton illustrates this technique in Examples 3 and 4. In Example 4 he redoes the problem we discussed above: minimize

$$S = f(x,y,z) = xy + 2xz + 2yz$$

subject to the constraint

$$g(x,y,z) = xyz - 32 = 0$$

The four equations which must be solved are

$$
\begin{cases}
y + 2z = \lambda\, yz \\[4pt]
x + 2z = \lambda\, xz \\[4pt]
2x + 2y = \lambda\, xy \\[4pt]
xyz = 32 \qquad \longleftarrow \boxed{\text{DON'T FORGET THE CONSTRAINT!}}
\end{cases}
$$

Solving the first three equations for λ and then equating the resulting expressions lead essentially to two distinct equations in x, y and z ; combined with the final constraint equation we then obtain three equations in the three unknowns x, y and z. These will yield $x = 4$, $y = 4$ and $z = 2$, in agreement with the earlier solution.

4. Optimization with two constraints (optional). Suppose we wish to solve the following type of problem:

> maximize or minimize $w = f(x, y, z)$ relative
>
> to the two constraints $g(x, y, z) = 0$
>
> and $h(x, y, z) = 0$

If we suppose that f, g and h all have continuous first partial derivatives, and that

$\nabla g \neq \bar{0}$ and $\nabla h \neq \bar{0}$ along the curve C of intersection of the two level surfaces $g = 0$

and $h = 0$,   then the following Lagrange Multiplier Rule is valid:

> If (x_0, y_0, z_0) is a constrained relative extreme
>
> point, then at (x_0, y_0, z_0) we have
>
> $$\nabla f = \lambda_1 \nabla g + \lambda_2 \nabla h$$
>
> for some numbers λ_1 and λ_2 .

The geometric idea behind this rule is that

all three vectors $\nabla f, \nabla g$ and ∇h will

be perpendicular to the curve C at (x_0, y_0, z_0) ,

and hence must be coplanar. This means that

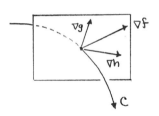

∇f will be a sum of multiples of the vectors ∇g and ∇h .

To apply this rule we must solve the system of equations

$$\nabla f = \lambda_1 \nabla g + \lambda_2 \nabla h$$
$$g = 0 \quad \text{and} \quad h = 0$$

for x, y, z, λ_1 and λ_2 . Since the first equation is a vector equation in   3-space,   then we

really have five scalar equations in five unknowns. This can be a mess to work with! We refer

the interested student to a course in advanced vector calculus.

Chapter 17: Multiple Integrals ·

Section 17. 1:   Double Integrals

§1.   The definition of the double integral.   The definition of the integral of a one variable function

was motivated by the concept of area:

for a nonnegative function   f(x)

the integral of   f   over an interval

[a , b] ,

$$\int_a^b f(x)\, dx$$

is the area above   [a , b]   and

under the graph   y = f(x) .

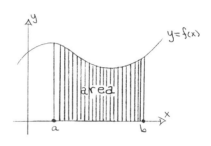

In a similar fashion we will motivate the double integral of a two variable function by the use of

volume:

for a nonnegative function   f(x, y) ,

the double integral of   f   over a

plane region   R ,

$$\iint_R f(x, y)\, dA$$

will be the volume above   R   and

under the graph   z = f(x, y) .

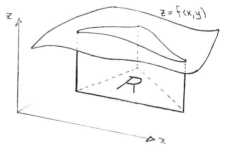

It is important to realize, however, that the computation of volumes will be only one small

application of double integration, just as the computation of areas was only one small application

of ordinary integration. Other uses of double integration will be developed in later sections.

Anton begins this section by recalling that the formal definition of the one variable

integral $\displaystyle\int_a^b f(x)\,dx$ involved limits of Riemann sums for f over the interval [a, b]:

Definition
of
$$\int_a^b f(x)\,dx$$

divide [a, b] into n subintervals

of lengths $\Delta x_1, \Delta x_2, \dots, \Delta x_n$, choose

a point x_k^* in each subinterval, and

form the Riemann sum

$$\sum_{k=1}^n f(x_k^*)\,\Delta x_k$$

This Riemann sum can be thought

of as a sum of areas of rectangles,

and hence as an approximation to

the total area.

Then $\displaystyle\int_a^b f(x)\,dx = \lim_{n \to \infty} \sum_{k=1}^n f(x_k^*)\,\Delta x_k$

where $\Delta x_k \to 0$ as $n \to \infty$

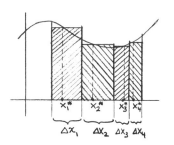

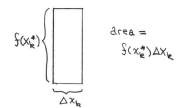

It is important to remember this because, in parallel fashion, the formal definition of the double

integral $\displaystyle\iint_R f(x,y)\,dA$ uses limits of Riemann sums for f over the region R :

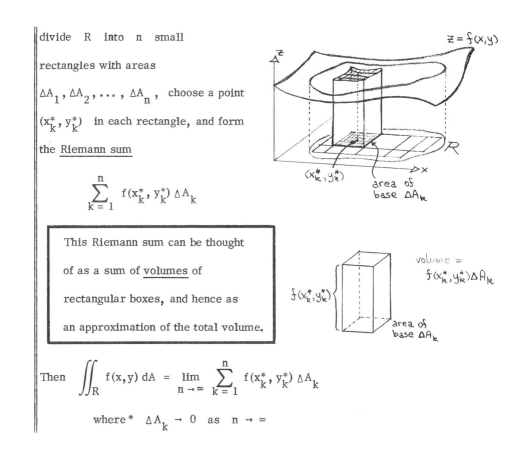

$z = f(x,y)$

$(x_k^\#, y_k^\#)$

area of base ΔA_k

volume $=$ $f(x_k^*, y_k^*)\Delta A_k$

$f(x_k^*, y_k^*)$

area of base ΔA_k

divide R into n small rectangles with areas ΔA_1, ΔA_2, ..., ΔA_n, choose a point (x_k^*, y_k^*) in each rectangle, and form the Riemann sum

$$\sum_{k=1}^{n} f(x_k^*, y_k^*)\, \Delta A_k$$

Definition of
$$\iint_R f(x,y)\, dA$$

This Riemann sum can be thought of as a sum of <u>volumes</u> of rectangular boxes, and hence as an approximation of the total volume.

Then $\displaystyle\iint_R f(x,y)\, dA = \lim_{n \to \infty} \sum_{k=1}^{n} f(x_k^*, y_k^*)\, \Delta A_k$

where* $\Delta A_k \to 0$ as $n \to \infty$

Thus, the formal definition of the double integral is just a "higher dimensional" version of the formal definition of the single integral. The basic procedure in both is to approximate by Riemann sums and then make the approximations become better and better by taking the limit.

As you can probably guess, this complicated definition (yes, even we admit that it's pretty complicated!) makes the proofs of the properties of double integrals technical and difficult. You will notice that Anton does not include most of these proofs, and we won't either - for a very

---

* $\Delta A_k \to 0$ is to mean that the length and width of the rectangle both go to zero.

good reason: slogging through them adds very little to your understanding and it uses time and space which can be spent more productively on other things.

In view of this, you may be asking yourself "so why should I learn the Riemann sum definition of the double integral?" That question has two good answers:

1. The precise, formal definition is essential to a solid theory. A study of double integrals based only on a loose, intuitive idea of "volume under a surface" will lead to misunderstandings and mistakes.

2. Double integrals arise in applications through their Riemann sums definition. If you don't know this definition (and know it well) you will be completely befuddled when the double integral "magically" appears in physics or economics. We will give such examples later in the chapter, especially in §§17.4 and 17.6.

So be sure to study carefully the definition of the double integral.

2. <u>Existence and basic properties of the double integral.</u> The greater complexity of a region R in the xy-plane (as compared with a closed interval [a, b] on the x-axis) gives rise to a complexity in double integrals that was not present in single integrals. Some closed intervals are bigger than others, some smaller; some are shifted more to the left, some more to the right. But that's as much difference as there can be.

However, even relatively "well-behaved" regions in the plane can exhibit a very complex structure.

One example of the added difficulty this complexity causes is with the existence question

("When does the integral exist?"). In the one variable situation our answer is simple:

$$\text{If } y = f(x) \text{ is continuous on } [a, b]$$

$$\text{then } \int_a^b f(x)\, dx \text{ exists.}$$

Unfortunately the answer to the existence questions is much more complex for double integrals.

The best we can do at this level is the following loose rule:

<table>
<tr><td>The Existence
Principle</td><td>If z = f(x, y) is continuous on a "well-behaved" closed and bounded region R in the xy-plane, then $\iint_R f(x, y)\, dA$ exists.</td></tr>
</table>

Recall that from §16.3 a region R in the xy-plane is closed if it contains all its boundary

points; a region is bounded if it can be placed inside some suitably large rectangle.

There is no simple definition for "well-behaved." However, two important classes of

"well-behaved" regions are referred to as regions of type I and type II, and will be defined in

§17.2. In particular, any rectangle of the form

$$\{(x, y) : a \le x \le b, \ c \le y \le d\} = [a, b] \times [c, d]$$

is a region of both type I and type II. Another important class of "well-behaved" regions are

those of polar type, to be defined in §17.3.

Aside from the existence question, most of the basic algebraic properties for double

integrals are the same as for their one variable counterparts! For example, "the integral of a

sum is the sum of the integrals." Anton lists these properties just prior to Example 1, and

you should memorize them.

3. <u>Evaluation of double integrals.</u> In the next section we will discuss methods for evaluating double integrals over any region R of type I or type II; in this section we restrict ourselves to the case where R is a rectangle of the form

$$R = \{(x,y) : a \le x \le b, \ c \le y \le d\} = [a,b] \times [c,d]$$

We do not evaluate double integrals by using their Riemann sums definition. Instead we reduce them to <u>iterated</u> (or <u>repeated</u>) integrals via Theorem 17.1.1:

Suppose $z = f(x,y)$ is continuous on the rectangle $R = [a,b] \times [c,d]$

Iterated
Integrals
on a
rectangle

Then $\iint_R f(x,y) \, dA = \int_a^b \left[\int_c^d f(x,y) \, dy \right] dx$

$$= \int_c^d \left[\int_a^b f(x,y) \, dx \right] dy$$

Said in words:

To evaluate a double integral over a rectangle

- hold one variable fixed and integrate with respect to the other;

- integrate the result with respect to the previously "fixed" variable.

Anton gives a good geometric justification for the result when $f(x,y)$ is non-negative: it is just an application of the method of slicing for the computation of volume (as discussed in §6.2). You should carefully study this justification.

Theorem 17.1.1 reduces the computation of a double integral to the computation of two one-variable integrals! However, the inner integral is an example of <u>partial integration</u>: one variable is held constant while we integrate with respect to the other. It is an analogous concept to that of partial differentiation.

Anton's Examples 2, 3 and 4 illustrate Theorem 17. 1. 1. Here are some further illustrations:

<u>Example A.</u>   Evaluate the integral $\iint_R y\,e^{xy}\,dA$ where

$$R = [1,2] \times [0,3]$$

<u>Solution.</u>   Using Theorem 17. 1. 1 we have

$$\iint_R y\,e^{xy}\,dA = \int_0^3 \left[\int_1^2 y\,e^{xy}\,dx \right] dy \qquad\qquad (A)$$

We first evaluate the inner integral (remember: during this partial integration y is considered to be a constant):

$$\int_1^2 y\,e^{xy}\,dx = y \int_1^2 e^{xy}\,dx \qquad\qquad (y \text{ is a constant!})$$

$$= y \left[\frac{e^{xy}}{y} \Big|_{x=1}^{x=2} \right] \qquad\qquad \begin{array}{l}\text{(ordinary integration ...}\\ \text{... with } y \text{ a constant!)}\end{array}$$

$$= e^{xy} \Big|_{x=1}^{x=2}$$

$$= e^{2y} - e^{y}$$

Thus the full integral is evaluated by replacing the inner integral in (A) by $e^{2y} - e^{y}$:

$$\iint_R y\,e^{xy}\,dA = \int_0^3 \left[e^{2y} - e^y \right] dy$$

$$= \left(\frac{e^{2y}}{2} - e^y \right) \Bigg|_0^3$$

$$= \left(\frac{e^6}{2} - e^3 \right) - \left(\frac{1}{2} - 1 \right)$$

$$= \frac{e^6}{2} - e^3 + \frac{1}{2}$$

Alternate Solution.   Theorem 17. 1. 1 allows for iterated integrals in either order: x   then

y ,   or   y   then   x . Here's the other order:

$$\iint_R y\,e^{xy}\,dA = \int_1^2 \left[\int_0^3 y\,e^{xy}\,dy \right] dx \qquad\qquad (B)$$

The inner integral in this case is not as simple as in the first solution; we need to use integration

by parts to evaluate it:

$$\int_0^3 y\,e^{xy}\,dy = uv \Bigg|_{y=0}^{y=3} - \int_0^3 v\,du$$

u = y	dv = e^{xy} dy
du = dy	v = e^{xy}/x
(x is a constant !)	

$$= y\,\frac{e^{xy}}{x} \Bigg|_{y=0}^{y=3} - \int_0^3 \frac{e^{xy}}{x}\,dy$$

$$= \frac{3\,e^{3x}}{x} - \left(\frac{e^{xy}}{x^2} \Bigg|_{y=0}^{y=3} \right)$$

$$= \frac{3\,e^{3x}}{x} - \frac{e^{3x}}{x^2} + \frac{1}{x^2}$$

Whew! Perhaps the worst is over!? The full integral is now evaluated by replacing the inner integral in (B) by $\dfrac{3\,e^{3x}}{x} - \dfrac{e^{3x}}{x^2} + \dfrac{1}{x^2}$:

$$\iint_R y\,e^{xy}\,dA = \int_1^2 \left[\frac{3\,e^{3x}}{x} - \frac{e^{3x}}{x^2} + \frac{1}{x^2}\right] dx$$

$$= 3\int_1^2 \frac{e^{3x}}{x}\,dx - \int_1^2 \frac{e^{3x}}{x^2}\,dx + \int_1^2 \frac{dx}{x^2} \qquad\qquad (C)$$

$$= \dots\ \ !\#?!*\$\%!?!\ \dots\ \text{this looks horrendous!}$$

... The point of this example should be evident: Theorem 17. 1. 1 gives us two ways to attempt

the evaluation of any double integral on a rectangle. However, that does not mean that the two

methods are equally easy! If you get into trouble with one order of integration, <u>then try the</u>

<u>other!</u>...

... Returning to our alternate solution, with some care and patience Equation (C) can

be evaluated. The trick is to use an integration by parts on the middle integral. Try it!   □

<u>Example B.</u>   Find the volume in the first octant, bounded by the coordinate planes, the plane   x = 2 ,

and the surface $z + y^2 = 4$.

<u>Solution.</u>   In order to apply the techniques of this section we should somehow express the

specified volume as the volume <u>over a rectangle   R</u>

<u>and under the graph of a function   z = f(x, y)</u>. To

do this it is helpful to sketch the volume we are

considering. First draw a set of positive   xyz-axes

(positive since we only are considering the first

octant). Then draw in the plane   x = 2 .

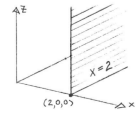

17.1.10

Finally add in the surface $z + y^2 = 4$ (i.e., draw

the parabola $z = 4 - y^2$ in the yz-plane and

extend it out parallel to the x-axis).

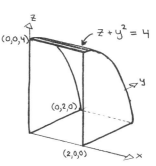

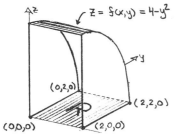

Ahh...! We now can see how the volume is

expressed in the form we want! We need to

calculate the volume over the rectangle

$R = \{(x,y) : 0 \le x \le 2, 0 \le y \le 2\}$ and under the graph of $z = f(x,y) = 4 - y^2$. Our

volume is thus

$$V = \iint_R (4 - y^2)\, dA$$

$$= \int_0^2 \left[\int_0^2 (4 - y^2)\, dx \right] dy \qquad \text{by Theorem 17.1.1}$$

$$= \int_0^2 \left[(4 - y^2)x \, \Big|_{x=0}^{x=2} \right] dy$$

$$= \int_0^2 (8 - 2y^2)\, dy$$

$$= 8y - \frac{2}{3}y^3 \Big|_0^2 = 16 - \frac{16}{3} = \boxed{\frac{32}{3}} \qquad \qquad \Box$$

As Example B demonstrates, being able to sketch reasonable pictures in three dimensions is

important in solving certain types of double integral problems. If you are still weak in this area,

review the graphing procedures in The Companion, §§15.8 and 16.1.

§4. <u>Volumes below the xy-plane</u>. The double integral $\iint_R f(x,y)\,dx\,dy$ was motivated as a

volume <u>when   f(x,y)   is a nonnegative function.</u> If   f(x,y)   is not always positive, then the

situation is just slightly more complicated:

$$\iint_R f(x,y)\,dx\,dy = \text{Volume above the   xy-plane}$$

<u>minus</u> the volume below the   xy-plane

$$= V_a - V_b$$

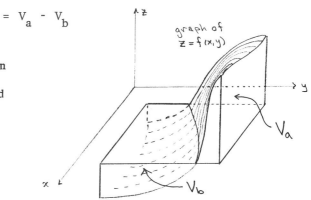

where V_a and V_b are shown

to the right.   This should remind

you of the similar property of

ordinary integrals:

$$\int_a^b f(x)\,dx = \text{area above the   x-axis}$$

<u>minus</u> the area below

the   x-axis

$$= A_a - A_b$$

where A_a and A_b are shown to the right.

<u>Example C.</u>   Find the volume   V   of the two-piece solid bounded by $z^3 + y = 2$,   the   xy-plane ,

and the planes $x = \pm 1$, $y = 1$ and $y = 4$.

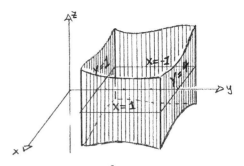

Solution. First we sketch the solid. On a

set of xyz-axes we place the four planes

$x = \pm 1$, $y = 1$ and $y = 4$.

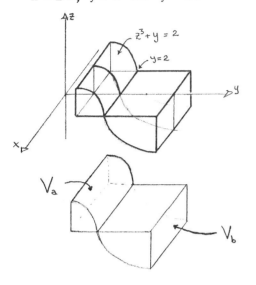

Then add in the surface $z^3 + y = 2$ (i.e., draw

the cube root function $z = \sqrt[3]{2 - y}$ in the yz-plane

and extend it out parallel to the x-axis). Notice

that the surface intersects the xy-plane in the

line $y = 2$.

We can now see what is happening. The

desired volume V is the sum of two volumes,

$V_a + V_b$, each of which can be obtained from

a double integral:

$$V_a = \iint_{R_a} \sqrt[3]{2 - y} \, dA = \int_{-1}^{1} \int_{1}^{2} (2 - y)^{1/3} \, dy \, dx$$

$$= \int_{-1}^{1} \left[-\frac{3(2 - y)^{4/3}}{4} \Big|_{y = 1}^{y = 2} \right] dx = \int_{-1}^{1} \left[0 + \frac{3}{4} \right] dx$$

$$= \int_{-1}^{1} \frac{3}{4} \, dx = \frac{3}{4} x \Big|_{-1}^{1} = \frac{3}{2}$$

We might be tempted to write

$$V_b = \iint_{R_b} \sqrt[3]{2 - y} \, dA \quad \ldots \quad \text{but this is WRONG!}$$

This is incorrect because $z = \sqrt[3]{2 - y}$ is below the xy-plane on R_b, and hence the double integral denotes <u>negative</u> V_b! To correct for this we take a "negative of a negative":

$$V_b = -\iint_{R_b} \sqrt[3]{2-y} \; dA = -\int_{-1}^{1} \int_{2}^{4} (2-y)^{1/3} \; dy \; dx$$

$$= -\int_{-1}^{1} \left[-\frac{3(2-y)^{4/3}}{4} \Big|_{y=2}^{y=4} \right] dx = -\int_{-1}^{1} \left[-\frac{3 \cdot 2^{4/3}}{4} + 0 \right] dx$$

$$= -\int_{-1}^{1} \left(-\frac{3}{2} \sqrt[3]{2} \right) dx = \frac{3}{2} \sqrt[3]{2} \; x \Big|_{-1}^{1} = 3\sqrt[3]{2}$$

Thus $V = V_a + V_b = \frac{3}{2} + 3\sqrt[3]{2}$ □

Section 17.2: Double Integrals over Nonrectangular Regions

1. <u>Regions of Type I and Type II.</u> A region R in the xy-plane is said to be of <u>Type I</u> if it appears as shown to the right: it is bounded on the top and bottom by two smooth * functions

$y = g_1(x)$ and $y = g_2(x)$, and (if necessary) on the sides by two vertical lines x = a and

x = b . Many times the top and bottom functions will intersect at one or both of the x-values

x = a or x = b , thereby making one or both of the lines x = a or x = b unnecessary;

the resulting region is still of Type I :

---

* $y = g(x)$ is a "smooth" function if g' is continuous.

17.2.2

Examples
of Regions
of Type I

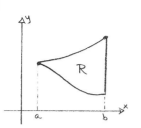
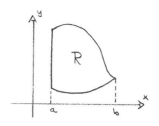
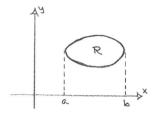

A region of Type II simply reverses the roles of x and y : it is bounded on the

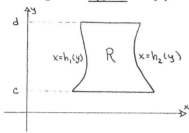

right and left by two smooth functions $x = h_1(y)$

and $x = h_2(y)$, and (if necessary) on the top

and bottom by two horizontal lines $y = c$ and

$y = d$. As in the Type I case, the lines might

not actually appear.

Examples
of Regions
of Type II

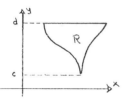
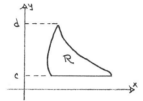
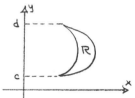

There are three reasons for singling out regions of Types I and II for special

consideration:

1. Regions of Type I and Type II are easy to deal with ;

2. A majority of regions occurring in applications are of Type I or II , or unions

of such regions.

3. If $z = f(x,y)$ is continuous on a region R of Type I or Type II , then

$$\iint_R f(x,y) \, dA$$

exists and can be evaluated by iterated integrals using Theorem 17.2.1 :

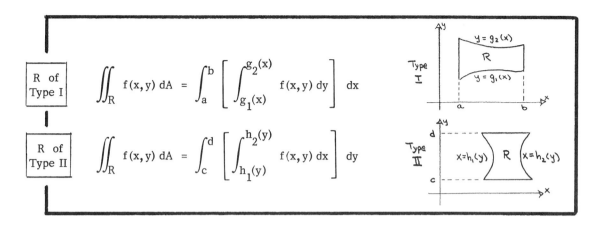

You should recognize this result as a generalization of Theorem 17. 1. 1, the

corresponding theorem for <u>rectangular</u> regions R . As in the case where R is a rectangle,

the present theorem can be justified geometrically by using the method of slicing for the compu-

tation of volume (§6. 2). Anton writes out this justification very carefully following the theorem

statement.

2.   <u>Applying Theorem 17. 2. 1.</u>   Theorems 17. 1. 1 and 17. 2. 1 enable us to <u>reduce a double</u>

<u>integral into two one-variable integrals.</u>

> The only major difficulty is in determining
>
> the correct <u>limits of integration!</u>

For this reason you should give careful attention to Anton's two-step method which follows

Theorem 17. 2. 1. We illustrate this method in ...

<u>Example A.</u>   Compute $\iint_R (xy + 1)\, dA$ where   R   is the region bounded by the   y-axis ,   the

line $y = 1$, and the curve $y = x^2$.

17.2.4

Solution. As a first step you should graph the

region. In this case you see from the graph

that the three corner points are $(0,0)$,

$(0,1)$ and $(1,1)$, and that R can be

considered as a region of <u>either</u> Type I or

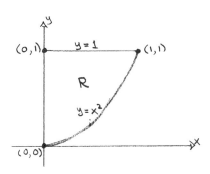

Type II. We will choose to regard it now as a region of Type I, while in Example C we will

consider it again as a region of Type II. We'll follow Anton's two-step procedure:

<u>Step 1.</u> Fix an arbitrary value of x and observe how

y varies. This amounts to cutting R by the vertical

line through $(x,0)$, and noting the lowest and highest

values for y. From the picture we see

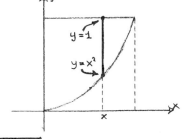

$$\boxed{\text{(i) for } x \text{ fixed, } x^2 \le y \le 1}$$

<u>Step 2.</u> We now need to find the lowest and highest values of x in the region R. This

amounts to moving the line drawn in Step 1 first

as far to the left as possible, and then as far to

the right as possible. Since our corner points are

$(0,0)$, $(0,1)$ and $(1,1)$, we obtain

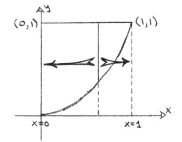

$$\boxed{\text{(ii) } 0 \le x \le 1}$$

The bounds in (i) are now our <u>inner</u> limits of integration, while the bounds in (ii) are our outer limits. Hence

$$\iint_R (xy + 1)\, dA = \int_0^1 \left[\int_{x^2}^1 (xy + 1)\, dy \right] dx$$

We first evaluate the inner integral (remember: this is a partial integration with respect to y ,

so that x is considered to be a constant):

$$\int_{x^2}^1 (xy + 1)\, dy = \left[\frac{xy^2}{2} + y \right] \Bigg|_{y = x^2}^{y = 1}$$

$$= \left(\frac{x}{2} + 1 \right) - \left(\frac{x^5}{2} + x^2 \right)$$

$$= - \frac{x^5}{2} - x^2 + \frac{x}{2} + 1$$

We now return to the original iterated integral and substitute in our value for the inner integral:

$$\iint_R (xy + 1)\, dA = \int_0^1 \left[- \frac{x^5}{2} - x^2 + \frac{x}{2} + 1 \right] dx$$

$$= \left[- \frac{x^6}{12} - \frac{x^3}{3} + \frac{x^2}{4} + x \right] \Bigg|_0^1$$

$$= - \frac{1}{12} - \frac{1}{3} + \frac{1}{4} + 1 = \boxed{\frac{5}{6}} \qquad \Box$$

Example B.   Find the volume of the solid under the surface $z = x^2 + y^2$ and above the region R

in the first quadrant bounded by $y = x^2$ and $x = y^2$.

<u>Solution.</u> Graphing R as shown to the right, we find that the curves intersect at $(0,0)$ and $(1,1)$. We will consider R as a region of Type I (it is also a region of Type II).

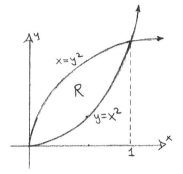

Analyzing our sketch, we see

> (i) for x fixed, $x^2 \leq y \leq \sqrt{x}$
>
> (ii) $0 \leq x \leq 1$

Thus $\displaystyle \iint_R (x^2 + y^2)\, dA = \int_0^1 \int_{x^2}^{\sqrt{x}} (x^2 + y^2)\, dy\, dx$

Inner integral: $\displaystyle \int_{x^2}^{\sqrt{x}} (x^2 + y^2)\, dy = \left[x^2 y + \frac{y^3}{3} \right] \Bigg|_{y = x^2}^{y = \sqrt{x}}$

$$= x^{5/2} + \frac{x^{3/2}}{3} - x^4 - \frac{x^6}{3}$$

Full integral: $\displaystyle \iint_R (x^2 + y^2)\, dA = \int_0^1 \left[x^{5/2} + \frac{x^{3/2}}{3} - x^4 - \frac{x^6}{3} \right] dx$

$$= \frac{2x^{7/2}}{7} + \frac{2x^{5/2}}{15} - \frac{x^5}{5} - \frac{x^7}{21} \Bigg|_0^1$$

$$= \frac{2}{7} + \frac{2}{15} - \frac{1}{5} - \frac{1}{21} = \boxed{\frac{6}{35}}$$

□

Our procedure for setting up the iterated integral over a region R of Type I can be summarized as follows:

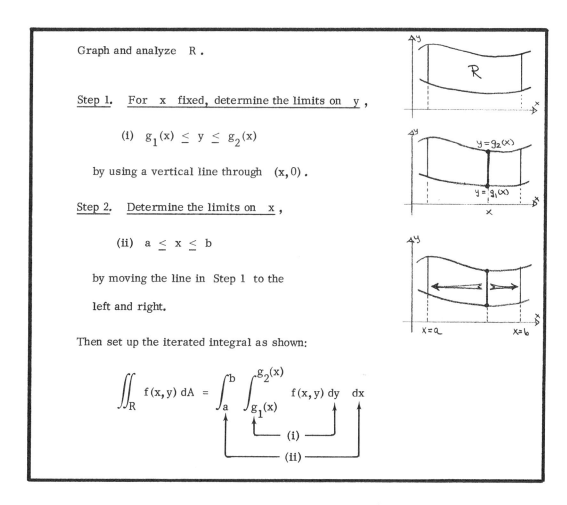

Graph and analyze R .

Step 1. For x fixed, determine the limits on y ,

(i) $g_1(x) \le y \le g_2(x)$

by using a vertical line through $(x, 0)$.

Step 2. Determine the limits on x ,

(ii) $a \le x \le b$

by moving the line in Step 1 to the

left and right.

Then set up the iterated integral as shown:

$$\iint_R f(x, y)\, dA = \int_a^b \int_{g_1(x)}^{g_2(x)} f(x, y)\, dy \; dx$$

(i)

(ii)

Regions of Type II are handled in the same way except that in Steps 1 and 2 the roles of x and y are reversed (and vertical lines become horizontal lines, etc.). Anton outlines the procedure following Example 3 . We'll illustrate it by redoing Example A as a region of Type II :

Example C.   Compute $\displaystyle\iint_R (xy + 1)\, dA$ where   R   is the region bounded

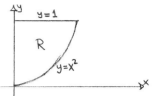

by the   y-axis,   the line $y = 1$,   and

the curve $y = x^2$.

Solution.   The graph of   R   was determined in   Example A,   and is shown above.

Step 1.   Fix an arbitrary value of   y   and observe

how   x   varies.   This amounts to cutting   R   by

the horizontal line through $(0, y)$ and noting the

lowest and highest values for   x.   From our

picture we see

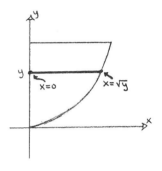

> (i)   for   y   fixed, $0 \le x \le \sqrt{y}$

Step 2.   We now need to find the lowest and highest values of   y

in the region   R.   This amounts to moving the line drawn in

Step 1   first as far down as possible and then as far up as

possible.   From this we obtain

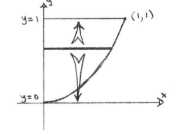

> (ii) $0 \le y \le 1$

Thus $\displaystyle\iint_R (xy + 1)\, dA \; = \; \int_0^1 \left[\int_0^{\sqrt{y}} (xy + 1)\, dx \right] dy$

notice the   x   and   y   reversal
from the   Type I   solution.

Inner integral: $\displaystyle\int_0^{\sqrt{y}} (xy + 1)\, dx = \left(\dfrac{x^2 y}{2} + x\right)\Bigg|_{x = 0}^{x = \sqrt{y}}$

$$= \dfrac{y^2}{2} + y^{1/2}$$

Full integral: $\displaystyle\iint_R (xy + 1)\, dA = \int_0^1 \left[\dfrac{y^2}{2} + y^{1/2}\right] dy$

$$= \left(\dfrac{y^3}{6} + \dfrac{2 y^{3/2}}{3}\right)\Bigg|_0^1$$

$$= \dfrac{1}{6} + \dfrac{2}{3} = \boxed{\dfrac{5}{6}}$$

This agrees with the answer originally obtained in Example A.   □

3.   Regions which are of both Type I and Type II.   As we saw in Examples A and C above,

integrals over regions which are of both Type I and Type II can be found by "integrating in

either order," i. e. , first integrate y , then x   (Type I), or first x , then y (Type II).

In Examples A and C , there was little reason to prefer one order of integration over the other.

However, this is not always the case; the choice of order of integration can at times be quite

crucial ! We illustrate this in ...

Example D.   Evaluate $\displaystyle\iint_R \sin(\pi y^2)\, dA$ for the region R bounded by the lines y = x ,

x = 0   and   y = 1/2 .

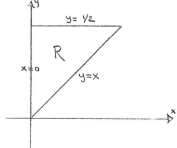

Solution.   The region R is drawn to the right.

It is easy to see that R is both of Type I and of

Type II since it can be described in two ways:

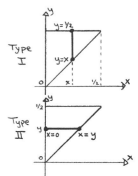

R as Type I: (i) for x fixed, $x \leq y \leq 1/2$

    (ii) $0 \leq x \leq 1/2$

R as Type II: (i) for y fixed, $0 \leq x \leq y$

    (ii) $0 \leq y \leq 1/2$

Suppose we try to evaluate using the Type I limits. Then

$$\iint_R \sin(\pi y^2)\, dA = \int_0^{1/2} \left[\int_x^{1/2} \sin(\pi y^2)\, dy \right] dx = \dots \quad ? \quad \text{Oops!}$$

The trouble here is that we need to integrate $\sin(\pi y^2)$ with respect to y, and there is <u>no</u> <u>simple antiderivative for</u> $\sin(\pi y^2)$. We could expand $\sin(\pi y^2)$ as a series in x and then integrate term-by-term, but that looks like no fun at all. Instead let's examine the Type II integration.

Using the Type II limits we obtain

$$\iint_R \sin(\pi y^2)\, dA = \int_0^{1/2} \left[\int_0^y \sin(\pi y^2)\, dx \right] dy$$

The inner integral is now a snap since the integrand is only a "constant" with respect to integration by x :

$$\int_0^y \sin(\pi y^2)\, dx = x \sin(\pi y^2) \Big|_{x=0}^{x=y} = y \sin(\pi y^2)$$

The full integral becomes

$$\iint_R \sin(\pi y^2)\, dA = \int_0^{1/2} y \sin(\pi y^2)\, dy$$

⌐ ahh ... the extra y
will save the day

$$= \int_0^{\pi/4} \sin u \left(\frac{du}{2\pi}\right)$$

$$
\boxed{
\begin{array}{l}
u = \pi y^2 \\[6pt]
du = 2\pi y\, dy \\[6pt]
\dfrac{du}{2\pi} = y\, dy
\end{array}
}
$$

$$= -\frac{1}{2\pi} \cos u \Big|_0^{\pi/4}$$

$$= -\frac{1}{2\pi}\left(\frac{\sqrt{2}}{2} - 1\right)$$

$$= \boxed{(2 - \sqrt{2})/4\pi}$$ □

The moral: When integrating over a region which is of both Type I and Type II, use the order which makes integration easier.

4. **Unions of regions of Type I and II.** Sometimes a region will not be a simple Type I or Type II region, as shown in...

Example E. Determine'the volume undér the surface $z = 2x^2 + 2$ and above the triangle R with vertices $(0,0)$, $(1,2)$ and $(2,1)$.

Solution. The region R is pictured to the right; the equations for the three lines were all determined by the two-point form of a line (see The Companion, §1.5.1).

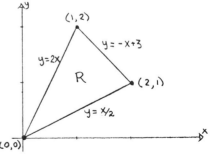

Suppose we wish to consider R as a region of Type I. Well, taking an arbitrary x, we draw the vertical line to determine the y limits and find ...

... oops! Depending on where we pick our x-value we have two different sets of limits for y, i.e., y = 2x is the upper limit when x ≤ 1, but y = -x + 3 is the upper limit when x ≥ 1. Thus R is not a region of Type I, since the upper y limit is not a smooth function of x (it has a "corner" at (1, 2)).

So we try a Type II integration. Taking an arbitrary y we draw the horizontal line to determine the x limits and find ...

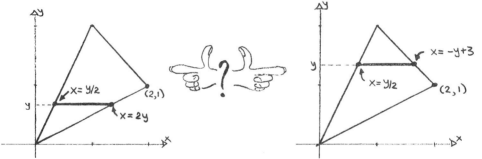

... oops again! Depending on where we pick our y value we have two different sets of limits for x, i.e., x = 2y is the upper limit when y ≤ 1, but x = -y + 3 is the upper limit when y ≥ 1. Thus R is not a region of Type II!

Are we at a dead end? Of course not ...

... because R is easily broken up into a union of two regions of Type I, or ...

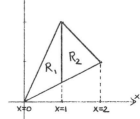

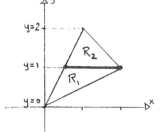

... into a union of two regions

of Type II.

We then determine the double integrals over R_1 and R_2 separately, and add them

together. Here is the Type I approach:

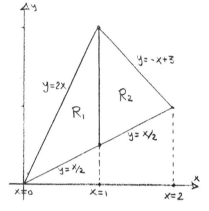

For R_1 :   (i) for x fixed, $\dfrac{x}{2} \le y \le 2x$

(ii) $0 \le x \le 1$

For R_2 :   (i) for x fixed, $\dfrac{x}{2} \le y \le -x + 3$

(ii) $1 \le x \le 2$

Thus $\displaystyle\iint_{R_1} (2x^2 + 1)\, dA \; = \; \int_0^1 \left[\int_{x/2}^{2x} (2x^2 + 2)\, dy \right] dx$

$$= \int_0^1 (2x^2 y + 2y) \Big|_{y=x/2}^{y=2x} dx \; = \; \int_0^1 (4x^3 + 4x - x^3 - x)\, dx$$

$$= \int_0^1 (3x^3 + 3x)\, dx \; = \; \frac{3}{4}\, x^4 + \frac{3}{2}\, x^2 \Big|_0^1 \; = \; \frac{9}{4}$$

and $\displaystyle\iint_{R_2} (2x^2 + 2)\, dA \; = \; \int_1^2 \left[\int_{x/2}^{-x+3} (2x^2 + 2)\, dy \right] dx$

$$= \int_1^2 \left[2x^2 y + 2y \right] \Big|_{y=x/2}^{y=-x+3} dx \; = \; \int_1^2 (-3x^3 + 6x^2 - 3x + 6)\, dx$$

$$= \left(-\frac{3}{4}\, x^4 + 2x^3 - \frac{3}{2}\, x^2 + 6x \right) \Big|_1^2 \; = \; \frac{17}{4}$$

17.2.14

Thus

$$\iint_R (2x^2 + 2)\, dA \;=\; \frac{9}{4} \;+\; \frac{17}{4} \;=\; \frac{26}{4} \;=\; \boxed{\frac{13}{2}}$$

The same answer will also be obtained by using the Type II approach. We'll leave that as an exercise for you. □

5. **When the region is given implicitly** .... There are double integral problems in which the region R is not given in the exact Type I or Type II form. In such a situation you must determine the equations for the boundary curves of R and put them into the Type I/Type II form (... or unions of such forms).

We have already done one example like this. In Example E we were given only boundary points for the triangle R, and we had to come up with equations for the lines ourselves. *

But things can get even more complicated: many times it is not apparent at first what is to be used as the region R, let alone whether it is Type I or Type II! In these cases R is determined **implicitly** by the problem and you must figure out what it is. Most often this is done with a sketch in three dimensions. Anton's Examples 7 and 8 are typical illustrations of this procedure, as is Example B of §17.1.3 of The Companion. Here's another:

Example F. Find the volume of the solid enclosed by the surfaces $x = 2y^2$, $z = 0$ and $x + 2z = 2$.

Solution. Somehow we must express the specified volume as the volume over a "nice" region R and under the graph of a function $z = f(x,y)$. To do this we must sketch the solid we are

_____

* Study that example well. It is a common type of problem, and it tends to give some people a lot of trouble.

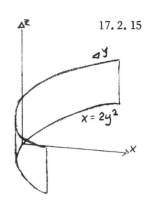

considering. On a set of xyz axes we draw the parabolic

surface $x = 2y^2$ (i. e. , draw the parabola $x = 2y^2$ in

the xy-plane and extend it parallel to the z-axis).

Then draw the plane x + 2z = 2 (i. e. , start with the

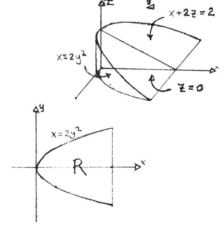

line x + 2z = 2 in the xz-plane and

extend it parallel to the y-axis). Closing

up the bottom with the plane z = 0 (the

xy-plane) we see that our solid is a parabolic-

shaped wedge, and we have the form we

want: we wish to compute the volume over

the parabolic region R (shown to the

right) and under the graph of

$z = f(x, y) = (2 - x)/2$ (i. e. , the

equation x + 2z = 2 rewritten as

$z = (2 - x)/2$).

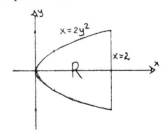

We have only to determine the x-value of the line which bounds R: this is given

by the intersection of the planes x + 2z = 2 and z = 0 . Solving these two equations

simultaneously we obtain x = 2 . Hence we have reworked Example F into:

Find $\iint_R (1 - \frac{x}{2})\, dA$

when R is the region to the right.

The corner points are found to be $(2, \pm 1)$ by equating $x = 2y^2$ and $x = 2$. Thus R is a region of Type II whose limits of integration are

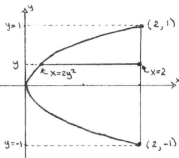

    (i) for y fixed, $2y^2 \le x \le 2$

    (ii) $-1 \le y \le 1$

The integral is thus

$$\iint_R (1 - \tfrac{x}{2}) \, dA = \int_{-1}^{1} \left[\int_{2y^2}^{2} (1 - \tfrac{x}{2}) \, dx \right] dy$$

$$= \int_{-1}^{1} \left(x - \frac{x^2}{4} \right) \Bigg|_{x = 2y^2}^{2} \, dy = \int_{-1}^{1} (y^4 - 2y^2 + 1) \, dy$$

$$= \left(\frac{1}{5} y^5 - \frac{2}{3} y^3 + y \right) \Bigg|_{-1}^{1} = \boxed{\frac{16}{15}} \qquad \square$$

6.   <u>Some final remarks.</u> Anton shows how the <u>area</u> of a region R can be computed by a double integral using the formula

Area from a double integral

$$[\text{area of } R] = \iint_R 1 \, dA = \iint_R dA$$

This is simply a double integral of one of the simplest functions possible: $f(x, y) = 1$. If you remember the volume formula

$$\text{volume} = (\text{area of base}) \times \text{height}$$

then it is easy to remember that the area of the base equals the volume when the height is 1.

Hence the area can be found by using what is normally a volume-finding approach.   Anton's

Example 6 is an illustration of this technique.

Finally, Anton has a number of exercises that ask you to take a given iterated integral

and write it with the order of integration reversed (Exercises 49 - 56). We'll do one to indicate

the method:

Example G.   Express the iterated integral

$$\int_0^2 \int_1^{e^y} f(x, y)\, dx\, dy$$

as an equivalent integral with the order of integration reversed.

Solution.   We will take the given limits and use them to sketch the region   R   in question.   Then

we will read off the limits in the other order.

The given iterated integral is written in Type II form (since the inner integral is with

respect to   x).   Our limits are thus

     (i)   For   y   fixed, $1 \le x \le e^y$

     (ii) $0 \le y \le 2$

So we plot the two boundary curves $x = 1$ and $x = e^y$,   and take the region between them

bounded by the lines $y = 0$

and $y = 2$:

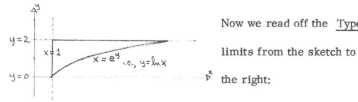

Now we read off the Type I

limits from the sketch to

the right:

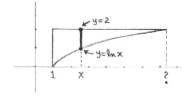

(i)   For   x   fixed,   ln x $\leq$ y $\leq$ 2

(ii)   1 $\leq$ x $\leq$?

To find the upper x-limit   we must determine the   x-value   at the intersection of $x = e^y$
with   y = 2 : that's clearly $x = e^2$.   Hence our integral becomes

$$\int_1^{e^2} \int_{\ln x}^{2} f(x, y)\, dy\, dx \qquad \Box$$

Section 17.3:   Double Integrals in Polar Coordinates

1.   <u>Regions of polar type.</u>   A region   R   in the   xy-plane   is said to be of <u>polar type</u> if it can be
described as follows: it is enclosed by the
two polar coordinate rays $\theta = \alpha$ and
$\theta = \beta$ and is between the two smooth polar
curves $r = r_1(\theta)$ and $r = r_2(\theta)$,   where
$r_1(\theta) \leq r_2(\theta)$ for all $\alpha \leq \theta \leq \beta$. (See

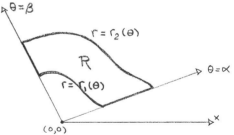

the diagram to the right.) Examples are shown in Anton's Figure 17.3.2.

Regions of polar type form a third class of regions (in addition to those of Type I and II)
over which it is common to have double integrals.   The major theorem of this section (Theorem
17.3.1) tells how to evaluate the double integral of a continuous function $z = f(r, \theta)$ on a
region   R   of polar type:

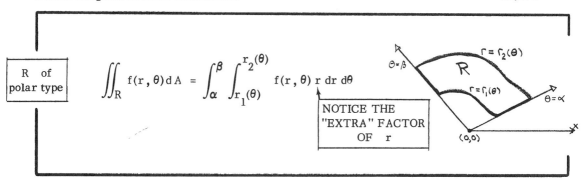

$$\iint_R f(r,\theta)\, dA = \int_\alpha^\beta \int_{r_1(\theta)}^{r_2(\theta)} f(r,\theta)\, r\, dr\, d\theta$$

R of polar type

NOTICE THE "EXTRA" FACTOR OF r

$r = r_2(\theta)$

R

$r = r_1(\theta)$

$\theta = \beta$

$\theta = \alpha$

$(0,0)$

Remember the "extra" factor of r !! It is needed in the computation of area in polar coordinates. Anton justifies this in his discussion following Theorem 17.3.1.

Your should recognize this result as a variant of Theorems 17.1.1 and 17.2.1, the corresponding theorems for rectangular regions and regions of Type I or II.

2. Applying Theorem 17.3.1. Theorem 17.3.1 (like the two earlier results, 17.1.1 and 17.2.1) enables us to reduce a double integral (in this case a polar double integral) into two one-variable integrals. And, as with our earlier results,

> the only major difficulty is in determining
>
> the correct limits of integration!

Just before Example 1 Anton gives a two-step method for determining these limits. He then uses his method in Examples 1-4. Here is another example in which the process is described in greater detail:

Example A. Evaluate $\displaystyle\iint_R \sin 3\theta\, dA$

where R is the region in the first quadrant that is outside the circle $r = \sqrt{2}/2$ but inside the three-petal rose $r = \sin 3\theta$.

17.3.3

Solution. We begin, as always, by drawing the region; the result is shown to the right. We will certainly need to know the points of intersection of the two curves, so we solve their equations simultaneously:

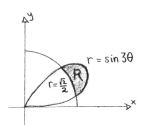

$$\sin 3\theta \;=\; r \;=\; \sqrt{2}\,/\,2$$

$$3\theta \;=\; \pi/4 \quad \text{or} \quad 3\pi/4$$

$$\theta \;=\; \pi/12 \quad \text{or} \quad \pi/4$$

Step 1. Fix an arbitrary value of θ and see how r varies. This amounts to cutting R by the ray of polar angle θ, and noting the lowest and highest values for r. From our diagram we see

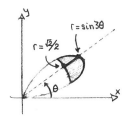

> (i) for θ fixed, $\sqrt{2}/2 \le r \le \sin 3\theta$

Step 2. We now need to find the lowest and highest values of θ in the region R. This amounts to rotating the ray drawn in Step 1 first as far clockwise as possible, and then as far counterclockwise as possible. Since we have already determined the intersection angles to be $\pi/12$ and $\pi/4$, we obtain

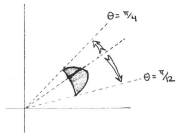

> (ii) $\pi/12 \le \theta \le \pi/4$

As usual, the bounds in (i) are the <u>inner</u> limits of integration, while the bounds in (ii) are the

<u>outer</u> limits. Hence

$$
* \quad \iint_R \sin 3\theta \, dA \;=\; \int_{\pi/12}^{\pi/4} \left[\int_{\sqrt{2}/2}^{\sin 3\theta} (\sin 3\theta) \, r \, dr \right] d\theta
$$

> DON'T FORGET THE
> "EXTRA"   r !!

$$
= \int_{\pi/12}^{\pi/4} (\sin 3\theta) \left. \frac{r^2}{2} \right|_{r=\sqrt{2}/2}^{r=\sin 3\theta} d\theta
$$

$$
= \int_{\pi/12}^{\pi/4} \left(\frac{\sin^3 3\theta}{2} - \frac{\sin 3\theta}{4} \right) d\theta
$$

$$
= \int_{\pi/4}^{3\pi/4} \frac{\sin^3 u}{6} \, du \;-\; \int_{\pi/4}^{3\pi/4} \frac{\sin u}{12} \, du
$$

> $u = 3\theta$
> $du = 3 \, d\theta$
> $\dfrac{1}{3} \, du = d\theta$

> from (6) §9.3

$$
= \frac{1}{6} \left. \left(-\cos u + \frac{1}{3}\cos^3 u \right) \right|_{\pi/4}^{3\pi/4} + \frac{1}{12} \left. \cos u \right|_{\pi/4}^{3\pi/4}
$$

$$
= \frac{1}{6} \left(\frac{\sqrt{2}}{2} - \frac{\sqrt{2}}{12} \right) - \frac{1}{6} \left(-\frac{\sqrt{2}}{2} + \frac{\sqrt{2}}{12} \right) + \frac{1}{12} \left(-\frac{\sqrt{2}}{2} - \frac{\sqrt{2}}{2} \right)
$$

$$
= \boxed{\sqrt{2} \, / \, 18} \qquad\qquad\qquad \square
$$

The procedure for setting up the iterated polar integral over a region   R   of polar type
can be summarized as follows:

Graph and analyze R.

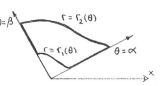

Step 1. For θ fixed, determine the limits on r,

(i) $r_1(\theta) \le r \le r_2(\theta)$

using a ray with polar angle θ.

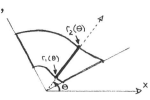

Step 2. Determine the limits on θ,

(ii) $\alpha \le \theta \le \beta$

by rotating the ray in Step 1 clockwise
and counterclockwise.

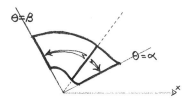

Then set up the iterated integral as shown:

DON'T FORGET
THE "EXTRA" r

$$\iint_R f(r,\theta)\,dA = \int_\alpha^\beta \int_{r_1(\theta)}^{r_2(\theta)} f(r,\theta)\,r\,dr\,d\theta$$

——— (i) ———

——————— (ii) ———————

3. **Volumes and areas.** Polar double integrals can be used to compute volumes and areas just

as we did with ordinary double integrals:

$$\iint_R f(r,\theta)\,dA = \text{the volume above R and}$$
$$\text{under the graph of } z = f(r,\theta)$$
$$\text{whenever } f(r,\theta) \ge 0$$

$$\iint_R 1\,dA = \text{the area of the region R}$$

Example B.   Find the area of the region   R   outside the circle

$$x^2 + y^2 = 1$$

but inside the circle

$$x^2 + (y - 1)^2 = 1$$

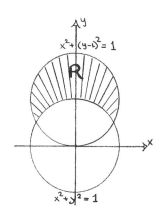

Solution.   The region   R   is as shown to the right.

It is easy to see that   R   is not a region of Type I

or II; however, it does have a certain symmetry

about the origin, which should make you suspect that

it is a region of <u>polar type</u> (and hence we should

integrate using polar coordinates).   We therefore convert the two   xy   equations into $r\theta$

equations by using $x = r \cos \theta$ and $y = r \sin \theta$; this will yield

$$r = 1 \qquad \text{and} \qquad r = 2 \sin \theta$$

Now we can find our limits of integration:

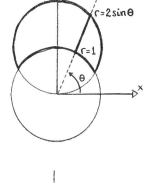

Step 1.   For θ fixed we see that

> (i) $1 \leq r \leq 2 \sin \theta$

Step 2.   To determine the θ limits we first need

to determine the θ values of the intersections of

the curves.   By solving simultaneously we find

$$1 = r = 2 \sin \theta$$

$$\sin \theta = \frac{1}{2}$$

$$\theta = \frac{\pi}{6} \quad \text{or} \quad \frac{5\pi}{6}$$

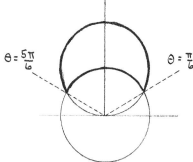

Hence rotating the θ-ray as far clockwise as possible $(\theta = \pi/6)$ and then as far counter-clockwise as possible $(\theta = 5\pi/6)$, we find

$$(ii) \quad \frac{\pi}{6} \le \theta \le \frac{5\pi}{6}$$

Now we can compute the area by using the polar double integral area formula:

$$\text{area (R)} = \iint_R 1 \, dA$$

$$= \int_{\pi/6}^{5\pi/6} \int_1^{2\sin\theta} r \, dr \, d\theta$$

DON'T FORGET THE "EXTRA" r

$$= \int_{\pi/6}^{5\pi/6} \frac{r^2}{2} \Big|_{r=1}^{r=2\sin\theta} d\theta$$

$$= \int_{\pi/6}^{5\pi/6} \left(\frac{4\sin^2\theta - 1}{2} \right) d\theta$$

$$= \int_{\pi/6}^{5\pi/6} \left(\frac{1 - 2\cos 2\theta}{2} \right) d\theta$$

using
$$\cos 2\theta = \cos^2\theta - \sin^2\theta$$
$$= 1 - 2\sin^2\theta$$

$$= \frac{\theta - \sin 2\theta}{2} \Big|_{\pi/6}^{5\pi/6} = \frac{\pi}{3} + \frac{\sqrt{3}}{2} \approx \boxed{1.9132} \qquad \square$$

Anton's Example 2 is an example of volume computation using double polar integrals. We'll give another in the following subsection (Example C).

4.   Conversion from rectangular to polar coordinates.   Notice that in the statement of Example B

there was no mention of polar coordinates. In doing the problem, however, we found that the

region   R   in question was more conveniently described with polar coordinates than with

rectangular coordinates, and that the double integral was more conveniently evaluated using

polar coordinates.   This is often the case with a double integral

$$\iint_R f(x, y)\, dA$$

when   R   is of polar form and when the integrand $f(x, y)$ involves an expression like

$x^2 + y^2$ (which converts to r^2 in polar coordinates).   You must learn when it is convenient

to convert an ordinary double integral into a polar double integral.

When   R   is a region of polar form*,   conversion of $\iint_R f(x, y)\, dA$ into a polar

double integral is easy

<table>
<tr><td rowspan="2">Polar
Change of
Coordinates</td><td>

Suppose $f(x, y)$ is a continuous function on a region   R

of polar type.   Then the double integral of   f   over   R

may be expressed in terms of polar coordinates as follows:

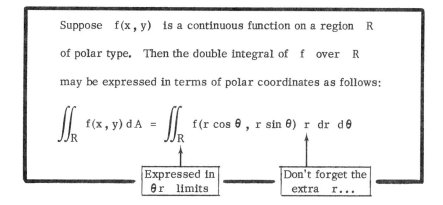

</td></tr>
</table>

The polar double integral on the right is then evaluated by using iterated integrals as in

Theorem 17.3.1.

---

*
   We would never convert $\iint_R f(x, y)\, dA$ to polar form unless   R   is a region of polar form.

The evaluation of ordinary double integrals by conversion to polar double integrals is the most common and important use of polar double integrals!

Example C.   Find the volume of the solid bounded below by the xy-plane and above by the paraboloid $z = 1 - (x^2 + y^2)$.

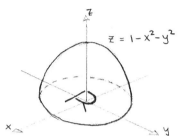

Solution.   When we sketch our solid, we see that its volume is

$$V = \iint_R (1 - x^2 - y^2)\, dA$$

where R is the unit disk in the xy-plane.

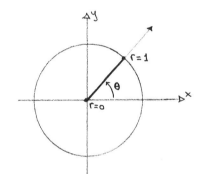

Ahh... a prime candidate for conversion to polar coordinates! Using $x = r \cos \theta$ and $y = r \sin \theta$, the equation $x^2 + y^2 = 1$ for the boundary curve becomes $r = 1$, and hence our limits are given by:

  (i)   for θ fixed, $0 \le r \le 1$

  (ii) $0 \le \theta \le 2\pi$

Hence $\displaystyle V = \iint_R (1 - x^2 - y^2)\, dA$

$$= \iint_R (1 - r^2)\, dA \qquad \text{converting to a polar integral by using}$$
$$r^2 = x^2 + y^2$$

$$= \int_0^{2\pi} \int_0^1 (1 - r^2)\, r\, dr\, d\theta$$

WE DIDN'T FORGET THE "EXTRA" r

$$= \int_0^{2\pi} \left[\frac{r^2}{2} - \frac{r^4}{4} \right]_{r=0}^{r=1} d\theta = \int_0^{2\pi} \frac{d\theta}{4} = \boxed{\frac{\pi}{2}} \qquad \square$$

Section 17.4:   Surface Area

§1.   The basic law: surface area on a plane. Suppose we are given a (possibly tilted) plane

$z = ax + by + c$, and wish to compute

the area of that portion R_0 of the plane

that lies directly over a rectangle R with

side lengths Δx and Δy . Anton proves

in Theorem 17.4.2 that the area of R_0

(which Anton denotes as S) is simply the

area of R multiplied by the factor

$\sqrt{a^2 + b^2 + 1}$,   i.e.,

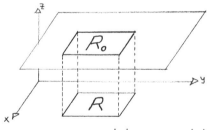

Surface area on a plane

$$\text{area } (R_0) \;=\; \sqrt{a^2 + b^2 + 1}\ \ \text{area } (R)$$

where   area $(R) = \Delta x\, \Delta y$.

Notice that the formula does make intuitive sense. If the plane is horizontal (i.e., if it

is not tilted), then $a = b = 0$, so that

area (R_0) = area (R) . This is to be expected

since in this case R_0 is an exact copy of R ,

as shown to the right. However, if the plane is

extremely tilted, then $|a|$ and/or $|b|$

are large, making   area (R_0) much larger

than   area (R) , as shown to the left.

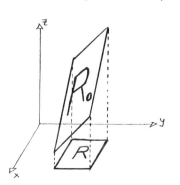

§2. <u>The general surface area formula.</u> The Riemann sums definition for the double integral

supplies the critical link between the surface area on a plane formula

$$S = \sqrt{a^2 + b^2 + 1} \; \Delta x \, \Delta y$$

and the general formula in terms of double integrals,

General Surface Area Formula

$$S = \iint_R \sqrt{\left(\frac{\partial z}{\partial x}\right)^2 + \left(\frac{\partial z}{\partial y}\right)^2 + 1} \; dA$$

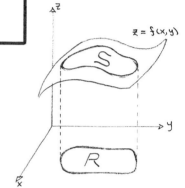

(Definition 17.4.3). Here $z = f(x, y)$ is a function

with continuous first partial derivatives over the closed

region R , and S is the surface area of that portion

of the graph of $z = f(x, y)$ which lies over R .

To help take the mystery out of the general formula

we will restate Anton's four-step motivation in the language of <u>infinitesimals.</u> This is less

precise than Anton's careful Riemann sums approach, but it clearly illustrates all the key

elements in the argument and is less burdensome notationally. You may find it easier to under-

stand and remember:

<u>Step 1.</u>   Divide R into infinitesimal rectangles;

let d R be one of these rectangles, with

area $dA = dx \, dy$, and center point

(x_0, y_0)

<u>Step 2.</u>   Let d S be the infinitesimal piece of surface

area over d R .

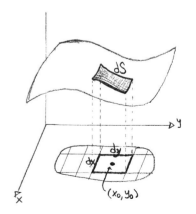

<u>Step 3.</u> (The key approximation!) Since dS is

infinitesimal, it can be closely approxi-

mated by the corresponding portion of

the <u>tangent plane</u> to $z = f(x, y)$ at

(x_0, y_0). From Theorem 16.6.2 the

equation of this plane is:

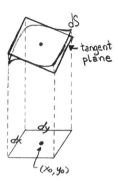

$$z = \frac{\partial z}{\partial x} (x - x_0) + \frac{\partial z}{\partial y} (y - y_0) + z_0$$

$$= \left(\frac{\partial z}{\partial x}\right) x + \left(\frac{\partial z}{\partial y}\right) y + c \qquad \text{for some constant} \quad c$$

$$\left(\frac{\partial z}{\partial x} \quad \text{and} \quad \frac{\partial z}{\partial y} \quad \text{are evaluated at} \quad (x_0, y_0) \, .\right)$$

Hence, by the <u>surface area on a plane formula</u> $\left(a = \frac{\partial z}{\partial x}, \, b = \frac{\partial z}{\partial y} \right)$:

$$dS \approx \sqrt{\left(\frac{\partial z}{\partial x}\right)^2 + \left(\frac{\partial z}{\partial y}\right)^2 + 1} \quad dA$$

<u>Step 4.</u> Now "add up" all the infinitesimal surface areas, i.e., integrate over all

possible (x, y) values in R , to get the total surface area:

$$S = \iint dS = \iint_R \sqrt{\left(\frac{\partial z}{\partial x}\right)^2 + \left(\frac{\partial z}{\partial y}\right)^2 + 1} \quad dA$$

as desired.

17.4.4

In Anton's more rigorous development, Step 4 comes down to evaluating a limit:

$$S = \lim_{n \to \infty} \underbrace{\sum_{k=1}^{n} \sqrt{\left[\frac{\partial f}{\partial x}(x_k, y_k)\right]^2 + \left[\frac{\partial f}{\partial y}(x_k, y_k)\right]^2 + 1} \; \Delta A_k}$$

But this is a <u>Riemann sum</u> for the function

$$z = \sqrt{\left(\frac{\partial f}{\partial x}\right)^2 + \left(\frac{\partial f}{\partial y}\right)^2 + 1} \quad \text{over the region} \quad R$$

Hence

$$S = \iint_R \sqrt{\left(\frac{\partial f}{\partial x}\right)^2 + \left(\frac{\partial f}{\partial y}\right)^2 + 1} \; dA$$

by the Riemann sums definition for a double integral as given in §17.1. As we said, it is the critical link!

*   *   *   *   *   *   *   *   *

In the applications to be developed in §17.6 we will give further illustrations of the technique:

Basic non-calculus formula

--- transformed via
Riemann sum
approximations into ---

General integration formula

In all these cases the "infinitesimal approach" can be used to clarify the basic procedure.

§3. Calculation of surface areas. Keep the following points in mind regarding the use of the surface area formula in Definition 17.4.3:

    1. Because of the square root in the integrand, many surface area integrals cannot be evaluated except by numerical techniques. The problem is similar to that for the arclength integral of §13.4.

    2. The regions R will commonly be of Types I or II, or of polar type. Hence in many instances (such as Anton's Example 2) you will need to convert to a polar double integral.

Example A. Find the surface area of the portion of the surface $z = \dfrac{\sqrt{2}}{3}(x+y)^{3/2}$ in the first octant that is bounded by the plane $x+y = 2$.

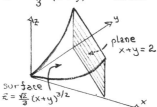

Solution. When we sketch the surface and the plane, it is apparent that the region R over which we need to integrate is a triangle. This triangle is shown to the right. As can be seen, the limits of integration are

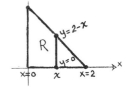

> (i) for x fixed, $0 \le y \le 2 - x$
>
> (ii) $0 \le x \le 2$

Thus
$$S = \iint_R \sqrt{\left(\frac{\partial z}{\partial x}\right)^2 + \left(\frac{\partial z}{\partial y}\right)^2 + 1}\; dA$$

$$= \int_0^2 \int_0^{2-x} \sqrt{\left(\frac{\partial z}{\partial x}\right)^2 + \left(\frac{\partial z}{\partial y}\right)^2 + 1}\; dy\, dx$$

We now must compute the partial derivatives:

$$\frac{\partial z}{\partial x} = \frac{\sqrt{2}}{3} \left(\frac{3}{2}\right) (x + y)^{1/2}, \quad \text{so} \quad \left(\frac{\partial z}{\partial x}\right)^2 = \frac{1}{2} (x + y)$$

$$\frac{\partial z}{\partial y} = \frac{\sqrt{2}}{3} \left(\frac{3}{2}\right) (x + y)^{1/2}, \quad \text{so} \quad \left(\frac{\partial z}{\partial y}\right)^2 = \frac{1}{2} (x + y)$$

Thus

$$S = \int_0^2 \int_0^{2-x} \sqrt{x + y + 1} \; dy \, dx$$

Inner integral:

$$\int_0^{2-x} \sqrt{x + y + 1} \; dy = \int_{x+1}^3 u^{1/2} \, du$$

$$\boxed{\begin{array}{l} u = x + y + 1 \\ du = dy \end{array}} \qquad = \left. \frac{2 u^{3/2}}{3} \right|_{x+1}^3$$

$$= 2\sqrt{3} - \frac{2}{3} (x + 1)^{3/2}$$

Full integral:

$$S = \int_0^2 \left[2\sqrt{3} - \frac{2}{3} (x + 1)^{3/2} \right] dx$$

$$= \left. \left[2\sqrt{3} \, x - \frac{4}{15} (x + 1)^{5/2} \right] \right|_0^2$$

$$= 4\sqrt{3} - \frac{12}{5} \sqrt{3} + \frac{4}{15} \approx 3.0379 \qquad \qquad \square$$

Example B. Find the surface area of the portion of the surface $x^2 + y^2 + z = 4$ that is above the region $1 \le x^2 + y^2 \le 4$.

Solution.   Writing our surface as $z = 4 - x^2 - y^2$ and graphing it, we see that it's a down-

ward-turning paraboloid. The region   R   is

the region between the two circles $x^2 + y^2 = 1$

and $x^2 + y^2 = 4$. Thus   R   is a region of

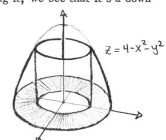

$z = 4-x^2-y^2$

polar type (the region between the circles

$r = 1$ and $r = 2$),   and we convert the

integral to a polar double integral with limits

(i)   for θ fixed, $1 \leq r \leq 2$

(ii) $0 \leq \theta \leq 2\pi$

The partial derivatives are

$$\frac{\partial z}{\partial x} = -2x \quad \text{and} \quad \frac{\partial z}{\partial y} = -2y$$

so that

$$S = \iint_R \sqrt{\left(\frac{\partial z}{\partial x}\right)^2 + \left(\frac{\partial z}{\partial y}\right)^2 + 1} \; dA$$

$$= \iint_R \sqrt{4x^2 + 4y^2 + 1} \; dA$$

$$= \iint_R \sqrt{4r^2 + 1} \; dA \qquad \text{conversion to polar coordinates}$$

$$= \int_0^{2\pi} \int_1^2 \sqrt{4r^2 + 1} \; \; r \; dr \, d\theta$$

REMEMBER TO INCLUDE THE "EXTRA" r

Inner integral: $\qquad \displaystyle\int_1^2 \sqrt{4r^2 + 1}\ \ r\,dr\ =\ \int_5^{17} u^{1/2}\left(\frac{du}{8}\right)$

$$\boxed{\begin{aligned} u &= 4r^2 + 1 \\ du &= 8r\,dr \\ \frac{du}{8} &= r\,dr \end{aligned}} \qquad = \frac{1}{8}\left(\frac{2u^{3/2}}{3}\right)\Bigg|_5^{17}$$

$$= \frac{1}{12}\left(17\sqrt{17}\ -\ 5\sqrt{5}\right)$$

Full integral: $\qquad S\ =\ \displaystyle\int_0^{2\pi} \frac{1}{12}\left(17\sqrt{17}\ -\ 5\sqrt{5}\right) d\theta$

$$= \frac{\pi}{6}\left(17\sqrt{17}\ -\ 5\sqrt{5}\right)\ \approx\ \boxed{30.8465} \qquad\qquad \square$$

Section 17.5: Triple Integrals

§1. <u>The definition of the triple integral.</u> As with ordinary (i. e. , single) and double integrals, the <u>triple integral</u> is defined via Riemann sums. As we have seen, the ordinary integral has a natural interpretation as <u>area,</u> and the double integral has a natural interpretation as <u>volume.</u> That raises the question: what natural interpretation can be attached to the triple integral?

One natural interpretation is as the <u>mass</u> of a solid. Let's develop this carefully. Given a point (x,y,z) in a solid G , we define the <u>mass density</u> of G at (x,y,z) to be

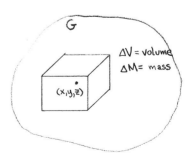

$$f(x,y,z)\ =\ \lim_{\Delta V \to 0}\ \frac{\Delta M}{\Delta V}$$

where ΔV is a small volume about (x,y,z)

with mass ΔM . If ΔV is very small, then

$$f(x,y,z) \approx \frac{\Delta M}{\Delta V}$$

or

$$\boxed{\Delta M \approx f(x,y,z) \, \Delta V}$$

In words this says "the mass of a small piece of G is approximately equal to its volume times the mass density at an interior point. " Using this idea we can justify the following:

> For a non-negative function $f(x,y,z)$, the triple integral
>
> of f over a solid G
>
> $$\iiint_G f(x,y,z) \, dV$$
>
> will be the <u>mass</u> of the solid G with mass density
>
> function $f(x,y,z)$.

To see this, note that the formal definition of $\iiint_G f(x,y,z) \, dV$ using limits of Riemann sums

(Anton's (1)) now can be interpreted in the

following <u>physical</u> sense:

divide G into n small

rectangular boxes with volumes

$\Delta V_1, \Delta V_2, \ldots, \Delta V_n$; choose a point

(x_k^*, y_k^*, z_k^*) in each box; and form

the <u>Riemann sum</u>

$$\sum_{k=1}^{n} f(x_k^*, y_k^*, z_k^*) \, \Delta V_k$$

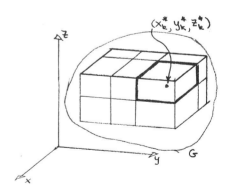

This Riemann sum can be thought of as a sum of masses of rectangular boxes, and hence as an approximation of the total mass. Then

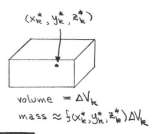

volume $= \Delta V_k$

mass $\approx f(x_k^*, y_k^*, z_k^*) \Delta V_k$

Definition of the Triple Integral

$$\iiint_G f(x,y,z)\, dV = \lim_{n \to \infty} \sum_{k=1}^{n} f(x_k^*, y_k^*, z_k^*)\, \Delta V_k$$

where* $\Delta V_k \to 0$ as $n \to \infty$.

Thus $\iiint_G f(x,y,z)\, dV$ is the limit of <u>approximations</u> for the mass of G as the approximations get better and better. This justifies our claim that $\iiint_G f(x,y,z)\, dV$ is <u>the (exact) mass of G</u>.

One other important use of triple integrals is easily obtained from our mass interpretation. If the mass density of a solid G is always 1 , then (since volume times density equals mass),

$$\text{volume of } G = \text{mass of } G$$

which shows

$$\text{volume of } G = \iiint_G 1\, dV$$

This formula corresponds nicely with equation (7) of §17.2:

$$\text{area of } R = \iint_R 1\, dA$$

---

* $\Delta V_k \to 0$ is to mean that the length, width and depth of each box all go to zero.

Other applications of triple integration will be discussed in §17.6. As usual, these applications will arise through Riemann sums.

2. <u>Existence and evaluation of the triple integral.</u>   The enormous complexity and diversity of solids in space make it difficult to state existence theorems for the triple integral which are both simple and broadly applicable (see the discussion for double integrals in §17.1.2 of <u>The Companion</u>).   Nonetheless, our basic principle is the same for triple integrals as it was for double integrals:

<table>
<tr><td>The Existence
Principle</td><td>If $w = f(x, y, z)$ is continuous on a "well-behaved"

closed and bounded* solid G in space,

then $\iiint_G f(x, y, z)\, dV$ exists.</td></tr>
</table>

Most of the "well-behaved" solids which Anton discusses are what he calls simple solids:

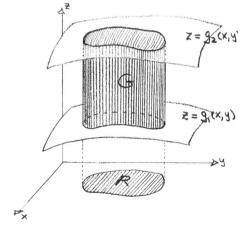

G   is a <u>simple solid</u> if it is the solid which lies

   i.   above a closed and bounded

       region   R   in the   xy-plane , and

  ii. between the graphs of two

       continuous functions

$$g_1(x, y) \leq g_2(x, y)$$

---

* A solid   G   is <u>closed</u> if it contains all its boundary points, and it is <u>bounded</u> if it can be placed inside some suitably large rectangular box.

In Theorem 17. 5. 2 Anton says how to evaluate the triple integral over a simple solid by iteration into one ordinary integral and one double integral:

$$\iiint_G f(x,y,z)\, dV = \iint_R \left[\int_{g_1(x,y)}^{g_2(x,y)} f(x,y,z)\, dz \right] dA$$

The idea is to use our knowledge of double integrals to transform the double integral

$\iint_R [\, - \,]\, dA$ into an iterated integral of two ordinary integrals. <u>The form this iteration</u>

<u>takes will depend upon the nature of the region R .</u> In most situations   R   will be either of

Type I, Type II or polar type.

For instance, if   R   is of Type I

as shown to the right, then Theorem 17. 5. 2

will become

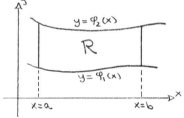

$$\iiint_G f(x,y,z)\, dV = \int_a^b \int_{\varphi_1(x)}^{\varphi_2(x)} \int_{g_1(x,y)}^{g_2(x,y)} f(x,y,z)\, dz\, dy\, dx$$

Thus our triple integral has been written as an iterated integral consisting of three ordinary integrals.

<u>Example A.</u>   Suppose   G   is the solid bounded by the   xy-plane , the   xz-plane , x = 1 , x = 2 , $z = \dfrac{1}{x}$ and $y = x^2$. Calculate the mass of   G   if the mass density of   G   is given by $f(x,y,z) = xy^2 z$.

<u>Solution.</u>   We first attempt to sketch   G ,   labelling the various surfaces. With six boundary

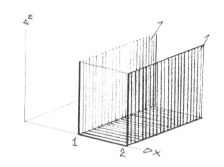

surfaces to draw, we first start with the easy

ones: the four planes, as shown to the left.

Then cut these planes with each of the

remaining surfaces, first $z = \dfrac{1}{x}$,   and

then $y = x^2$.

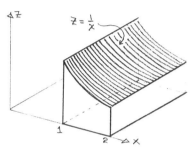

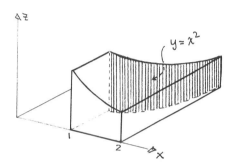

We now wish to determine if   G   is a "simple solid, " and (more importantly) to deter-

mine what the limits of integration should be.   To accomplish both of these purposes we

proceed as follows:

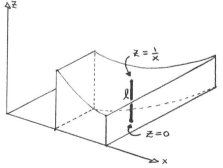

<u>Step 1.</u>   Fix arbitrary values of   x   and   y

and observe how   z   varies. This amounts

to cutting   G   by the vertical line ℓ through

$(x, y, 0)$ and noting the lowest and highest values

for   z .   From our picture we see

$$(1) \quad \text{for   x   and   y   fixed,} \quad 0 \leq z \leq \frac{1}{x}$$

Ahh...! From this we see that our solid G is indeed a "simple solid:" it is the solid which

lies

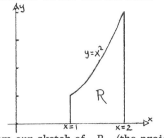

i. above the region R shown to the left, and

ii. between the graphs of the two continuous

functions z = 0 and z = 1/x .

<u>Step 2.</u> From our sketch of R (the <u>projection</u>

of G onto the xy-plane) we can obtain our last

two limits of integration. Clearly R is a region

of Type I ; we thus cut R by a vertical line to

obtain

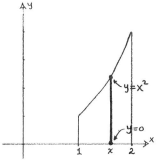

(2)   for x fixed, $0 \le y \le x^2$

(3) $1 \le x \le 2$

Our triple integral can now be written as a triple iterated integral, with the bounds in

(1) becoming the <u>innermost</u> limits of integration, and the bounds in (3) becoming the <u>outermost</u>

limits:

$$\text{mass of } G = \iiint_R xy^2z \, dx \, dy \, dz = \int_1^2 \int_0^{x^2} \int_0^{\frac{1}{x}} xy^2z \, dz \, dy \, dx$$

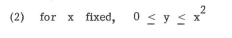

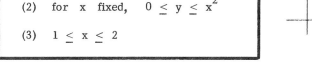

Inner integral:

$$\int_0^{1/x} xy^2 z \, dz = \frac{xy^2 z^2}{2} \Bigg|_{z=0}^{z=\frac{1}{x}} = \frac{y^2}{2x}$$

Middle integral:

$$\int_0^{x^2} \frac{y^2}{2x} \, dy = \frac{y^3}{6x} \Bigg|_{y=0}^{y=x^2} = \frac{x^6}{6x} = \frac{x^5}{6}$$

Full integral:

$$\text{mass of } T = \int_1^2 \frac{x^5}{6} \, dx = \frac{x^6}{36} \Bigg|_1^2 = \boxed{\frac{63}{36}} \qquad \square$$

The procedure for setting up the iterated integral over a simple solid G can be summarized as follows:

Carefully sketch and analyze

  the given solid.

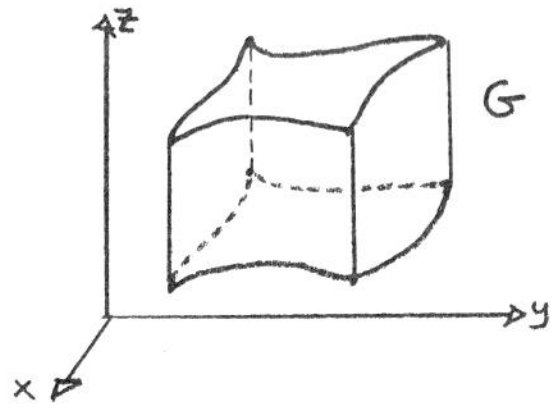

Step 1. For x and y fixed, determine

  the limits on z,

  (1) $g_1(x, y) \le z \le g_2(x, y)$

  by using a vertical line through $(x, y, 0)$.

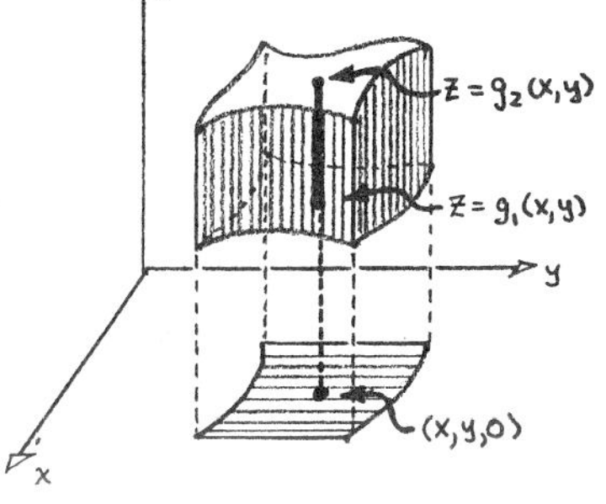

Step 2. Sketch the projection R of the

solid G on the xy-plane and

determine the remaining two limits,

(2) ....

(3) ....

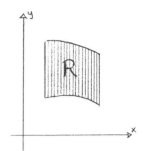

The form these limits take will depend on the type of R .

Then (1) gives the innermost limits, while (2) and (3) give the

outer (double integral) limits:

$$\iiint_G f(x,y,z) \, dV = \iint_R \left[\int_{g_1(x,y)}^{g_2(x,y)} f(x,y,z) \, dz \right] dA$$

3. Various Examples. We will illustrate our triple integral procedure by examples which

complement those in Anton. Each problem will also feature some additional technique or trick

that you should know about.

Our first example deals with solids determined by two surfaces. It is like Anton's

Example 4 .

Example B. Determine the volume between the paraboloids $2x^2 + y^2 + z - 4 = 0$ and

$2x^2 + 3y^2 - z = 0$.

Solution. We first make a rough sketch for

the solid G in question. It is then

relatively easy to see that

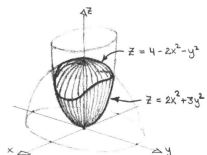

> **(1) for x and y fixed, $2x^2 + 3y^2 \le z \le 4 - 2x^2 - y^2$**

Hence we have a simple solid since the z-values are all bounded between two continuous

functions

$$z = 2x^2 + 3y^2 \quad \text{and} \quad z = 4 - 2x^2 - y^2$$

However, there is one problem: <u>how do we determine the region R which is the projection of</u>

<u>G onto the xy-plane</u>? Well, studying our

sketch again should convince you that if   C

is the intersection curve of the two paraboloids,

then the boundary of   R   is the projection of

C onto the xy-plane . And how do we

compute this projection curve? Easy! Any

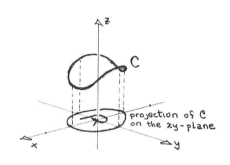

point (x, y, z) on C satisfies the two paraboloid equations, so use the two equations to

eliminate z . The resulting equation in x and y will be the equation of the projection

curve:

$$\begin{cases} 2x^2 + y^2 + z - 4 = 0 \\ 2x^2 + 3y^2 - z = 0 \end{cases}$$

Addition yields $4x^2 + 4y^2 = 4$

or $x^2 + y^2 = 1$

     This is the unit circle in the xy-plane , and thus R can be treated as a region of

<u>polar type</u>:

17.5.11

$$\boxed{\begin{array}{ll} (2) & \text{for} \quad \theta \quad \text{fixed}, \quad 0 \le r \le 1 \\[2ex] (3) & 0 \le \theta \le 2\pi \end{array}}$$

Thus

$$\text{volume of } G = \iiint_G 1 \, dV$$

$$= \int_0^{2\pi} \int_0^1 \left[\int_{2x^2+3y^2}^{4-2x^2-y^2} 1 \, dz \right] r \, dr \, d\theta$$

$\boxed{\text{REMEMBER THE EXTRA } r!}$

Inner integral:

$$\int_{2x^2+3y^2}^{4-2x^2-y^2} 1 \, dz = z \Big|_{z=2x^2+3y^2}^{z=4-2x^2-y^2}$$

$$= (4 - 2x^2 - y^2) - (2x^2 + 3y^2) = 4 - 4x^2 - 4y^2$$

$$= 4 - 4r^2 \qquad (\text{using } r^2 = x^2 + y^2)$$

Middle integral:

$$\int_0^1 (4 - 4r^2) \, r \, dr = \int_0^1 (4r - 4r^3) \, dr = 2r^2 - r^4 \Big|_0^1 = 1$$

Full integral:

$$\int_0^{2\pi} 1 \, d\theta = \boxed{2\pi} \qquad\qquad \square$$

Another type of triple integral problem that sometimes gives people trouble occurs when the solid G is specified as bounded by planes passing through a given set of points. These problems tend to be much easier than they appear at first.

Example C. Let G be the tetrahedron determined by the four points $(0,0,0)$, $(1,0,0)$, $(0,\frac{1}{2},0)$

and $(0,0,\frac{1}{3})$. If the mass density of G is given by $f(x,y,z) = x$, determine the total mass of G.

Solution. A sketch of the solid G appears

to the right. First we should determine

the equation for the top plane of the solid

(we will certainly need this piece of

information when we look for the limits

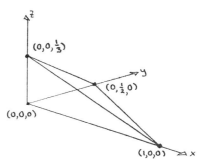

of integration). One method is to recall the general equation for a plane P :

$$ax + by + cz = d$$

Plug in the three points $(1,0,0)$, $(0,\frac{1}{2},0)$ and $(0,0,\frac{1}{3})$ to obtain the three equations

$$\left\{ \begin{array}{c} a = d \\ \dfrac{b}{2} = d \\ \dfrac{c}{3} = d \end{array} \right\}$$

Letting d = 1 gives $x + 2y + 3z = 1$

$$\text{or} \qquad \boxed{z = \frac{1 - 2y - x}{3}}$$

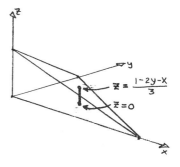

From our sketch to the right we now see

$$\boxed{(1) \quad \text{for x and y fixed,} \quad 0 \le z \le \frac{1 - 2y - x}{3}}$$

G is therefore a simple solid, and its projection R onto the xy-plane is the triangle

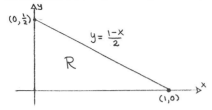

shown to the left. The equation for the top line

is obtained by setting z = 0 in the plane equation

x + 2y + 3z = 1. From this diagram we see that

R is a region of Type I with limits

$$(2) \quad \text{for } x \text{ fixed,} \quad 0 \le y \le \frac{1-x}{2}$$

$$(3) \quad 0 \le x \le 1$$

Thus

$$\text{mass of } G = \iiint_G x \, dV$$

$$= \int_0^1 \int_0^{\frac{1-x}{2}} \int_0^{\frac{1-2y-x}{3}} x \, dz \, dy \, dx$$

Inner integral:
$$\int_0^{\frac{1-2y-x}{3}} x \, dz = x z \Big|_{z=0}^{z=\frac{1-2y-x}{3}}$$

$$= \left(\frac{x - x^2}{3}\right) - \left(\frac{2x}{3}\right) y$$

Middle integral:
$$\int_0^{\frac{1-x}{2}} \left[\left(\frac{x - x^2}{3}\right) - \left(\frac{2x}{3}\right) y \right] dy = \left(\frac{x - x^2}{3}\right) y - \left(\frac{x}{3}\right) y^2 \Big|_{y=0}^{y=\frac{1-x}{2}}$$

$$= \left(\frac{x - x^2}{3}\right) \left(\frac{1-x}{2}\right) - \left(\frac{x}{3}\right) \left(\frac{1-x}{2}\right)^2$$

$$= \frac{x^3 - 2x^2 + x}{12}$$

Full integral:
$$\int_0^1 \frac{x^3 - 2x^2 + x}{12}\, dx = \frac{x^4}{48} - \frac{x^3}{18} + \frac{x^2}{24}\Big|_0^1 = \boxed{\frac{1}{144}}\qquad \square$$

Finally, it should be made clear that there is nothing sacred about always having z as

the innermost limit of integration. If we started with x instead, then we would project G

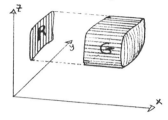

back onto the yz-plane to get the region R as

shown to the left; if we start with y , then we

would project onto the xz-plane .

Example D.   Consider the triple iterated integral

$$\int_0^1 \int_x^1 \int_0^{\sqrt{x}} f(x,y,z)\, dz\, dy\, dx$$

Express this as an iterated integral with the   y-integration   as the innermost integral and the

z-integration as the outermost integral.

Solution.   We must first use the limits of integration from the given integral to sketch the region

G . From the outer two limits,

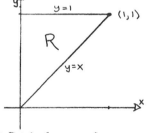

(3) $0 \le x \le 1$

(2)   for x fixed, $x \le y \le 1$

we can draw the region R shown to the right. Then laying this out flat in the xy-plane , we

use the innermost limits,

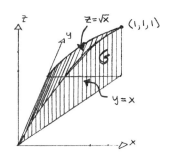

$$(1) \quad \text{for} \quad x, y \quad \text{fixed}, \quad 0 \leq z \leq \sqrt{x}$$

to draw in the top and bottom surfaces

$$z = g_1(x, y) = 0 \quad \text{and} \quad z = g_2(x, y) = \sqrt{x}$$

G   is thus as shown to the right.

To obtain the limits in the yxz order we first

hold x and z constant to determine the y limits:

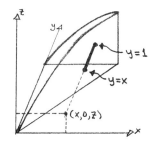

$$(1) \quad \text{for} \quad x \quad \text{and} \quad z \quad \text{fixed}, \quad x \leq y \leq 1$$

The projection R of G onto the xz-plane then

appears as shown to the right. Since we want x to

be the next integration, we treat R as a region of

Type II and obtain

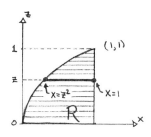

$$(2) \quad \text{for} \quad z \quad \text{fixed}, \quad z^2 \leq x \leq 1$$

$$(3) \quad 0 \leq z \leq 1$$

Thus $\displaystyle\int_0^1 \int_{z^2}^1 \int_x^1 f(x, y, z) \, dy \, dx \, dz$ is the desired answer.   □

Section 17. 6:   Centroids, Centers of Gravity, Theorem of Pappus

1.   Riemann sums and the applications of integration.   As we have said time and again, integration

(single, double or triple) almost always appears in applications through its definition in terms of

Riemann sums. The usual technique is:

Basic non-calculus formula	--- transformed via Riemann sum approximations into ---

General integration formula

We saw this with surface area in §17.4, and we see it again with mass and moments about lines in this section.

Mass. Suppose R is a two-dimensional lamina with mass M and area A. The basic formula occurs when the mass density δ is a constant throughout all of R :

$$\boxed{\text{Basic formula.} \qquad M = \delta A}$$

When $\delta = \delta(x,y)$ varies with position (x,y) , then approximation via Riemann sums yields Anton's Definition 17.6.1:

$$\boxed{\text{General formula.} \qquad M = \iint_R \delta(x,y)\, dA}$$

Moments about lines. Suppose m is a point mass located at (x,y). Then the moment about the line $x = a$ is given by

$$\begin{bmatrix} \text{moment} \\ \text{about} \\ x = a \end{bmatrix} = (x - a)\,m$$

Now suppose that m is not just a point mass, but is a lamina R with small area A   and

<u>constant</u> density δ . This yields:

$$
\text{Basic formula.} \qquad \begin{bmatrix} \text{moment} \\ \text{about} \\ x = a \end{bmatrix} \approx (x - a) \, \delta A
$$

When $\delta = \delta(x,y)$ <u>varies</u> with position (x,y) , then approximation via Riemann sums yields Anton's equation (10) :

$$
\text{General formula.} \qquad \begin{bmatrix} \text{moment} \\ \text{about} \\ x = a \end{bmatrix} = \iint_R (x - a) \, \delta(x,y) \, dA
$$

In the same way we can compute the moment about $y = c$:

$$
\text{General formula.} \qquad \begin{bmatrix} \text{moment} \\ \text{about} \\ y = c \end{bmatrix} = \iint_R (y - c) \, \delta(x,y) \, dA
$$

* * * * *

All the above formulas and procedures generalize to solids G ; we have only to replace areas A with volumes V and double integrals $\iint_R (\text{---}) \, dA$ with triple integrals $\iiint_R (\text{---}) \, dV$ (e.g., the formula for the mass of G is Anton's equation (17)). A detailed discussion of the mass of a solid G was given in §17. 5. 1 of <u>The Companion.</u>

* * * * *

When given a lamina R or solid G with a mass density function δ , two quantities which are almost always desired are the mass M and the center of gravity $(\bar{x}, \bar{y})$ or

$(\bar{x}, \bar{y}, \bar{z})$. We have just discussed the mass. The center of gravity is defined to be that point

$(\bar{x}, \bar{y})$ in the lamina or $(\bar{x}, \bar{y}, \bar{z})$ in the solid for which all the moments are zero. For a

lamina this means

$$\iint_R (x - \bar{x})\, \delta(x,y)\, dA \; = \; 0 \qquad \text{and} \qquad \iint_R (y - \bar{y})\, \delta(x,y)\, dA \; = \; 0$$

(Similar equations can be obtained for a solid.) Solving these equations for $\bar{x}$ and $\bar{y}$ yields:

Center of gravity $(\bar{x}, \bar{y})$ of R	$$\bar{x} \; = \; \frac{1}{M} \iint_R x\, \delta(x,y)\, dA \qquad \text{and} \qquad \bar{y} \; = \; \frac{1}{M} \iint_R y\, \delta(x,y)\, dA$$ where $M \; = \; \iint_R \delta(x,y)\, dA$ is the mass of R

Anton's equations (18) give the corresponding formulas for a solid.

$$* \qquad * \qquad * \qquad * \qquad *$$

As far as calculation of these quantities goes, there is very little new to learn. If you

learned the techniques for evaluating double and triple integrals in the previous sections, there

is little more to do for mass and center of gravity than memorize the appropriate formulas.

Anton's Examples 1, 3 and 4 are illustrations.

There is just one small point which can cause trouble: in mass and center of gravity

problems it is common to have the density function defined via a proportionality. Here are two

examples:

Example A.   Translate into an equation for $\delta(x,y)$:

"The density $\delta(x,y)$ of a lamina

is proportional to the distance

to the origin. "

<u>Solution.</u>   To say that $\delta(x,y)$ "is proportional to" another quantity Q means there exists

a constant k (the "constant of proportionality") such that

$$\delta(x,y) \;=\; kQ$$

In this case our quantity Q is the distance

from (x,y) to $(0,0)$, i.e.,

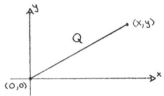

$$Q \;=\; \sqrt{x^2 + y^2}$$

Thus $\delta(x,y) \;=\; k\sqrt{x^2 + y^2}$.   □

<u>Example B.</u>   Translate into an equation for $\delta(x,y,z)$:

"The density $\delta(x,y,z)$ of a solid

is inversely proportional to the

distance to the   xy-plane. "

<u>Solution.</u>   To say that $\delta(x,y,z)$ "is inversely proportional to" another quantity Q means

there exists a constant k such that

$$\delta(x,y,z) \;=\; \frac{k}{Q}$$

(No, it <u>DOES NOT MEAN</u> $\delta(x,y,z) = \frac{Q}{k}$, although this is a common mistake!)

In this case our quantity Q is the distance from

(x, y, z) to $(x, y, 0)$ (the point in the

xy-plane directly below or above (x, y, z)),

which is merely

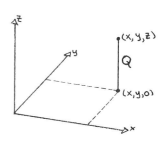

$$Q = |z|$$

Thus $\delta(x, y, z) = \dfrac{k}{|z|}$ □

Anton's Example 5 shows a complete calculation of mass and center of gravity starting with

"density is proportional to .... "

2. The Theorem of Pappus. Suppose a plane region R

is rotated about a non-intersecting line L contained

in the same plane. The solid so formed will have a

volume which is related to the centroid * of R

as follows:

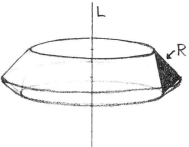

Theorem of Pappus

$$\begin{bmatrix} \text{volume} \\ \text{of solid} \end{bmatrix} = \begin{bmatrix} \text{area} \\ \text{of R} \end{bmatrix} \times \begin{bmatrix} \text{distance traveled} \\ \text{by centroid of R} \end{bmatrix}$$

This curious result tends to be used most often in the following way:

---

* Remember: The centroid of an object is simply its center of gravity when the density
    is taken to be the constant 1 .

If R is a reasonably symmetric region, then the computation of its area and centroid will often be easy. Once they are obtained, the Theorem of Pappus makes computing the volume of any solid generated by revolution of R quite simple!

Anton's Example 6 illustrates this procedure. Before giving another such example we should say a word or two about the use of symmetry in computing the centroid $(\bar{x}, \bar{y})$ of a region R .

Suppose R is symmetric about the line x = a .

Then clearly the centroid $(\bar{x}, \bar{y})$ lies on this line, so that $\bar{x} = a$.

Symmetry and centroids

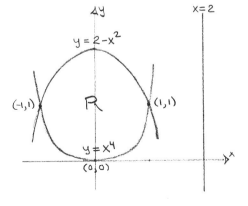

Suppose R is symmetric about the line y = c .

Then clearly the centroid $(\bar{x}, \bar{y})$ lies on this line, so that $\bar{y} = c$.

The following example shows the value of these symmetry principles:

<u>Example C.</u>   Suppose R is the region enclosed by the curves $y = x^4$ and $y = 2 - x^2$. Find the volume of the solid generated when R is revolved about the line x = 2 .

<u>Solution.</u>   The region R is pictured to the right. Let $(\bar{x}, \bar{y})$ be the centroid of R . Since R is symmetric about the y-axis (x = 0), then $\underline{\bar{x} = 0}$!

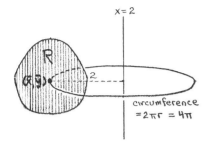

Thus the distance of the centroid to the line   x = 2

is   2 ,   and the distance traveled by the centroid

will be 4π (the circumference of a circle of

radius   2).

The area of   R   is easy to compute:

$$\begin{bmatrix} \text{area of} \\ R \end{bmatrix} = \int_{-1}^{1} \left[(2 - x^2) - x^4 \right] dx = 2x - \frac{x^3}{3} - \frac{x^5}{5} \Big|_{-1}^{1}$$

$$= \left(2 - \frac{1}{3} - \frac{1}{5} \right) - \left(-2 + \frac{1}{3} + \frac{1}{5} \right) = \frac{44}{15}$$

Hence, using the Theorem of Pappus:

$$\begin{bmatrix} \text{volume} \\ \text{of solid} \end{bmatrix} = \begin{bmatrix} \frac{44}{15} \end{bmatrix} \times [4\pi] = \boxed{\frac{176}{15}\pi} \qquad \square$$

This particular example can also be solved by the method of cylindrical shells (see §6.3),
with about the same degree of effort. Such is not always the case! Anton's Exercise 33 is a
case in point: it is easy to do with the Theorem of Pappus, but it is quite unpleasant to do by
cylindrical shells. Try it!

3.   Food for thought .... (optional).   We close this section with one last application of "multiple"
integration, one we hope will cause you to think a bit. Probabilty and statistics rely heavily on
multiple integration. Our example, with obvious relevance to air traffic control, is right out of
probability theory.

Suppose D_1 and D_2 are two "nice" disjoint regions in space, and let $\overline{x}$ and $\overline{y}$ be two "randomly chosen" points in D_1 and D_2 respectively. What is the "average" or "expected" distance between $\overline{x}$ and $\overline{y}$?

Answer:

$$\left(\frac{1}{\text{volume } D_1}\right)\left(\frac{1}{\text{volume } D_2}\right) \underset{D_1 \times D_2}{\iiint\iiint} \|\overline{x} - \overline{y}\| \; dx_1 \, dx_2 \, dx_3 \, dy_1 \, dy_2 \, dy_3$$

i.e., a "six-fold" multiple integral! We refer the interested reader to Vector Calculus for the relevant integration theory, and to Probability for the relevant probability theory.

Section 17.7: Triple Integrals in Cylindrical and Spherical Coordinates

1. **Cylindrical coordinates.** As discussed in §15.9, cylindrical coordinates are nothing more than polar coordinates in the xy-plane with an added z-coordinate. Hence, if you understand polar coordinates, you will not find cylindrical coordinates difficult to master.

First make sure that you understand the basic cylindrical coordinate surfaces $r = r_0$, $\theta = \theta_0$ and $z = z_0$ as shown in Anton's Figure 15.9.2 and §15.9.1 of The Companion. Of central importance is that $r = r_0$ is a cylinder with the z-axis as its axis of symmetry. This is why cylindrical coordinates are useful in triple integration problems involving solids having symmetry about the z-axis. In particular,

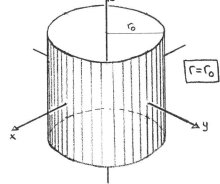

a simple solid G whose projection R on the xy-plane

is a region of polar type (see Anton's Figure 17.7.4)

is a prime candidate for the use of cylindrical coordinates.

Our primary concern in this subsection is the conversion of ordinary xyz-triple integrals into cylindrical triple integrals for easier evaluation. Here is the major conversion theorem:

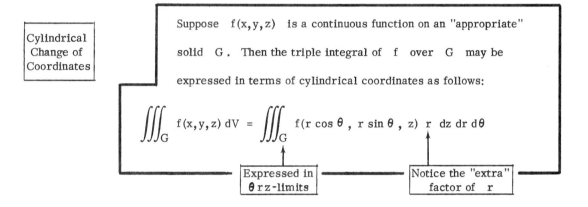

Cylindrical
Change of
Coordinates

Suppose $f(x,y,z)$ is a continuous function on an "appropriate"

solid G . Then the triple integral of f over G may be

expressed in terms of cylindrical coordinates as follows:

$$\iiint_G f(x,y,z)\, dV = \iiint_G f(r \cos \theta ,\ r \sin \theta ,\ z)\ r\ dz\, dr\, d\theta$$

Expressed in
θ r z-limits

Notice the "extra"
factor of r

The procedure for using this result is exactly the same as for the evaluation of an ordinary xyz-integral (see §17.5.2 of The Companion) on a simple solid G . The use of cylindrical coordinates is dictated when R , the projection of G on the xy-plane, is of polar type. In fact, we did such an example in §17.5.3 of The Companion (Example B) --- and we hadn't even studied cylindrical triple integrals yet! (See! We told you that cylindrical coordinates are easy if you've mastered polar coordinates.) Anton's Examples 1 and 2 are also illustrations. Here's another:

Example A.   Consider the solid G which is that portion of the cone $4(z-1)^2 = x^2 + y^2$, $z \le 1$, which lies in the first octant (i.e., $x \ge 0$, $y \ge 0$, $z \ge 0$). Find the mass of G if its mass density function is $f(x,y,z) = z\sqrt{x^2 + y^2}$.

17.7.3

Solution. We first make a rough sketch of G,
as shown to the right. (We do this by
obtaining the traces of $4(z - 1)^2 = x^2 + y^2$
on the three xyz-coordinate planes.)

Then we express the top surface (the cone) with

z as a function of x and y :

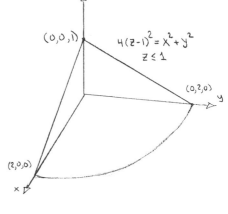

$$4(z - 1)^2 = x^2 + y^2$$

yields $z = 1 \pm \dfrac{1}{2}\sqrt{x^2 + y^2}$

However, since $z \leq 1$, we must use the <u>negative</u> sign in our expression for z, i.e.,

$$z = 1 - \frac{1}{2}\sqrt{x^2 + y^2}$$

We can now find the limits of integration for G using the two-step procedure Anton gives

prior to Example 2 in §17.5:

Step 1. Fix arbitrary values of x and y and

observe how z varies. This amounts to cutting

G by the vertical line ℓ through $(x, y, 0)$ and

noting the lowest and highest values for z.

From our picture we see

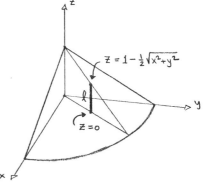

(1) for x and y fixed, $0 \leq z \leq 1 - \dfrac{1}{2}\sqrt{x^2 + y^2}$

This proves that   G   is a simple solid whose

projection   R   is one-quarter of the disk

bounded by $x^2 + y^2 = 4$ (just set   z = 0

in the equation for the cone).   But wait a minute ...

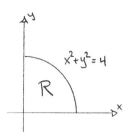

R   is a region of polar type, which should

immediately encourage us to use cylindrical coordinates.   We get further encouragement by

noting that the bounds for   z   and the expression $f(x, y, z)$ both involve $x^2 + y^2$ terms,

which become r^2!   We thus convert our   z   limits into cylindrical coordinate form:

$$\text{(1)   for   r   and   } \theta \text{   fixed,   } 0 \leq z \leq 1 - \frac{1}{2} r$$

(Notice that fixing   r   and θ is the same as fixing   x   and   y.)

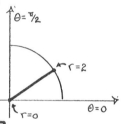

Step 2.   From our sketch of   R   (the projection of

G   on the   xy-plane)   we can obtain our polar

coordinate limits of integration:

$$\text{(2)   for   } \theta \text{   fixed,   } 0 \leq r \leq 2$$

$$\text{(3)   } 0 \leq \theta \leq \pi/2$$

The mass of   G   can now be written as a cylindrical triple integral:

$$\begin{bmatrix} \text{mass} \\ \text{of } G \end{bmatrix} = \iiint_G z \sqrt{x^2 + y^2} \; dV$$

$$= \iiint_G z\,r \; dV \qquad \text{since} \quad \sqrt{x^2 + y^2} = r$$

DON'T FORGET THE "EXTRA" r ...

$$= \int_0^{\pi/2} \int_0^2 \int_0^{1-r/2} z\,r^2 \; dz \; dr \; d\theta$$

(1)

(2)

(3)

Inner integral: $\displaystyle\int_0^{1-r/2} zr^2 \; dz = \left.\frac{z^2 r^2}{2}\right|_0^{1-r/2} = \frac{r^2}{2}\left(1 - \frac{r}{2}\right)^2 = \frac{r^2}{2} - \frac{r^3}{2} + \frac{r^4}{8}$

Middle integral: $\displaystyle\int_0^2 \left(\frac{r^2}{2} - \frac{r^3}{2} + \frac{r^4}{8}\right) dr = \left.\frac{r^3}{6} - \frac{r^4}{8} + \frac{r^5}{40}\right|_0^2 = \frac{2}{15}$

Full integral: $\displaystyle\int_0^{\pi/2} \frac{2}{15} \; d\theta = \boxed{\dfrac{\pi}{15}}$ $\qquad\qquad$ □

$$* \qquad * \qquad * \qquad * \qquad *$$

As you become more familar with the use of cylindrical coordinates in integration you will be able to recognize <u>at the beginning of a problem</u> when a solid G should be described via cylindrical coordinates. But remember:

> When converting from xyz to θrz coordinates,
> <u>every</u> x and y must be replaced by its
> corresponding expression in r and θ before
> evaluating a cylindrical triple integral !!!

In particular, don't forget that <u>all the</u> x and y's <u>in the limits for</u> z must be converted.

Here's one further cylindrical coordinate example, one in which we choose to convert to the $\theta r z$ variables immediately:

<u>Example B.</u>   Find the volume of the solid   G   bounded by the paraboloid $x^2 + y^2 = 2z$, the xy-plane , and the cylinder $x^2 + y^2 = 4x$.

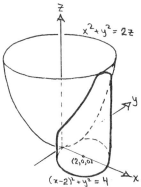

<u>Solution.</u>   A sketch of   G   is shown to the right, where, by completing the square (Appendix D), we have rewritten the cylinder equation as

$$(x - 2)^2 + y^2 = 4$$

This makes it clear that the base region is a circle with radius 2 and center $(2,0)$.

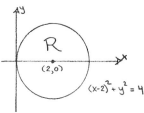

Because of the presence of so many $x^2 + y^2$ terms, the symmetry of the paraboloid about the z-axis , and the polar type of the region R , we decide to convert immediately into cylindrical coordinates. The equation for the paraboloid converts easily:

$$z = \frac{1}{2}(x^2 + y^2) = \frac{1}{2}r^2$$

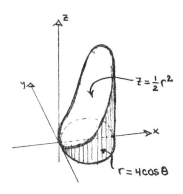

The equation for the cylinder also converts easily:

$$x^2 + y^2 = 4x$$
$$r^2 = 4r\cos\theta$$
$$r = 4\cos\theta$$

<u>Step 1.</u> Fixing arbitrary values of r and θ (that's the same as fixing arbitrary values of

x and y!) we see from our picture that

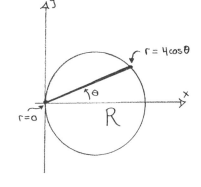

> (1) for r and θ fixed, $0 \leq z \leq \frac{1}{2} r^2$

<u>Step 2.</u> From our sketch of R we can obtain

our <u>polar coordinate</u> limits of integration:

> (2) for θ fixed, $0 \leq r \leq 4 \cos \theta$
>
> (3) $-\pi/2 \leq \theta \leq \pi/2$

The volume of G can now be written as a cylindrical triple integral:

$$\begin{bmatrix} \text{volume} \\ \text{of } G \end{bmatrix} = \iiint_G 1 \, dV$$

$$= \int_{-\pi/2}^{\pi/2} \int_0^{4 \cos \theta} \int_0^{r^2/2} r \; dz \; dr \; d\theta$$

DON'T FORGET THE r!!...

$$= \dots = 12\pi \qquad \text{(we'll leave the integration details as an exercise!)} \qquad \square$$

2. <u>Spherical coordinates.</u> These are a bit more confusing than cylindrical coordinates because

ρ and ϕ are probably much less familar to you than are z and r. Moreover, the important

conversion formulas

$$x = \rho \sin \phi \cos \theta$$

$$y = \rho \sin \phi \sin \theta$$

$$z = \rho \cos \phi$$

$$x^2 + y^2 + z^2 = \rho^2$$

are not easy to remember! We urge you to review the memory tricks introduced in §15.9.2 of

The Companion (the polar coordinate and spherical coordinate triangles).

You should also make sure that you understand the basic spherical coordinate surfaces

$\rho = \rho_0$, $\theta = \theta_0$ and $\phi = \phi_0$ as shown in Anton's Figure 15.9.5 and §15.9.2 of _The_

Companion.

Of central importance is that $\rho = \rho_0$ is a

sphere centered on the origin. This is why spherical

coordinates are useful in triple integration problems

involving solids <u>having symmetry about the origin.</u>

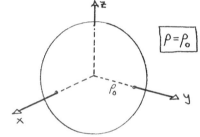

To convert a triple integral into spherical coordinate form we have the following result:

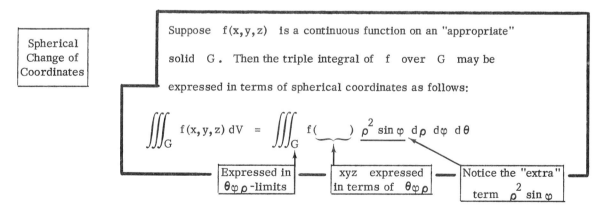

| Spherical Change of Coordinates |

Suppose $f(x,y,z)$ is a continuous function on an "appropriate"
solid G. Then the triple integral of f over G may be
expressed in terms of spherical coordinates as follows:

$$\iiint_G f(x,y,z)\, dV = \iiint_G f(\underline{\quad})\, \rho^2 \sin \phi \, d\rho \, d\phi \, d\theta$$

| Expressed in $\theta \phi \rho$-limits | — | xyz expressed in terms of $\theta \phi \rho$ | — | Notice the "extra" term $\rho^2 \sin \phi$ |

Before giving some examples, a few remarks may be helpful:

1.   The "integrating factor" $\rho^2 \sin \varphi$ may be more easily remembered as ρr (of course, that's helpful only if you remember $r = \rho \sin \varphi$).

2.   With very few exceptions (such as the fourth example in Anton's Table 17.7.1), the solids   G   which are "appropriate" for spherical coordinates are portions of spheres centered on the origin.

> When asked to integrate over a solid which
>
> is a portion of a sphere, you should always
>
> consider using spherical coordinates.

3.   Very often "appropriate" solids for spherical coordinates are not <u>simple solids</u>, i.e., they <u>cannot</u> be described as the volume above a region   R   in the   xy-plane   and between two surfaces $z = g_1(x,y)$ and $z = g_2(x,y)$.   For this reason we do not have the luxury of reading off our "last two" limits from a region   R   in the   xy-plane. This is one reason why determining spherical coordinate limits can be difficult:

> All three spherical coordinate limits of
>
> integration must be determined from
>
> the three dimensional sketch of the solid.

4.   The vast majority of the time we determine spherical coordinate limits of integration in the following order:

Step 1	for θ and φ fixed,	------- $\leq \rho \leq$ -------
Step 2	for θ fixed,	----- $\leq \varphi \leq$ -----
Step 3		--- $\leq \theta \leq$ ---

Anton's Table 17.7.1 works through this exact procedure for five commonly occurring

solids; you should look this table over very carefully!

Example C.   Consider the solid G which is the region in the 1-st octant $(x \geq 0 , y \geq 0 , z \geq 0)$

between the two concentric spheres $x^2 + y^2 + z^2 = 1$ and $x^2 + y^2 + z^2 = 4$. Find the mass

of G if its mass density at any point (x, y, z) is inversely proportional to its distance to the

origin.

Solution.   We first make a rough sketch of G

as shown to the right. Since it is a portion of

a sphere, we take heed of our Remark 3

above and decide to use spherical coordinates. *

The equations for our two spheres become $\rho = 1$

and $\rho = 2$, while the mass density becomes

$$f(x, y, z) = \frac{k}{\rho}$$ for some constant k ,

since ρ is the distance from (x, y, z) to the origin.

---

*
  Another indication that spherical coordinates should be used is that G is not a simple solid:
  the xy-expression for the lower z-bound is not the same over all of G . When
$x^2 + y^2 \leq 1$ we have $\sqrt{1 - x^2 - y^2} \leq z$, but when $1 \leq x^2 + y^2 \leq 4$ we have $0 \leq z$.

Step 1. Fixing θ and φ is equivalent to
fixing a ray ℓ out of the origin; we then have
only to determine the smallest and largest values
of ρ on this ray. From the picture we obtain

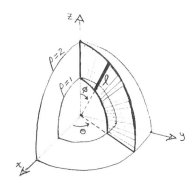

> (1) for θ and φ fixed, $1 \leq \rho \leq 2$

Step 2. Fixing only θ is equivalent to looking
at a slice S of G with constant θ value,
as shown to the right. On this surface we see
that φ varies from the z-axis (φ = 0)
to the xy-plane $(\varphi = \frac{\pi}{2})$. Thus

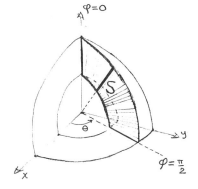

> (2) for θ fixed, $0 \leq \varphi \leq \pi/2$

Step 3. Determining the θ limits without
holding anything else constant is equivalent to
moving the slice S to its smallest possible
θ position (θ = 0) and then to its largest
possible θ position (θ = π/2). Thus

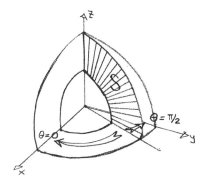

> (3) $0 \leq \theta \leq \pi/2$

So putting it all together:

$$\begin{bmatrix} \text{mass} \\ \text{of } G \end{bmatrix} = \iiint_G \frac{k}{\rho} \, dV$$

DON'T FORGET ... !

$$= \int_0^{\pi/2} \int_0^{\pi/2} \int_1^2 \left(\frac{k}{\rho}\right) \overbrace{\rho^2 \sin \varphi}^{} \, d\rho \, d\varphi \, d\theta$$

- (1) -
- (2) -
- (3) -

Inner integral: $\displaystyle\int_1^2 k\rho \sin \phi \, d\rho = (k \sin \phi) \left.\frac{\rho^2}{2}\right|_1^2 = \frac{3k}{2} \sin \phi$

Middle integral: $\displaystyle\int_0^{\pi/2} \frac{3k}{2} \sin \phi \, d\phi = \frac{3k}{2}(-\cos\phi)\Big|_0^{\pi/2} = \frac{3k}{2}$

Full integral: $\displaystyle\int_0^{\pi/2} \frac{3k}{2} \, d\theta = \frac{3\pi k}{4}$ □

*   *   *   *   *

As occurred in Example C, it is very common for all the limits of integration in a spherical triple integral to be constants. This is not always the case, however, as the next example shows:

<u>Example D.</u>   Determine the volume above the sphere $x^2 + y^2 + z^2 = 1$ but below the sphere $x^2 + y^2 + (z-1)^2 = 1$.

<u>Solution.</u>   We sketch the solid G to the right, and because of the presence of two spheres we decide to use spherical coordinates. The lower sphere then becomes $\rho = 1$, while the upper sphere converts as follows:

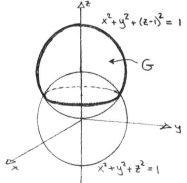

17. 7. 13

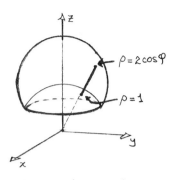

$$x^2 + y^2 + z^2 - 2z + 1 = 1$$

$$\rho^2 - 2\rho \cos \varphi = 0$$

$$\rho = 2 \cos \varphi$$

Step 1. Holding θ and φ fixed (i. e. , following a ray out from the origin) we see from our picture that

> (1)   for θ and φ fixed, $1 \leq \rho \leq 2 \cos \varphi$

Step 2. Fixing θ to find φ limits, we have a slice S that looks rather much like a rose thorn. The problem is: what is the φ value at the tip of the thorn? Some elementary geometry will save the day: φ is the corner of an equilateral triangle of side length one (two sides are radii of spheres of radius one, and the vertical side is the distance 1 between their centers). Hence $\varphi = \pi/3$, giving

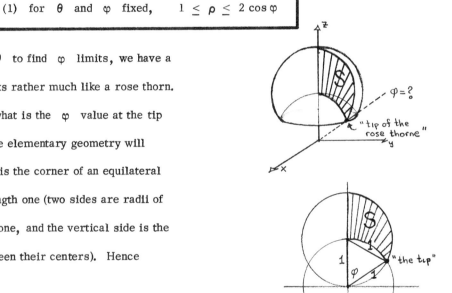

> (2)   for θ fixed, $0 \leq \varphi \leq \pi/3$

Step 3. The θ limits are easy since the slice S can rotate all the way around the z-axis , i. e. ,

> (3) $0 \leq \theta \leq 2\pi$

Hence our volume is given by

$$\begin{bmatrix} \text{volume} \\ \text{of } G \end{bmatrix} = \int_0^{2\pi} \int_0^{\pi/3} \int_1^{2\cos\varphi} \underbrace{\rho^2 \sin\varphi}\ d\rho\ d\varphi\ d\theta$$

DON'T FORGET ... !!

Inner integral: $\displaystyle\int_1^{2\cos\varphi} \rho^2 \sin\varphi\ d\rho = (\sin\varphi)\left.\frac{\rho^3}{3}\right|_{\rho=1}^{\rho=2\cos\varphi} = \frac{8}{3}\cos^3\varphi \sin\varphi - \frac{1}{3}\sin\varphi$

Middle integral: $\displaystyle\int_0^{\pi/3}\left(\frac{8}{3}\cos^3\varphi \sin\varphi - \frac{1}{3}\sin\varphi\right)d\varphi = \left.\left(-\frac{2}{3}\cos^4\varphi + \frac{1}{3}\cos\varphi\right)\right|_{\varphi=0}^{\varphi=\pi/3}$

$$= \left(-\frac{2}{3}\right)\left(\frac{1}{2}\right)^4 + \frac{1}{3}\left(\frac{1}{2}\right) + \left(\frac{2}{3}\right) - \frac{1}{3} = \frac{11}{24}$$

Full integral: $\displaystyle\int_0^{2\pi} \frac{11}{24}\ d\theta = \left.\frac{11}{24}\theta\right|_{\theta=0}^{\theta=2\pi} = \boxed{\frac{11}{12}\pi}$ □

Chapter 18:   Topics in Vector Calculus

Section 18. 1.   Line Integrals

Suppose C is a smooth, directed* plane curve, and

$$\overline{F}(x,y) = f(x,y)\,\overline{i} + g(x,y)\,\overline{j}$$

is a vector-valued function, where f and g are continuous real-valued functions defined

on some open region containing C. The purposes of this section are (1) to define the <u>line</u>

<u>integral of $\overline{F}$ along C</u>

$$\int_C \overline{F}(x,y) \cdot d\overline{r}$$

(2) to illustrate how it is computed, and (3) to describe its use in the calculation of work.

1.   <u>Work.</u>   The definition of the line integral looks pretty strange and unintuitive at first glance.

However, when derived via Riemann sums using the concept of <u>work</u>, the definition seems more

natural. So we will temporarily forget about line integrals and concentrate on learning how

to compute work:

Suppose the force vector at a point (x,y) in the plane is given by

$$\overline{F}(x,y) = f(x,y)\,\overline{i} + g(x,y)\,\overline{j}$$

---

*
A smooth curve is <u>directed</u> if it has a direction assigned to it
(designated by arrows along its path). Thus the curves shown
to the right are all "directed." If C is a directed curve,
then the same curve traced in the opposite direction will be
denoted by - C .

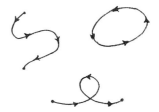

and a particle is moving through this force field along a smooth, directed

curve   C.   For computational purposes assume a set of parametric

equations for   C   is given by the position vector

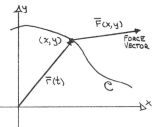

$$\bar{r}(t) = x(t)\bar{i} + y(t)\bar{j} , \qquad a \le t \le b$$

Using a straightforward approximation procedure, we will derive

a formula for the work done by the force when moving the particle along the curve   C :

1.   Divide the curve   C   into   n   pieces by

   partitioning the interval   [a, b]:

$$a = t_0 < t_1 < \ldots < t_n = b$$

   Let $\Delta t_k = t_{k+1} - t_k$

2.   Consider the piece of   C   from $\bar{r}(t_k)$

   to $\bar{r}(t_{k+1})$.   When Δt_k is small, then

   this piece of the curve is closely approximated

   by

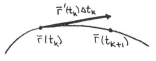

$$\bar{r}'(t_k) \Delta t_k$$

$$\left(\text{Reason:} \ \ \bar{r}'(t_k) = \lim_{t \to t_k} \frac{\bar{r}(t) - \bar{r}(t_k)}{t - t_k} \approx \frac{\bar{r}(t_{k+1}) - \bar{r}(t_k)}{\Delta t_k} \right)$$

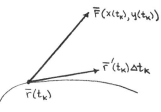

3.   Hence ΔW_k,   the work done in moving from

   $\bar{r}(t_k)$ to $\bar{r}(t_{k+1})$,   is approximated by

$$\Delta W_k \approx \bar{F}(x(t_k), y(t_k)) \cdot [\bar{r}'(t_k) \Delta t_k]$$

(Reason: This is an application of Equation (13) from §15. 3

$$\text{work} = \overline{F} \cdot \overline{v}$$

where $\overline{F}$ is a constant vector force and $\overline{v}$ is the displacement vector of the object being moved. Also see §15. 3. 3 of The Companion.)

4.   Hence W ,   the total work done, is the sum of all the ΔW_k :

$$W = \sum \Delta W_k \approx \sum_k (\overline{F}(x(t_k), y(t_k)) \cdot \overline{r}'(t_k)) \Delta t_k$$

But wait a minute ... . This is a Riemann sum for the function

$$t \longmapsto \overline{F}(x(t), y(t)) \cdot \frac{d\overline{r}}{dt}$$

over the interval $a \le t \le b$.

Hence, taking the limit as $\Delta t_k \to 0$ gives

$$\boxed{W = \int_a^b \left[\overline{F}(x(t), y(t)) \cdot \frac{d\overline{r}}{dt} \right] dt}$$

Example A.   Suppose Howard Ant is moving through the force field

$$\overline{F}(x, y) = e^x \overline{i} + e^y \overline{j}$$

along the curve   C   parameterized by

$$\overline{r}(t) = t\overline{i} + t^2 \overline{j} , \qquad 0 \le t \le 2$$

How much work does the force do on Howard?

18.1.4

<u>Solution.</u> From the formula we obtain

$$W = \int_a^b \left[\overline{F}(x(t), y(t)) \cdot \frac{d\overline{r}}{dt} \right] dt$$

where $\overline{F}(x(t), y(t)) = \overline{F}(t, t^2) = e^t \overline{i} + e^{t^2} \overline{j}$

$$\frac{d\overline{r}}{dt} = \frac{d}{dt}[t\overline{i} + t^2\overline{j}] = \overline{i} + 2t\overline{j}$$

Thus

$$W = \int_0^2 \left(e^t \overline{i} + e^{t^2} \overline{j} \right) \cdot (\overline{i} + 2t\overline{j}) \, dt$$

$$= \int_0^2 \left(e^t + 2t e^{t^2} \right) dt$$

$$= e^t + e^{t^2} \Big]_0^2 = e^2 + e^4 - 2 \qquad \text{units of work.} \qquad \square$$

2. <u>The definition and various formulations of a line integral.</u> We will use our formula for work

to define the line integral: *

Suppose

1. C is a smooth, directed curve given parametrically by the position

vector

$$\overline{r}(t) = x(t)\overline{i} + y(t)\overline{j}, \qquad a \le t \le b$$

The
Line
Integral

---
*
We have chosen a different definition for the line integral than Anton uses. Since the two
definitions are equivalent, it is only a matter of taste as to which one you prefer.

2. $\overline{F}(x,y)$ is a vector-valued function given by

$$\overline{F}(x,y) \; = \; f(x,y)\,\overline{i} \; + \; g(x,y)\,\overline{j}$$

where $f(x,y)$ and $g(x,y)$ are continuous on some open region containing C.

Then the line integral of $\overline{F}$ along the curve C is defined to be

$$\int_C \overline{F}(x,y) \cdot d\overline{r} \; = \; \int_a^b \left[\overline{F}(x(t), y(t)) \cdot \frac{d\overline{r}}{dt} \right] dt$$

Hence, if $\overline{F}(x,y)$ is a force field, and C is the path of a particle in $\overline{F}$, then

$W = \int_C \overline{F}(x,y) \cdot d\overline{r}$ is the work done by the force on the particle. Work is the basic

interpretation you should always use for a line integral!! This adds the fourth entry to the

following table:

Integral	Basic Interpretation
$\int_a^b f(x)\,dx$	Area
$\iint_R f(x,y)\,dA$	Volume
$\iiint_G f(x,y,z)\,dV$	Mass
$\int_C \overline{F}(x,y) \cdot d\overline{r}$	Work

Example B. Compute $\displaystyle\int_C \overline{F}(x,y) \cdot d\overline{r}$ if $\overline{F}(x,y) = \sqrt{x}\,\sin y\,\overline{i} - \cos y\,\overline{j}$ and C is

parameterized by $\overline{r}(t) = t^2\,\overline{i} + t^3\,\overline{j}$, $0 \le t \le \pi^{1/3}$.

Solution. $\displaystyle\int_C \overline{F}(x,y) \cdot d\overline{r} = \int_a^b \left[\overline{F}(x(t), y(t)) \cdot \frac{d\overline{r}}{dt} \right] dt$

where $\overline{F}(x(t), y(t)) = \overline{F}(t^2, t^3) = t\,\sin t^3\,\overline{i} - \cos t^3\,\overline{j}$

$$\frac{d\overline{r}}{dt} = \frac{d}{dt}\,[t^2\,\overline{i} + t^3\,\overline{j}] = 2t\,\overline{i} + 3t^2\,\overline{j}$$

Thus $\displaystyle\int_C \overline{F}(x,y) \cdot d\overline{r} = \int_0^{\pi^{1/3}} (t\,\sin t^3\,\overline{i} - \cos t^3\,\overline{j}) \cdot (2t\,\overline{i} + 3t^2\,\overline{j})\,dt$

$$= \int_0^{\pi^{1/3}} (2t^2\,\sin t^3 - 3t^2\,\cos t^3)\,dt$$

$$= -\frac{2}{3}\,\cos t^3 - \sin t^3 \Big]_0^{\pi^{1/3}} = \frac{2}{3} + \frac{2}{3} = \frac{4}{3} \qquad \square$$

$$* \quad * \quad * \quad * \quad *$$

The notational symbol

$$\int_C \overline{F}(x,y) \cdot d\overline{r}$$

was invented because it captures so well the essence of the computational formula

$$\int_a^b \left[\overline{F}(x(t), y(t)) \cdot \frac{d\overline{r}}{dt} \right] dt$$

If you "cancel" the dt's, suppress (i. e., erase) the t's, and replace $\displaystyle\int_a^b$ with $\displaystyle\int_C$,

then the computational formula becomes the notational symbol. Hence, when asked to compute a line integral, you simply "add back" all these deleted or replaced symbols:

<u>Example C.</u>   Compute $\displaystyle\int_C \overline{F}(x,y) \cdot d\overline{r}$ if $\overline{F}(x,y) = 5x^2 y\,\overline{i} + \dfrac{24}{x}\,\overline{j}$ and C is parameterized

by $\overline{r}(t) = t^2\overline{i} + \dfrac{1}{t}\,\overline{j}, \quad 1 \le t \le 2.$

<u>Solution.</u> $\displaystyle\int_C \overline{F}(x,y) \cdot d\overline{r} = \int_a^b \left[\overline{F}(x(t), y(t)) \cdot \dfrac{d\overline{r}}{dt} \right] dt$

> The arrows indicate symbols which have been "added back."

where $\overline{F}(x(t), y(t)) = \overline{F}(t^2, \dfrac{1}{t})$

$$= 5(t^2)^2 \, (\dfrac{1}{t})\overline{i} + \dfrac{24}{t^2}\,\overline{j}$$

$$= 5t^3\overline{i} + \dfrac{24}{t^2}\,\overline{j}$$

$$\dfrac{d\overline{r}}{dt} = \dfrac{d}{dt}\,[\, t^2\overline{i} + \dfrac{1}{t}\,\overline{j}\,] = 2t\,\overline{i} - \dfrac{1}{t^2}\,\overline{j}$$

Hence

$$\int_C \overline{F}(x,y) \cdot d\overline{r} = \int_1^2 [\, 5t^3\overline{i} + \dfrac{24}{t^2}\,\overline{j}\,] \cdot [\, 2t\,\overline{i} - \dfrac{1}{t^2}\,\overline{j}\,]\,dt$$

$$= \int_1^2 [\, 10t^4 - \dfrac{24}{t^2}\,]\,dt$$

$$= 2t^5 + \dfrac{8}{t^3}\,\Bigg]_1^2$$

$$= 64 + 1 - 2 - 8 = 55 \qquad\qquad \square$$

18. 1. 8

There is another notational symbol for the line integral which corresponds to a second computational formula. This second formula is easy to derive:

$$\int_C \overline{F}(x,y) \cdot d\overline{r} = \int_a^b \left[\overline{F}(x(t), y(t)) \cdot \frac{d\overline{r}}{dt} \right] dt$$

$$= \int_a^b [f(x(t), y(t)) \overline{i} + g(x(t), y(t)) \overline{j}] \cdot \left[\frac{dx}{dt} \overline{i} + \frac{dy}{dt} \overline{j} \right] dt$$

$$= \int_a^b \left[f(x(t), y(t)) \frac{dx}{dt} + g(x(t), y(t)) \frac{dy}{dt} \right] dt$$

This last expression is a perfectly good computational formula for the line integral. Now "cancel" the dt's, suppress the t's, and replace $\int_a^b$ with $\int_C$, and you're left with

$$\int_C f(x,y) \, dx + g(x,y) \, dy$$

which is the notational symbol for the line integral that Anton starts with! Here is a convenient summary:

Notational symbol		Computational formula
Vector Form	$\int_C \overline{F}(x,y) \cdot d\overline{r}$	$= \int_a^b \left[\overline{F}(x(t), y(t)) \cdot \frac{d\overline{r}}{dt} \right] dt$
Component Form	$\int_C f(x,y)\,dx + g(x,y)\,dy$	$= \int_a^b \left[f(x(t), y(t)) \frac{dx}{dt} + g(x(t), y(t)) \frac{dy}{dt} \right] dt$

where $\overline{F}(x,y) = f(x,y)\overline{i} + g(x,y)\overline{j}$, and C is parameterized

by $\overline{r}(t) = x(t)\overline{i} + y(t)\overline{j}$, $a \le t \le b$

The component form symbol $\displaystyle\int_C f(x,y)\,dx + g(x,y)\,dy$ is the symbol most often used in

applications. Sometimes it is further abbreviated to

$$\int_C f\,dx + g\,dy$$

<u>Example D.</u>   Compute $\displaystyle\int_C xy\,dx + x\,dy$ if C is parameterized

by $\overline{r}(t) = (1-t)\overline{i} + t^2\overline{j}$, $\ 0 \le t \le 1$.

<u>Solution.</u>

$$\int_C xy\,dx + x\,dy = \int_0^1 \left[x(t)\,y(t)\,\frac{dx}{dt} + x(t)\,\frac{dy}{dt} \right] dt$$

$$\text{where}\quad x(t) = 1-t \qquad y(t) = t^2$$

$$\frac{dx}{dt} = -1 \qquad \frac{dy}{dt} = 2t$$

$$= \int_0^1 \left[(1-t)\,t^2\,(-1) + (1-t)\,(2t) \right] dt$$

$$= \int_0^1 (t^3 - 3t^2 + 2t)\,dt$$

$$= \left. \frac{t^4}{4} - t^3 + t^2 \right]_0^1 = \frac{1}{4} - 1 + 1 = \frac{1}{4} \qquad \square$$

3.   <u>Parameterizations of the curve C</u>.   All the previous examples have one aspect in common:

we specified a parameterization for the directed curve C in the problem statement itself. In

applications this will rarely happen; instead, the directed curve will be described geometrically, or perhaps by an equation in x and y, and a suitable parameterization $\overline{r}(t) = x(t)\,\overline{i} + y(t)\,\overline{j}$

will have to be found. Notice that Theorem 18. 1. 2 (Independence of Parameterization) and

Theorem 18. 1. 3 (Reversal of Orientation) guarantee that <u>the value of a line integral</u>

$$\int_C \overline{F}(x,y) \cdot d\overline{r}$$ <u>depends only on the direction of the parameterization of C, i. e., if C is</u>

given two parameterizations $\overline{r}_1(t)$ and $\overline{r}_2(t)$, then

$$\int_C \overline{F}(x,y) \cdot d\overline{r}_1 \;=\; \int_C \overline{F}(x,y) \cdot d\overline{r}_2 \qquad \text{if } \overline{r}_1(t) \text{ and } \overline{r}_2(t)$$

have the same direction

$$\int_C \overline{F}(x,y) \cdot d\overline{r}_1 \;=\; -\int_C \overline{F}(x,y) \cdot d\overline{r}_2 \qquad \text{if } \overline{r}_1(t) \text{ and } \overline{r}_2(t)$$

have the opposite direction

<u>Example E.</u>   Compute $\displaystyle\int_C x\,y\,dx + x\,dy$ where C is the arc of the unit circle joining $(1,0)$

to $(0,1)$ in a counterclockwise direction.

<u>Solution.</u>   A picture of C is shown to the right.

In order to demonstrate different ways to obtain

parameterizations, we will parameterize C twice:

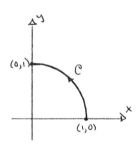

1.   Starting with the relationship

$$x^2 + y^2 = 1$$

which must be satisfied by every point (x,y) on C , we can solve for y in terms of x :

$$y = \sqrt{1 - x^2}$$

Thus, <u>choosing the parameter t to equal x</u> , we obtain parametric equations

$$\begin{cases} x(t) = 1 \\ y(t) = \sqrt{1 - t^2} \end{cases}$$

The starting point $(1, 0)$ corresponds to $t = 1$ and the terminal point $(0, 1)$ corresponds

to ... $t = 0$? ... Oops! This parameterization

is going in the <u>wrong direction.</u> However, that's

easily fixed by <u>choosing the parameter t to</u>

<u>equal $-x$!</u> Then

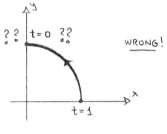

$$\begin{cases} x(t) = -t \\ y(t) = \sqrt{1 - t^2} \end{cases}$$

the starting point $(1, 0)$ corresponds to $t = -1$, and the terminal point $(0, 1)$ corresponds

to $t = 0$. Thus our parameterization for C will be

$$\bar{r}_1(t) = -t\,\bar{i} + \sqrt{1 - t^2}\,\bar{j} \quad \text{for} \quad -1 \le t \le 0$$

Computing the line integral gives

$$\int_C xy\,dx + x\,dy = \int_{-1}^{0} \left[x(t)\,y(t)\frac{dx}{dt} + x(t)\frac{dy}{dt} \right] dt$$

$$\text{where} \quad x(t) = -t \qquad y(t) = (1 - t^2)^{1/2}$$

$$\frac{dx}{dt} = -1 \qquad \frac{dy}{dt} = -t(1 - t^2)^{-1/2}$$

$$= \int_{-1}^{0} \left[(-t)(1 - t^2)^{1/2}(-1) + (-t)\left(-t(1 - t^2)^{-1/2}\right) \right] dt$$

$$= \int_{-1}^{0} t(1 - t^2)^{1/2} \, dt + \int_{-1}^{0} \frac{t^2 \, dt}{(1 - t^2)^{1/2}}$$

$$= \boxed{-\frac{1}{3} + \frac{\pi}{4}} \qquad \text{where the substitution} \quad u = 1 - t^2 \quad \text{is}$$

used to evaluate the first integral, and $t = \sin \theta$

is used in the second integral.

2.   If you were very alert you might have noticed a bit of a problem with the parameterization $\bar{r}_1$: the derivative of $\bar{r}_1(t)$ is not defined at $t = -1$, and so $\bar{r}_1(t)$ is not a "smooth" parameterization of C! Since it does give a smooth parameterization on any restricted interval $[-1 + \epsilon, 0]$, where $\epsilon > 0$, then we could use improper integrals to give a rigorous justification for our previous calculations. However, all of these problems can be cleared up by choosing a parameter t which is not equal to either $\pm x$ or $\pm y$:

how about $t = \theta$, the polar

coordinate angle?

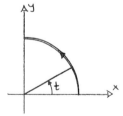

Then

$$\begin{cases} x(t) = \cos t \\ y(t) = \sin t \end{cases}$$

the starting point $(1, 0)$ corresponds to $t = 0$, and the terminal point $(0, 1)$ corresponds to $t = \pi/2$. Thus our parameterization for C will be

$$\bar{r}_2(t) = \cos t \, \bar{i} + \sin t \, \bar{j} \qquad \text{for} \qquad 0 \le t \le \pi/2$$

Computing the line integral gives

$$\int_C xy\,dx + x\,dy = \int_0^{\pi/2} \left[x(t)\,y(t)\,\frac{dx}{dt} + x(t)\,\frac{dy}{dt} \right] dt$$

where $x(t) = \cos t$ $y(t) = \sin t$

$$\frac{dx}{dt} = -\sin t \qquad \frac{dy}{dt} = \cos t$$

$$= \int_0^{\pi/2} [\cos t \sin t\,(-\sin t) + \cos t \cos t\,]\,dt$$

$$= -\int_0^{\pi/2} \sin^2 t \cos t\,dt + \int_0^{\pi/2} \cos^2 t\,dt$$

$$= \boxed{-\frac{1}{3} + \frac{\pi}{4}} \qquad \text{where the substitution} \quad u = \sin t \quad \text{is}$$

used in the first integral and $\cos^2 t = \frac{1}{2}(1 + \cos 2t)$

is used in the second.

As expected, our two different parameterizations $\bar{r}_1$ and $\bar{r}_2$ do yield the same answer for

the line integral.   □

*   *   *   *   *

One final aspect of parameterization needs to be stressed:

<u>many</u> times the curve   C   will not be smooth, but

will be a union of finitely many smooth curves

$C_1, C_2, \ldots, C_n$ joined end-to-end.   Then to

compute $\int_C \bar{F}(x,y) \cdot d\bar{r}$ you simply compute the line integral over each curve C_k

separately (each with its own parameterization $\bar{r}_k(t)$)   and add the integrals together:

$$\int_C = \int_{C_1} + \int_{C_2} + \cdots + \int_{C_n}$$

Anton illustrates this procedure in Example 3.

4. **Line integrals in three dimensions.** Integrating a vector function $\overline{F}$ along a curve C in three-dimensional space uses all the same procedures and equations that are used in two dimensions, except that z-components must also be used:

Notational symbol   Computational formula

$$\int_C \overline{F}(x,y,z) \cdot d\overline{r} = \int_a^b \left[\overline{F}(x(t), y(t), z(t)) \cdot \frac{d\overline{r}}{dt} \right] dt$$

$$\int_C f(x,y,z)\, dx + g(x,y,z)\, dy + h(x,y,z)\, dz$$

$$= \int_a^b \left[f(x(t), y(t), z(t)) \frac{dx}{dt} + g(x(t), y(t), z(t)) \frac{dy}{dt} + h(x(t), y(t), z(t)) \frac{dz}{dt} \right] dt$$

where $\overline{F}(x,y,z) = f(x,y,z)\overline{i} + g(x,y,z)\overline{j} + h(x,y,z)\overline{k}$, and C is

parameterized by $\overline{r}(t) = x(t)\overline{i} + y(t)\overline{j} + z(t)\overline{k}$, $a \le t \le b$

Don't let the length of these formulas bother you; the computational formulas follow easily from the notational symbols as in the two-dimensional case, and the notational symbols are the obvious generalizations of their two-dimensional analogues.

Example F.   Compute $\displaystyle\int_C y e^x\, dx + xz\, dy + 2e^{x-z}\, dz$ along the line segment C joining

$P_1(1, -1, 0)$ to $P_2(2, 1, 3)$.

Solution. We must first parameterize C. However,

parameterizing a line segment joining a point P_1 to

a point P_2 is easy: Let

$$\bar{v} = \overrightarrow{P_1 P_2} = (2 - 1)\bar{i} + (1 - (-1))\bar{j} + (3 - 0)\bar{k}$$

$$\boxed{\bar{v} = \bar{i} + 2\bar{j} + 3\bar{k}}$$

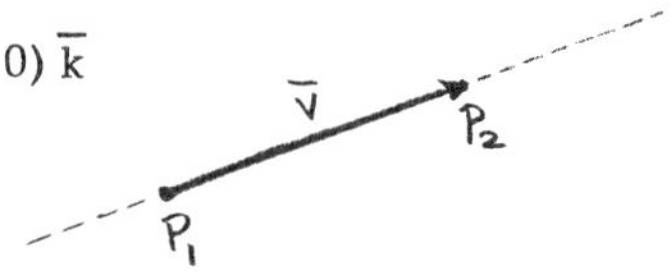

be the direction vector for the line. Then, starting at

$$P_1 (1, -1, 0)$$

parametric equations for the line are given by:

$$\begin{cases} x(t) = 1 + t \\ y(t) = -1 + 2t \\ z(t) = 0 + 3t \end{cases}$$

(Theorem 15. 2. 1 ; also see §15. 2. 5 of The Companion). The starting point $P_1(1, -1, 0)$

corresponds to $t = 0$ and the terminal point $P_2(2, 1, 3)$ corresponds to $t = 1$. Thus

our parameterization for C will be

$$\bar{r}(t) = (1 + t)\bar{i} + (-1 + 2t)\bar{j} + 3t\bar{k} \quad \text{for} \quad 0 \le t \le 1$$

Computing the integral gives

$$\int_C y e^x \, dx + xz \, dy + 2 e^{x-z} \, dz$$

$$= \int_0^1 [(-1+2t) e^{1+t} (1) + (1+t)(3t)(2) + 2 e^{1-2t} (3)] \, dt$$

$$= \int_0^1 (- e^{1+t} + 2t e^{1+t} + 6t + 6t^2 + 6 e^{1-2t}) \, dt$$

$$= - e^{1+t} + (2t e^{1+t} - 2 e^{1+t}) + 3t^2 + 2t^3 - 3 e^{1-2t} \Big]_0^1$$

$$= - e^2 - 6e + 5 - 3/e \qquad \square$$

Section 18.2.   Line Integrals Independent of Path

Conservative vector-valued functions

$$\overline{F}(x,y) = f(x,y) \, \overline{i} + g(x,y) \, \overline{j}$$

(also called conservative vector fields) lie at the heart of fluid mechanics and electromagnetic theory; in fact, they have more than just a name in common with conservation of energy laws. We will carefully summarize and (when it aids understanding) expand upon Anton's treatment of the subject; the connections with physics will be discussed briefly in an optional final subsection.

The main result is the following:

The Conservative Vector Fields Theorem

Suppose $\overline{F}(x,y)$ is a vector-valued function * on an open region U.

Then $\overline{F}$ is called <u>conservative</u> if it possesses any one (hence all)

of the following equivalent properties:

1. $\overline{F}$ is <u>circulation-free</u> on U, i.e., $\oint_C \overline{F} \cdot d\overline{r} = 0$ for

    every closed curve * C in U;

2. $\overline{F}$ has <u>path-independent</u> line integrals for every curve C

    in U, i.e., $\int_C \overline{F} \cdot d\overline{r}$ depends only on the endpoints

    of C;

3. $\overline{F}$ is a <u>gradient field</u> on U, i.e., $\overline{F} = \nabla\varphi$ for some

    real-valued function φ defined on U.

Moreover, if $\overline{F}(x,y)$ has continuous first partial derivatives on U,

and U is an open, <u>simply-connected</u> region, then there is a fourth

equivalence:

4. $\overline{F}$ is <u>irrotational</u> on U, i.e., $\dfrac{\partial g}{\partial x} = \dfrac{\partial f}{\partial y}$ at every point in U.

We will discuss these four properties in the order in which they are mentioned in the

theorem:

---

* In this and subsequent sections, curves C will all be <u>piecewise smooth</u> and vector-valued
functions $\overline{F}$ will all be <u>continuous</u>, i.e., if $\overline{F}(x,y) = f(x,y)\,\overline{i} + g(x,y)\,\overline{j}$, then $f(x,y)$
$g(x,y)$ will be continuous (as defined in §16.3).

1. Circulation-free vector-valued functions. A vector-valued function $\overline{F}(x,y)$ is said to be circulation-free on an open region U if

$$\oint_C \overline{F} \cdot d\overline{r} = 0 \quad \text{for any closed curve* } C \text{ in } U$$

Example A. Evaluate $\oint_C (2xy + e^x) dx + x^2 dy$ where C is the unit circle parameterized in the counterclockwise direction.

Brute force solution. Use the parameterization

$$\overline{r}(t) = \cos t \, \overline{i} + \sin t \, \overline{j} , \quad 0 \le t \le 2\pi$$

and evaluate the line integral as in the previous section. Since

$$x(t) = \cos t \qquad y(t) = \sin t$$

$$\frac{dx}{dt} = - \sin t \qquad \frac{dy}{dt} = \cos t$$

then

$$\oint_C (2xy + e^x) dx + x^2 dy$$

$$= \int_0^{2\pi} [(2 \cos t \sin t + e^{\cos t}) (- \sin t) + \cos^2 t (\cos t)] dt$$

$$= \dots \quad \text{well, it's "doable, " but we'll leave the details to you!}$$

Intelligent solution. As we will verify in Examples C and E, the vector-valued function $\overline{F}(x,y) = (2xy + e^x)\overline{i} + x^2\overline{j}$ is conservative on the whole plane, and hence is circulation-free. Thus

$$\oint_C (2xy + e^x) dx + x^2 dy = \oint_C \overline{F} \cdot d\overline{r} = 0 \qquad \square$$

---

* The symbol $\oint_C$ is commonly used when C is a closed curve, i.e., a curve whose initial and terminal points are the same.

As this example illustrates, we verify that $\overline{F}$ is circulation-free by verifying one of the other equivalences for conservative vector-valued functions.

2. <u>Path-independent line integrals.</u>   A line integral $\displaystyle\int_C \overline{F} \cdot d\overline{r}$ is said to be <u>independent of path</u> in an open region U if:

> the value of $\displaystyle\int_C \overline{F} \cdot d\overline{r}$ depends only on the
>
> endpoints of C and not on the particular
>
> path in U joining these endpoints.

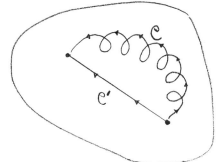

Thus, if C_1 and C_2 are curves in U with the same initial point (x_0, y_0) and terminal point (x_1, y_1) , then

$$\int_{C_1} \overline{F} \cdot d\overline{r} = \int_{C_2} \overline{F} \cdot d\overline{r} = \int_{(x_0, y_0)}^{(x_1, y_1)} \overline{F} \cdot d\overline{r}$$

> Notation indicating dependence
> only on the endpoints

Examples D and E in §18.1 of <u>The Companion</u> illustrate <u>non</u>-path-independent line integrals.

It is extremely convenient to know when a line integral is independent of path for the following reason: if we must compute $\displaystyle\int_C \overline{F} \cdot d\overline{r}$ for a complicated curve C , we can replace it with $\displaystyle\int_{C'} \overline{F} \cdot d\overline{r}$ where C' is a

simpler curve which has the same endpoints as C .

18.2.5

Example B. Evaluate $\int_C (2xy + e^x)\,dx + x^2\,dy$ where C is the curve parameterized by

$$\bar{r}(t) = (t + \sin t \ln(1+t))\,\bar{i} - t\cos t\,\bar{j}, \qquad 0 \le t \le \pi \ .$$

Brute force solution. Don't even try ... .

Intelligent solution. As we will verify in Examples C and E, the vector-valued function $\bar{F}(x,y) = (2xy + e^x)\bar{i} + x^2\bar{j}$ is <u>conservative</u> on the whole plane, and hence has path-independent line integrals. So let's change the path of integration. The endpoints of C are $(0,0)$ and (π,π), so we can replace C with the straight line segment C' parameterized by

$$\bar{r}(t) = t\bar{i} + t\bar{j}, \qquad 0 \le t \le \pi$$

Then $\quad x(t) = t \qquad y(t) = t$

$$\frac{dx}{dt} = 1 \qquad \frac{dy}{dt} = 1$$

$$\int_C (2xy + e^x)\,dx + x^2\,dy = \int_0^\pi [\,(2t^2 + e^t) + t^2\,]\,dt$$

$$= t^3 + e^t\Big]_0^\pi = \pi^3 + e^\pi - 1$$

(We will give yet another solution to this problem in Example D below.) ◻

* * * * *

Only elementary arguments are needed to prove the equivalence between "circulation-free" and "path-independent" in the conservative Vector Fields Theorem:

Suppose $\overline{F}$ is circulation-free in U, and C_1 and C_2 are

any two curves in U with the same initial and terminal

points. Let C be the closed curve which is the union

of C_1 and $-C_2$ (i.e., C_2 parameterized in the

opposite direction). Then

$$0 = \oint_C \overline{F} \cdot d\overline{r} = \int_{C_1} \overline{F} \cdot d\overline{r} - \int_{C_2} \overline{F} \cdot d\overline{r} \quad \text{(using Anton's Equation (4)}$$
$$\text{of } \S 18.1)$$

Thus $\int_{C_1} \overline{F} \cdot d\overline{r} = \int_{C_2} \overline{F} \cdot d\overline{r}$, proving that $\overline{F}$ has path-independent line

integrals in U.

Suppose $\overline{F}$ has path-independent line integrals in U, and C

is any closed curve in U. Choose any two points A

and B on C and write C as the union of two

curves C_1 and $-C_2$, where A is the initial point

of both C_1 and C_2, and B is the terminal point

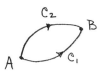

of both C_1 and C_2. Then, by Anton's Equation (4)

in $\S 18.1$

$$\oint_C \overline{F} \cdot d\overline{r} = \int_{C_1} \overline{F} \cdot d\overline{r} - \int_{C_2} \overline{F} \cdot d\overline{r} = 0$$

since $\overline{F}$ has path-independent line integrals in U.

Thus $\overline{F}$ is circulation-free in U.

3. <u>Gradient fields.</u> A vector-valued function $\overline{F}(x,y)$ is said to be a <u>gradient field</u> on an open

region U if

$\overline{F} = \nabla\varphi$ for some real-valued function φ defined on U

The function φ is called a <u>potential function</u> for $\overline{F}$. The importance of gradient fields lies

in the following generalization of the Fundamental Theorem of Calculus:

Fundamental Theorem of Line Integrals

Suppose $\overline{F}(x,y) = \nabla\varphi(x,y)$ on an open region U,

and C is a curve in U with initial point

(x_0, y_0) and terminal point (x_1, y_1).

Then $\displaystyle\int_C \overline{F} \cdot d\overline{r} = \varphi(x_1, y_1) - \varphi(x_0, y_0)$.

This is Anton's Theorem 18.2.1, and it makes the evaluation of line integrals for gradient

fields quite easy (... when the potential function is known, that is!). [See Anton's Example 2,

and Example D below.] Moreover, the theorem also shows immediately that a gradient field

has path-independent line integrals (because the quantity $\varphi(x_1, y_1) - \varphi(x_0, y_0)$ clearly

depends only on the endpoints (x_0, y_0) and (x_1, y_1) of C), and is thus circulation-free.

This is Anton's Theorem 18.2.2. The converse of this result (i.e., if $\overline{F}$ is circulation-free,

then it is a gradient field) is also true, although it requires some subtle analysis to construct

the potential function φ using path-independence.

This establishes the first three equivalences in the Conservative Vector Fields Theorem.

$$* \quad * \quad * \quad * \quad *$$

The property of being a gradient field, unlike the properties of being circulation-free and having path-independent line integrals, can be verified directly in many instances. The method is demonstrated in Anton's Example 3, and can be summarized as follows: Given

$$\overline{F}(x,y) = f(x,y)\,\overline{i} + g(x,y)\,\overline{j}$$

then $\overline{F}(x,y)$ will equal the gradient field

$$\nabla \varphi(x,y) = \frac{\partial \varphi}{\partial x}\,\overline{i} + \frac{\partial \varphi}{\partial y}\,\overline{j}$$

if and only if

$$\boxed{\frac{\partial \varphi}{\partial x} = f(x,y) \quad \text{and} \quad \frac{\partial \varphi}{\partial y} = g(x,y)}$$

To find φ, integrate the first equality with respect to x, treating y as a constant:

$$\varphi(x,y) = \int f(x,y)\,dx + k(y)$$

Here $k(y)$ is the "constant of integration," an unknown function of $\underline{y}$. Now take the partial with respect to y of this expression and equate it with $g(x,y)$:

$$g(x,y) = \frac{\partial \varphi}{\partial y} = \frac{\partial}{\partial y}\left[\int f(x,y)\,dx\right] + k'(y)$$

If $\overline{F}$ is in fact a gradient field, then all the x terms will cancel out of this equation, and you will be able to find $k(y)$ by an integration with respect to y. We illustrate in ...

<u>Example C.</u> Show that $\overline{F}(x,y) = (2xy + e^x)\,\overline{i} + x^2\,\overline{j}$ is conservative on the whole xy-plane.

<u>Solution.</u> We have only to prove that $\overline{F}$ is a gradient field. If $\nabla\varphi = F$, then to compute φ we must solve the partial differential equations

$$\frac{\partial \varphi}{\partial x} = 2xy + e^x \quad \text{and} \quad \frac{\partial \varphi}{\partial y} = x^2$$

Integrating the first equation with respect to x yields

$$\varphi(x,y) = x^2 y + e^x + k(y)$$

the "constant of integration" is a function $k(y)$ of y

Thus

$$x^2 = \frac{\partial \varphi}{\partial y} = x^2 + k'(y)$$

the derivative of the function $k(y)$ with respect to y

so that all the x-terms cancel leaving $k'(y) = 0$. This proves $k(y)$ is a constant C, and $\overline{F} = \nabla \varphi$ where

$$\varphi(x,y) = x^2 y + e^x + C$$

(We will solve this problem in another way in Example E below.) □

If you try this procedure with a non-conservative $\overline{F}$, then the equation for $k'(y)$ will involve non-cancelling x-terms, and hence cannot be solved. For example, with

$$\overline{F}(x,y) = (2xy + e^x)\overline{i} + x\overline{j}$$

you would obtain

$$x = \frac{\partial \varphi}{\partial y} = x^2 + k'(y)$$

which cannot be satisfied by any function $k(y)$ since $k(y)$ must be independent of x!

<u>Example D.</u>   Evaluate $\displaystyle\int_C (2xy + e^x)\,dx + x^2\,dy$ where C is the curve parameterized by

$$\overline{r}(t) = (t + \sin t \ln(1+t))\overline{i} - t \cos t\,\overline{j}, \quad 0 \le t \le \pi \ .$$

<u>Brilliant solution</u> (as opposed to the merely intelligent solution given above in Example B).

$$\int_C (2xy + e^x)\, dx + x^2\, dy = \int_C \nabla \varphi \cdot d\overline{r}$$

where $\varphi(x,y) = x^2 y + e^x$ from Example C.

Since the endpoints of C are $(0,0)$ and (π, π), then

$$\int_C (2xy + e^x)\, dx + x^2\, dy = \varphi(\pi, \pi) - \varphi(0,0) \quad \text{by the Fundamental Theorem}$$
of Line Integrals

$$= \pi^3 + e^\pi - 1 \qquad \square$$

4. <u>Simply connected regions.</u> Before considering irrotational vector-valued functions, it will be useful to expand upon Anton's discussion of simply connected regions. First we define what is meant by a connected region:

Connected Regions

Suppose U is an open region in the xy-plane.

Then U is said to be <u>connected</u> if any two points

(x_0, y_0) and (x_1, y_1) in U can be

joined by a smooth curve C lying

entirely in U.

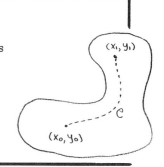

Loosely, a connected region comes in "just one piece;" it cannot be split up into the union of two open, disjoint subregions.

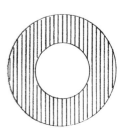

NOT A CONNECTED REGION

There is, however, one problem that connected sets can exhibit: they can have "interior holes." For example, the region between two concentric circles is connected, but it clearly has an "interior hole."

A simply connected region is just a connected region without any interior holes. It is rigorously defined as follows:

> Suppose U is an open region in the xy-plane.
>
> Then U is said to be simply connected if
>
> 1. U is connected, and
>
> 2. if C is a simple* closed curve
>
>   in U, then the interior of C
>
>   is also contained in U.

The second condition eliminates the possibility of interior holes in U. For example, consider the region U between two concentric circles. To show this is not simply connected, merely take C to be any simple closed curve which surrounds the inner circle, as shown to the right. The interior of C contains the "hole," and thus does not lie in U, proving U is not simply connected.

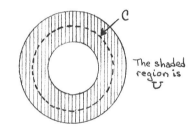

The shaded region is U

Anton's definition of simply connected region is more restrictive than the one we have given. While it is true that any region which is simply connected according to Anton's definition ("the boundary consists of one simple closed curve") is also simply connected according to our definition, the reverse is not the case. For example, the region consisting of the whole xy-plane and the region between two parallel lines are simply connected according to our definition, but not Anton's.

---

* A curve is simple if it does not intersect itself anywhere between its endpoints. See Anton's Figure 18.2.3 for some illustrations.

These and similar regions occur quite often in physics and engineering, and it is important to realize they are simply connected (so the results of the next subsection on irrotational vector-valued functions will apply to them)! This is why we have elaborated at such length on simply connectedness.

5.   Irrotational vector-valued functions.   There are numerous instances in which we wish to verify that a vector-valued function $\overline{F}(x,y)$ is conservative without going to the work of computing a potential function $\varphi(x,y)$. Moreover, in some cases the computation of $\varphi(x,y)$ might be too difficult to carry out.   In those cases we need a simple way to determine when $\overline{F}(x,y)$ is conservative!

The solution (at least when $\overline{F}$ has continuous partial derivatives and U is an open, simply connected region) lies with our fourth equivalent property of conservative vector-valued functions.   A vector-valued function

$$\overline{F}(x,y) = f(x,y)\,\overline{i} + g(x,y)\,\overline{j}$$

is said to be irrotational on an open region U if

$$\boxed{\dfrac{\partial g}{\partial x} = \dfrac{\partial f}{\partial y} \quad \text{at every point in } U}$$

If $\overline{F}$ is a conservative vector field (i. e., circulation-free; path-independent; a gradient field) and if the partials of f and g are continuous, then $\overline{F}$ is easily seen to be irrotational whether or not U is simply connected. Anton proves this in Theorem 18. 2. 3. The converse of the result (i. e., irrotational implies conservative) is also true provided that the open region U is simply connected! The proof is not elementary, requiring Green's Theorem, a result to be studied in the next section; the proof will be given in §18. 3. 4 of

The Companion. These statements finally complete the Conservative Vector Fields

Theorem.

Example E. Show that $\overline{F}(x,y) = (2xy + e^x)\overline{i} + x^2\overline{j}$ is conservative on the whole xy-plane.

Slick solution (as compared with the solution given in Example C). The functions

$$f(x,y) = 2xy + e^x$$

$$g(x,y) = x^2$$

certainly have continuous first partial derivatives, and the whole xy-plane is simply

connected. Thus we have only to check that $\overline{F}$ is irrotational:

$$\frac{\partial g}{\partial x} = 2x$$

$$\frac{\partial f}{\partial y} = 2x$$

BINGO! Thus $\overline{F}$ is irrotational, and hence it is also conservative.

6. A procedure and some examples. So ... when confronted by a horrendous-looking line

integral $\int_C \overline{F} \cdot d\overline{r}$ which needs evaluation, what should you do? No, you don't panic!

Instead, you use the following nifty procedure:

Step 1. Is $\overline{F}$ irrotational, i.e., is $\frac{\partial g}{\partial x} = \frac{\partial f}{\partial y}$?

Nifty procedure for evaluating $\int_C \overline{F} \cdot d\overline{r}$

If NO, then $\overline{F}$ is not conservative, so the techniques of this section

do not apply. You will either have to perform a direct evaluation

(as in §18. 1) or use Green's Theorem (to be studied in §18. 3).

If YES, then go to Step 2.

Step 2.   Is C contained in an open, <u>simply connected</u> region U in

the domain of $\overline{F}$?

If NO, then the irrotational property of $\overline{F}$ is not enough

to prove $\overline{F}$ is conservative on an appropriate region

U ; go to Step $2\frac{1}{2}$.

If YES, then $\overline{F}$ is conservative on an appropriate

region U ; go to Step 3 .

Step $2\frac{1}{2}$.   Is $\overline{F}$ a gradient field on an open set U containing C ?

Check this by trying to find a potential function φ .

If NO, then $\overline{F}$ is not conservative on an appropriate

region U , so the techniques of this section do not

apply. You will either have to perform a direct

evaluation (as in §18. 1) or use Green's Theorem

(to be studied in §18. 3).

If YES, then $\overline{F}$ is conservative on an appropriate

region U ; go to Step 3 .

Step 3. $\overline{F}$ is conservative on an open set U containing C.

Hence there are three possible ways to evaluate $\int_C \overline{F} \cdot d\overline{r}$:

1. Since $\overline{F}$ is _circulation-free_, then

$$\oint_C \overline{F} \cdot d\overline{r} = 0 \quad \underline{\text{if}} \quad C \quad \underline{\text{is closed}};$$

2. Since $\overline{F}$ is a _gradient field_, then finding a potential

function φ will allow $\int_C \overline{F} \cdot d\overline{r}$ to be evaluated

by the Fundamental Theorem for Line Integrals

3. Since $\overline{F}$ has _path-independent_ line integrals in U ,

then C can be replaced by a simpler curve C'

in U , and $\int_{C'} \overline{F} \cdot d\overline{r}$ can then be evaluated from the

definition.

Anton's Example 4 uses this method. Here are some other examples designed to illustrate a broad range of possibilities:

Example F. Evaluate $\int_C \dfrac{x}{x^2 + y^2} \, dx + \dfrac{y}{x^2 + y^2} \, dy$ where C is the curve parameterized

by $\overline{r}(t) = t^3 \overline{i} + e^t \overline{j}$, $-1 \le t \le 1$.

Solution.

Step 1.   Irrotational?

$$\frac{\partial g}{\partial x} = \frac{\partial}{\partial x}\left(\frac{y}{x^2 + y^2}\right) = \frac{(x^2 + y^2)(0) - y(2x)}{(x^2 + y^2)^2} = -\frac{2xy}{(x^2 + y^2)^2} \longleftarrow$$

equal

$$\frac{\partial f}{\partial y} = \frac{\partial}{\partial y}\left(\frac{x}{x^2 + y^2}\right) = \frac{(x^2 + y^2)(0) - x(2y)}{(x^2 + y^2)^2} = -\frac{2xy}{(x^2 + y^2)^2} \longleftarrow$$

Hence $\overline{F}$ is irrotational whenever $(x,y) \neq (0,0)$

Step 2.   Simply connected region? Since $y(t) = e^t > 0$, then C is contained in

the upper half-plane $U = \{(x,y) \mid y > 0\}$ which is a simply connected region in

the domain of $\overline{F}$. Thus $\overline{F}$ is conservative on U , an "appropriate" region in

the domain of $\overline{F}$.

Step 3.   Evaluate using conservative properties. Since C is not closed, we cannot

conclude that the line integral is zero. Instead we look for a potential function:

$$\frac{\partial \varphi}{\partial x} = \frac{x}{x^2 + y^2} \quad \text{and} \quad \frac{\partial \varphi}{\partial y} = \frac{y}{x^2 + y^2}$$

Integrating the first equation with respect to x gives

$$\varphi(x,y) = \int \frac{x\,dx}{x^2 + y^2} = \frac{1}{2}\int \frac{du}{u} = \frac{1}{2}\ln u + k(y)$$

$$\boxed{\begin{array}{l} u = x^2 + y^2 \\ du = 2x\,dx \end{array}} \qquad = \frac{1}{2}\ln(x^2 + y^2) + k(y)$$

Then

$$\frac{y}{x^2 + y^2} = \frac{\partial \varphi}{\partial y} = \frac{\partial}{\partial y}\left[\frac{1}{2}\ln(x^2 + y^2) + k(y)\right]$$

$$= \frac{y}{x^2 + y^2} + k'(y)$$

Thus $k'(y) = 0$, proving $k(y)$ is a constant. We'll take $k(y) = 0$ so that

$$\varphi(x,y) = \frac{1}{2} \ln (x^2 + y^2)$$

The line integral can now be evaluated by the Fundamental Theorem for Line Integrals:

$$\int_C \frac{x}{x^2 + y^2} \, dx + \frac{y}{x^2 + y^2} \, dy = \varphi(1, e) - \varphi(-1, 1/e)$$

since $(-1, 1/e)$ and $(1, e)$ are the endpoints of C

$$= \frac{1}{2} \ln (1 + e^2) - \frac{1}{2} \ln (1 + e^{-2}) \qquad \square$$

<u>Example G.</u>   Evaluate $\displaystyle\int_C \frac{x}{x^2 + y^2} \, dx + \frac{y}{x^2 + y^2} \, dy$ where C is the ellipse

$\dfrac{(x - 1)^2}{4} + y^2 = 1$ parameterized counterclockwise.

<u>Solution.</u>

<u>Step 1.</u>   <u>Irrotational?</u> Yes, since this is the same $\overline{F}$ as in the previous example.

<u>Step 2.</u>   <u>Simply connected region?</u> In this case, there is no simply connected region U

containing C which is in the domain of $\overline{F}$.

The reason is simple: If U is simply connected

and contains C , then U contains the interior

of C . But the interior of C contains $(0,0)$

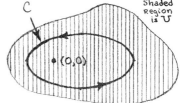

which is not in the domain of $\overline{F}$. So there is no appropriate <u>simply connected</u> region

U containing C .

Step $2\frac{1}{2}$.   Gradient field?   Even though the irrotational test for a conservative vector-valued

function failed, the computation in Example F shows that we do have a potential function

$$\varphi(x,y) \; = \; \frac{1}{2} \, \ln \, (x^2 + y^2)$$

which is valid on the full domain of $\overline{F}$, i.e., all $(x,y) \neq (0,0)$. Hence $\overline{F}$ is

conservative on an appropriate region containing C.

Step 3.   Evaluate using conservative methods.   Since C is a closed curve, then

$$\oint_C \frac{x}{x^2 + y^2} \, dx + \frac{y}{x^2 + y^2} \, dy \; = \; 0$$

$\square$

Example H.   Evaluate $\displaystyle\int_C \frac{y}{x^2 + y^2} \, dx - \frac{x}{x^2 + y^2} \, dy$ where C is the unit circle

parameterized counterclockwise.

Solution.   This line integral is simple enough to integrate directly. However, we will use our

procedure because we wish to make an important point.

Step 1.   Irrotational?

$$\frac{\partial g}{\partial x} \; = \; \frac{\partial}{\partial x} \left[- \frac{x}{x^2 + y^2} \right] \; = \; \frac{(x^2 + y^2)(-1) \, - \, (-x)(2x)}{(x^2 + y^2)^2} \; = \; \frac{x^2 - y^2}{(x^2 + y^2)^2} \; \longleftarrow$$

equal

$$\frac{\partial f}{\partial y} \; = \; \frac{\partial}{\partial y} \left[\frac{y}{x^2 + y^2} \right] \; = \; \frac{(x^2 + y^2)(1) \, - \, (y)(2y)}{(x^2 + y^2)^2} \; = \; \frac{x^2 - y^2}{(x^2 + y^2)^2} \; \longleftarrow$$

Thus $\overline{F}$ is irrotational.

Step 2.   Simply connected region?   For the same reasons as in Example G, there is no

appropriate simply connected region U containing C.

Step $2\frac{1}{2}$. __Gradient field?__ If there is a potential function φ, then

$$\frac{\partial \varphi}{\partial x} = \frac{y}{x^2 + y^2} \quad \text{and} \quad \frac{\partial \varphi}{\partial y} = -\frac{x}{x^2 + y^2}$$

Thus integrating the first equation with respect to x (treating y as a constant!) gives

$$\varphi(x,y) = \int \frac{y\, dx}{x^2 + y^2} = \int \frac{y^2\, du}{u^2 y^2 + y^2} = \int \frac{du}{u^2 + 1}$$

$$\boxed{\begin{array}{l} u = x/y \\ du = dx/y \\ \text{(assuming} \\ \quad y \neq 0) \end{array}} = \tan^{-1} u + k(y)$$

$$= \tan^{-1}\left(\frac{x}{y}\right) + k(y)$$

Then

$$-\frac{x}{x^2 + y^2} = \frac{\partial}{\partial y}\left[\tan^{-1}\left(\frac{x}{y}\right) + k(y)\right] = -\frac{x}{x^2 + y^2} + k'(y)$$

so as in Example H we can take $k(y) = 0$ to give

$$\varphi(x,y) = \tan^{-1}\left(\frac{x}{y}\right)$$

But wait a minute! This potential function is only valid when $y \neq 0$, i.e., off of the x-axis! Since C crosses the x-axis twice, then the potential function does not apply to C!

There are other potential functions for $\overline{F}$, but __all__ of them will suffer a defect just as $\tan^{-1}(x/y)$ does. For example, $-\tan^{-1}(y/x)$ is also a potential function ... but it is not valid on the y-axis!

This might make you suspect that $\overline{F}$ is not conservative on any appropriate region U. In fact, direct evaluation proves this as follows:

Using $\bar{r}(t) = \cos t\,\bar{i} + \sin t\,\bar{j}$, $0 \le t \le 2\pi$, gives

$$\int_C \frac{y}{x^2 + y^2}\,dx - \frac{x}{x^2 + y^2}\,dy = \int_0^{2\pi} \left[\frac{\sin t}{\sin^2 t + \cos^2 t}(-\sin t) - \frac{\cos t}{\sin^2 t + \cos^2 t}(\cos t) \right] dt$$

$$= \int_0^{2\pi} (-\sin^2 t - \cos^2 t)\,dt = -\int_0^{2\pi} dt = -2\pi$$

Since this integral is not zero, then $\bar{F}$ is not circulation-free on any open region

containing C, and hence is not conservative on any open region containing C.   □

The Moral of Example H: When a gradient function φ is computed, be sure
to check if it is valid on an open set containing the given curve C!

7.   Conservative vector fields and physics (Optional).   Suppose $\bar{F}(x,y)$ represents a force

field. Then the line integral $\displaystyle\int_C \bar{F} \cdot d\bar{r}$ represents the work done by the force in moving an

object along the curve C. Since a basic principle of physics says the "work done" on an

object equals the "change in kinetic energy" of the object, then in a conservative force field $\bar{F}$

it is not possible to extract any kinetic energy from the field by traversing a closed loop! In

other words, a circulation-free force field obeys a conservation of energy principle!

On the other hand, suppose $\bar{F}(x,y)$ is a "non-static electric field." Such a field is

not conservative, and hence it is possible to obtain a continuous or repeated exchange of energy

by traversing a closed loop. This is what puts the electric company into business!

The potential function φ of a gradient field $\bar{F} = \nabla\varphi$ contributes to a more general

conservation of energy principle. If $\bar{F}$ represents a force field, then $\bar{V}(x,y) = -\varphi(x,y)$

is defined to be the potential energy of the field $\bar{F}$ at the point (x,y). The total energy of an object moving under the influence of $\overline{F}$ is defined to be the sum of its kinetic and potential

energies, and it can be shown that this total energy is always a constant!

*   *   *   *   *

The terms "circulation-free" and "irrotational" both arise from applications of

conservative vector fields in fluid mechanics. Suppose $\overline{F}$ is the "fluid flow" of a two-

dimensional moving liquid, i. e. , at each point (x, y)

of the liquid, the vector $\overline{F}(x, y)$ points in the direction

of motion and has a magnitude equal to the rate at which

fluid is crossing the point (x, y). (Units for $\overline{F}(x, y)$

could be, for example, $(gm/cm^2)/sec.$) Take any

closed curve C in the fluid. Then the line integral

$$\oint_C \overline{F} \cdot d\overline{r}$$

can be interpreted as measuring the rate at which fluid circulates about C. For example,

if there is a whirlpool in the center of C, then the value of the line integral will probably

be very large. It is therefore natural to attach the name circulation-free to any vector field

$\overline{F}(x, y)$ for which $\oint_C \overline{F} \cdot d\overline{r} = 0$ for all closed curves C in a region U.

For similar reasons the quantity

$$curl \overline{F}(x, y) = \frac{\partial g}{\partial x} - \frac{\partial f}{\partial y}$$

(called the scalar curl of $\overline{F}$) can be interpreted as measuring the rate of rotation of the fluid

about the point (x, y). It is therefore natural to attach the name irrotational to any vector field

$\overline{F}(x, y)$ for which $curl \overline{F}(x, y) = 0$ at every point (x, y).

More will be said about fluid flow in the next section.

Section 18. 3.   Green's Theorem

1.   <u>A strange looking result ...</u>!   That's the typical **first reaction** to Green's Theorem (Theorem

18. 3. 1):

Suppose   R   is a region bounded by a simple closed curve*   C

traversed counterclockwise.   If $f(x,y)$ and $g(x,y)$ have

continuous first partial derivatives on an open set containing

R ,   then

$$\oint_C f(x,y)\, dx \; + \; g(x,y)\, dy \; = \; \iint_R \left(\frac{\partial g}{\partial x} \; - \; \frac{\partial f}{\partial y} \right) dA$$

Green's
Theorem

Thus, if $\overline{F}(x,y)$ is a vector-valued function on a simple closed curve   C   and $\overline{F}$ is

continuously differentiable on   C   <u>and its interior</u>, then the line integral $\oint_C \overline{F} \cdot d\overline{r}$

can be expressed as a certain double integral over the interior of the curve.

Green's Theorem is very similar to the Fundamental Theorem of Calculus (FTC)

in that an integral is "undoing" a differentiation: the <u>double</u> integral of the "derivative"

$\dfrac{\partial g}{\partial x} \; - \; \dfrac{\partial f}{\partial y}$ becomes a <u>single</u> integral (more precisely, a line integral) of the vector-valued

function $\overline{F}(x,y) = f(x,y)\,\overline{i} \; + \; g(x,y)\,\overline{j}$.   This similarity is not merely superficial, for Green's

Theorem, the FTC , and even the Fundamental Theorem for Line Integrals are all subcases

of a major theorem known as the <u>Generalized Stokes' Theorem</u>.   Unfortunately this theorem is

much too complicated to describe here; suffice it to say that its theme is an   "n-fold integral"

"undoes" a derivative to become an   "(n - 1)-fold integral."

---

*   As in the previous section, all curves in this section will be piecewise smooth.

2. **The proof of Green's Theorem for a rectangle (Optional).** The fundamental simplicity of the proof of Green's Theorem is hidden by notational complexity and, when C is a very general curve, by technical complications. If C is taken to be a rectangle as shown to the right, then all technical complications disappear, and the notational complexity is greatly reduced (... although not eliminated!). The only substantial part of the proof is the application of the FTC!

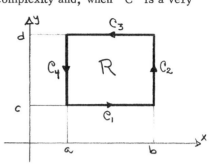

We start with the double integral

$$\iint_R \left(\frac{\partial g}{\partial x} - \frac{\partial f}{\partial y} \right) dA = \iint_R \frac{\partial g}{\partial x} dA - \iint_R \frac{\partial f}{\partial y} dA$$

Focusing on the first integral we obtain

$$\iint_R \frac{\partial g}{\partial x} dA = \int_c^d \left[\int_a^b \frac{\partial g}{\partial x} dx \right] dy \qquad \text{by writing the double integral as an iterated integral}$$

$$\boxed{\begin{array}{c} \text{THE} \\ \text{SUBSTANTIAL} \\ \text{STEP!!} \end{array}} \quad = \quad \int_c^d g(x,y) \Big]_{x=a}^{x=b} dy \qquad \begin{array}{l} \text{by using the FTC on} \\ \frac{\partial g}{\partial x}(x,y) \text{ while treating} \\ y \text{ as a constant} \end{array}$$

Thus

$$\boxed{\iint_R \frac{\partial g}{\partial x} dA = \int_c^d g(b,y) \, dy - \int_c^d g(a,y) \, dy}$$

We claim this quantity equals the sum of the two line integrals

$$\int_{C_2} [f(x,y)\, dx + g(x,y)\, dy] + \int_{C_4} [f(x,y)\, dx + g(x,y)\, dy]$$

To prove this we evaluate the line integrals, where

C_2 is parameterized by $\bar{r}_2(t) = b\bar{i} + t\bar{j}, \qquad c \le t \le d$

C_4 is parameterized by $\bar{r}_4(t) = a\bar{i} - t\bar{j}, \qquad -d \le t \le -c$

(See the diagram above.) Then

$$\int_{C_2} [f\, dx + g\, dy] + \int_{C_4} [f\, dx + g\, dy]$$

$$= \int_c^d [f(b,t) \cdot 0 + g(b,t) \cdot 1]\, dt + \int_{-d}^{-c} [f(a,-t) \cdot 0 + g(a,-t) \cdot (-1)]\, dt$$

$$= \int_c^d g(b,t)\, dt - \int_{-d}^{-c} g(a,-t)\, dt$$

$$= \int_c^d g(b,y)\, dy - \int_c^d g(a,y)\, dy$$

$\qquad$ (using $y = t$) $\qquad$ (using $y = -t$)

Hence, from the boxed equality above, we obtain

$$\boxed{\iint_R \frac{\partial g}{\partial x}\, dA = \int_{C_2} [f\, dx + g\, dy] + \int_{C_4} [f\, dx + g\, dy]}$$

Similar computations show

$$\boxed{-\iint_R \frac{\partial f}{\partial y}\, dA = \int_{C_1} [f\, dx + g\, dy] + \int_{C_3} [f\, dx + g\, dy]}$$

Adding these results together yields

$$\iint_R \left(\frac{\partial g}{\partial x} - \frac{\partial f}{\partial y} \right) dA = \left(\int_{C_1} + \int_{C_2} + \int_{C_3} + \int_{C_4} \right) (f\,dx + g\,dy)$$

$$= \oint_C f\,dx + g\,dy$$

which is the desired conclusion of Green's Theorem.

3. Calculations with Green's Theorem.   We will list a number of common ways in which Green's

Theorem is used in computational situations:

(1)   Many line integrals over closed curves

are more easily evaluated as the double integrals

that are obtained from Green's Theorem.   This

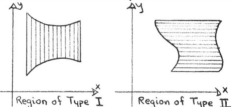

is particularly true if the curve   C   bounds a

region of Type I , Type II or polar type

(see §17.2 and the diagram to the right).

Anton illustrates this in Examples 1 and 2.

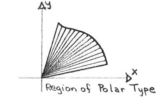

Here is another example:

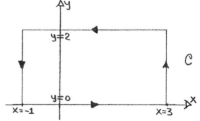

Example A.   Evaluate $\oint_C 3\,x\,y\,dx + x^2\,dy$

where   C   is the rectangle shown to the right.

Solution.   Checking to see if $\overline{F}(x,y) = 3\,x\,y\,\overline{i} + x^2\,\overline{j}$ is irrotational*, we find

$\frac{\partial g}{\partial x} - \frac{\partial f}{\partial y} = \frac{\partial}{\partial x}(x^2) - \frac{\partial}{\partial y}(3\,x\,y) = 2x - 3x = -x$.   Since this is not zero, then $\overline{F}$

---

* $\overline{F} = f\,\overline{i} + g\,\overline{j}$ is irrotational if $\frac{\partial g}{\partial x} - \frac{\partial f}{\partial y} = 0$ (see §18.2 of The Companion).

cannot be conservative, and hence we begin a direct evaluation of the line integral .... But

wait a minute ... might Green's Theorem help us out? The functions $3xy$ and x^2 clearly

have continuous first partial derivatives on the interior of C, and hence Green's Theorem

applies:

$$\oint_C 3xy\,dx + x^2\,dy = \iint_R \left[\frac{\partial}{\partial x}(x^2) - \frac{\partial}{\partial y}(3xy)\right]dA$$

$$= \iint_R (-x)\,dA = \int_{-1}^{3}\int_{0}^{2}(-x)\,dy\,dx = \int_{-1}^{3}(-2x)\,dx$$

$$= -x^2\Big|_{-1}^{3} = -9 + 1 = \boxed{-8}$$

A direct evaluation of the line integral in this example would not be difficult, but it would be

tedious: it would require four separate evaluations, one for each side of the rectangle.   □

Thus Green's Theorem provides us with a "fancy" tool to evaluate line integrals of

<u>non-conservative</u> vector-valued functions $\overline{F}$ over simple closed curves C ... at least

when $\overline{F}$ is differentiable on the interior of C !

(2)   More generally, suppose C_1 and C_2 are two simple closed curves with the same

direction, C_2 is inside of C_1,* and R is the region between them:

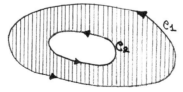

_____

*
   Actually we can allow C_1 and C_2 to cross over each other!

If $\dfrac{\partial g}{\partial x} - \dfrac{\partial f}{\partial y} = 0$ on an open set containing R, then Green's Theorem can be used to prove

$$\boxed{\int_{C_1} f\,dx + g\,dy = \int_{C_2} f\,dx + g\,dy}$$

The proof uses the trick of attaching C_1 to C_2 via a connecting curve C_3. We can now form a new closed curve as follows: trace around C_1, then C_3, then $-C_2$ (i.e., C_2 in the opposite direction) and then finally $-C_3$. The advantage of C over C_1 or C_2 is that <u>its interior is the region R</u> on which

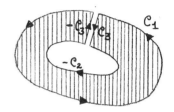

$\dfrac{\partial g}{\partial x} - \dfrac{\partial f}{\partial y} = 0$. Hence

$$0 = \iint_R 0\,dA = \iint_R \left(\dfrac{\partial g}{\partial x} - \dfrac{\partial f}{\partial y} \right) dA = \oint_C f\,dx + g\,dy$$

$$\underbrace{}_{\boxed{\text{Green's Theorem}}}$$

$$= \left(\int_{C_1} + \int_{C_3} - \int_{C_2} - \int_{C_3} \right)(f\,dx + g\,dy)$$

$$= \int_{C_1} f\,dx + g\,dy - \int_{C_2} f\,dx + g\,dy$$

which proves the boxed result above.

This result is just like path-independence for conservative vector-valued functions $\overline{F}$ except that we do not need $\overline{F}$ to be conservative on the interior of the inner curve C_2!

Example B.   Compute $\displaystyle\int_{C_1} \frac{y}{x^2 + y^2}\, dx - \frac{x}{x^2 + y^2}\, dy$ where C_1 is the circle

$(x - 1)^2 + y^2 = 9$ parameterized counterclockwise.

Solution.   The function $\overline{F}(x,y) = \dfrac{y}{x^2 + y^2}\, \overline{i} - \dfrac{x}{x^2 + y^2}\, \overline{j}$ was shown in Example H of

§18. 2 of The Companion to be irrotational everywhere except at $(0,0)$. Thus $\overline{F}$ is not

conservative on the interior of C_1 .

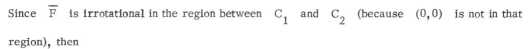

However, in that same example it was shown

that

$$\oint_{C_2} \frac{y}{x^2 + y^2}\, dx - \frac{x}{x^2 + y^2}\, dy = -2\pi$$

where C_2 is the unit circle parameterized counterclockwise.

Since $\overline{F}$ is irrotational in the region between C_1 and C_2 (because $(0,0)$ is not in that

region), then

$$\oint_{C_1} \frac{y}{x^2 + y^2}\, dx - \frac{x}{x^2 + y^2}\, dy = \oint_{C_2} \frac{y}{x^2 + y^2}\, dx - \frac{x}{x^2 + y^2}\, dy$$

$$= -2\pi \qquad \qquad \square$$

(3)   At times we can turn things around and use Green's Theorem to evaluate a double

integral by converting it into a line integral. This is, of course, a tricky proposition because,

given a double integral

$$\iint_R h(x,y)\, dA$$

you need to find a vector-valued function $\overline{F} = f\,\overline{i} + g\,\overline{j}$ for which

$$h(x,y) = \frac{\partial g}{\partial x} - \frac{\partial f}{\partial y}$$

This may be more trouble than it's worth unless R is a region which is described nicely as

the interior of a smooth simple closed curve C.

One application is in finding the area of a

region R bounded by a simple closed curve C.

As Anton shows,

$$\text{area (R)} = \iint_R 1\, dA = \frac{1}{2} \int_C -y\, dx + x\, dy$$

Anton illustrates this result in Example 3. Here is another illustration:

Example C. Find the area in the first quadrant enclosed by the curves $xy = 2$ and

$x^2 + y^2 = 5$.

Solution. We must first find the intersection points of the two curves:

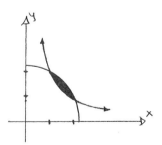

$$5 = x^2 + y^2 = x^2 + \left(\frac{2}{x}\right)^2$$

$$x^4 - 5x^2 + 4 = 0$$

$$(x^2 - 1)(x^2 - 4) = 0$$

$$x^2 = 1 \quad \text{or} \quad 4$$

$$x = 1 \quad \text{or} \quad 2 \quad \text{(since we are in} \\ \text{the first quadrant)}$$

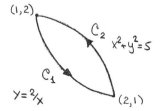

We now parameterize the two curves C_1 and

C_2 shown to the left. The parameterizing

functions may be taken to be

$$\bar{r}_1(t) = t\bar{i} + \frac{2}{t}\bar{j} \ , \quad 1 \le t \le 2$$

$$\bar{r}_2(t) = \sqrt{5} \ \cos t\,\bar{i} + \sqrt{5} \ \sin t\,\bar{j} \ , \quad a \le t \le b$$

$$\text{where} \quad a = \sin^{-1}\left(\frac{1}{\sqrt{5}}\right) \quad \text{and} \quad b = \sin^{-1}\left(\frac{2}{\sqrt{5}}\right)$$

With this parameterization, the area we desire is inside the curve formed by tracing C_1 and then tracing C_2. Thus

$$\text{Area} = \frac{1}{2}\int_{C_1} -y\,dx + x\,dy + \frac{1}{2}\int_{C_2} -y\,dx + x\,dy$$

$$= \frac{1}{2}\int_1^2 \left[-\left(\frac{2}{t}\right)(1) + t\left(-\frac{2}{t^2}\right)\right] dt$$

$$+ \frac{1}{2}\int_a^b \left[\,(-\sqrt{5}\ \sin t)(-\sqrt{5}\ \sin t) + (\sqrt{5}\ \cos t)(\sqrt{5}\ \cos t)\,\right] dt$$

$$= -2\int_1^2 \frac{dt}{t} + \frac{5}{2}\int_a^b dt$$

$$= -2\ln t\,\Big]_1^2 + \frac{5}{2}\,t\,\Big]_a^b = -2\ln 2 + \frac{5}{2}\,(b - a)$$

$$= -2\ln 2 + \frac{5}{2}\left(\sin^{-1}\left(\frac{2}{\sqrt{5}}\right) - \sin^{-1}\left(\frac{1}{\sqrt{5}}\right)\right) \approx .2225 \qquad \square$$

4. <u>Conservative vector fields revisited (Optional).</u> The Conservative Vector Fields Theorem

of §18.2 of <u>The Companion</u> is essentially a list of four equivalent properties of a conservative

vector field. Perhaps the most important part of that theorem is the following:

> Suppose $\overline{F}$ is a vector function with continuous first partial
>
> derivatives on a <u>simply-connected</u> open set U.
>
> If $\overline{F}$ is <u>irrotational</u> on U, then F is <u>circulation-free</u> on U.

(Recall that $\overline{F}(x,y) = f(x,y)\overline{i} + g(x,y)\overline{j}$ is irrotational on U if $\dfrac{\partial g}{\partial x} = \dfrac{\partial f}{\partial y}$ at every point

in U, and $\overline{F}$ is circulation-free if $\int_C \overline{F} \cdot d\overline{r} = 0$ for every closed curve C in U.)

This result is important because it gives an easy condition to check (Is $\dfrac{\partial g}{\partial x} = \dfrac{\partial f}{\partial y}$?) to test

whether a vector function $\overline{F}$ is conservative.

This result is not elementary, however: <u>it requires Green's Theorem,</u> as we now show.

With $\overline{F}$ and U as in the theorem statement (i.e., $\overline{F}$ is irrotational on U), take C

to be any simple closed curve in U. Since U is simply connected, the interior R of C

is contained in U (see §18.2.4 of <u>The Companion</u>). Hence Green's Theorem applies to

the line integral of $\overline{F}$ over C, and we obtain

$$\oint_C \overline{F} \cdot d r = \oint_C f(x,y)\,dx + g(x,y)\,dy$$

$$= \iint_R \left(\frac{\partial g}{\partial x} - \frac{\partial f}{\partial y} \right) dA$$

$$= \iint_R o\,dA \quad \text{since} \quad \frac{\partial g}{\partial x} = \frac{\partial f}{\partial y}$$

$$= 0$$

Thus $\overline{F}$ is circulation-free on U, at least with respect to all <u>simple</u> closed curves in U.

Using this result one can then show that $\overline{F}$ is a gradient field in U, which in turn will

show that $\overline{F}$ is circulation-free on U with respect to <u>all</u> closed curves in U. We will,

however, omit the details.

<u>Section 18.4</u>:   Introduction to Surface Integrals

1.   <u>The definition of the surface integral.</u>   Anton introduces the surface integral through

its simplest interpretation as the <u>mass of a lamina</u>

(a thin surface) in 3-space.

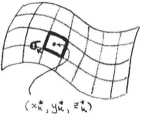

Suppose σ is a lamina in space with a

mass density (i.e., mass per unit area) of

$g(x,y,z)$ at each point (x,y,z) on σ. If

we subdivide σ into n parts $\sigma_1, \sigma_2, \cdots, \sigma_n$

with surface areas $\Delta S_1, \Delta S_2, \ldots, \Delta S_n$, respectively, and pick a point (x_k^*, y_k^*, z_k^*) in

each σ_k, then a good approximation to the total mass of σ will be

$$M = \begin{bmatrix} \text{mass} \\ \text{of } \sigma \end{bmatrix} = \sum_{k=1}^{n} \begin{bmatrix} \text{mass} \\ \text{of } \sigma_k \end{bmatrix} \approx \sum_{k=1}^{n} \underbrace{g(x_k^*, y_k^*, z_k^*) \Delta S_k}$$

i.e., the mass of σ_k
is approximately the
mass density at
(x_k^*, y_k^*, z_k^*) times
the surface area of σ_k.

As n gets larger and larger (and the σ_k get smaller and smaller) the summation should

become a better and better approximation for M. In the limit we should have equality:

$$M = \lim_{n \to +\infty} \sum_{k=1}^{n} g(x_k^*, y_k^*, z_k^*) \, \Delta S_k$$

Now forget the interpretation of $g(x,y,z)$ as the density of a lamina σ; instead, consider $g(x,y,z)$ as any continuous function defined on a surface σ. If the above approximation process produces a finite limit for M, we call this limit the surface integral of $g(x,y,z)$ over the surface σ, and denote it by $\iint_{\sigma} g(x,y,z) \, dS$:

Definition of a Surface Integral

$$\iint_{\sigma} g(x,y,z) \, dS = \lim_{n \to +\infty} \sum_{k=1}^{n} g(x_k^*, y_k^*, z_k^*) \, \Delta S_k$$

From this limiting formula we can see how to interpret the surface integral "infinitesimally."

Step 1. Let $d\sigma$ be an infinitesimal portion of the surface σ, with surface area dS.

Step 2. Pick a point (x,y,z) in $d\sigma$ and form the infinitesimal quantity

$$dM = g(x,y,z) \, dS$$

Step 3. The surface integral of g over σ is then the "sum" (i.e., integral) of all these infinitesimal quantities:

$$\iint_{\sigma} dM = \iint_{\sigma} g(x,y,z) \, dS$$

$d\sigma$

(x,y,z)

$dS = \begin{bmatrix} \text{surface} \\ \text{area of } d\sigma \end{bmatrix}$

This infinitesimal interpretation is useful in understanding the computational formulas

Anton lists in Theorem 18.4.2, as well as some other formulas we'll discuss in our (optional)

§18.4.3.

2. The calculation of surface integrals. There is a computational formula for the surface

integral $\iint_\sigma g(x,y,z)\, dS$ when the surface σ is a piece of the graph of $z = f(x,y)$:

Computation
of a Surface
Integral

Suppose σ is a portion of the graph of

$z = f(x,y)$, where f has continuous

first partial derivatives. Let R be

the projection of σ on the xy-plane.

If $g(x,y,z)$ is a continuous function

on σ, then

$$\iint_\sigma g(x,y,z)\, dS = \iint_R g(x,y,f(x,y))\, \sqrt{\left(\frac{\partial z}{\partial x}\right)^2 + \left(\frac{\partial z}{\partial y}\right)^2 + 1}\;\; dA$$

This is Anton's Theorem 18.4.2(a); it is significant in that it expresses a surface

integral $\iint_\sigma \ldots dS$ in terms of an ordinary double integral $\iint_R \ldots dA$. That's good

because, from Chapter 17, we know how to compute double integrals!

Although this formula may look strange at first, it is not hard to understand if you

"think infinitesimally":

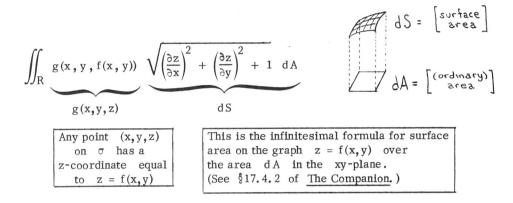

$$\iint_R \underbrace{g(x,y,f(x,y))}_{g(x,y,z)} \underbrace{\sqrt{\left(\frac{\partial z}{\partial x}\right)^2 + \left(\frac{\partial z}{\partial y}\right)^2 + 1}\ dA}_{dS}$$

$$dS = \begin{bmatrix} \text{surface} \\ \text{area} \end{bmatrix}$$

$$dA = \begin{bmatrix} \text{(ordinary)} \\ \text{area} \end{bmatrix}$$

Any point (x,y,z) on σ has a z-coordinate equal to $z = f(x,y)$

This is the infinitesimal formula for surface area on the graph $z = f(x,y)$ over the area dA in the xy-plane. (See §17.4.2 of The Companion.)

Using this formula for a double integral is straightforward, although it can be tedious.

Anton's Examples 4, 5 and 6 illustrate the procedure; here is one more example:

Example A. Evaluate the surface integral

$$\iint_\sigma y\sqrt{z-y}\ dS$$

where σ is that part of the surface $x^2 + y - z = 0$ for which $0 \le z \le 1$ and $0 \le x \le 1$.

Solution. In this example our surface σ is a portion of the graph of

$$z = f(x,y) = x^2 + y$$

and we are integrating the function

$$g(x,y,z) = y\sqrt{z-y}$$

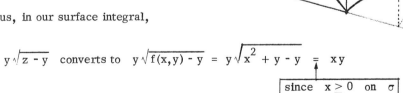

over σ. Thus, in our surface integral,

$$y\sqrt{z-y} \quad \text{converts to} \quad y\sqrt{f(x,y)-y} = y\sqrt{x^2+y-y} = xy$$

since $x \ge 0$ on σ

$$dS \quad \text{converts to} \quad \sqrt{\left(\frac{\partial z}{\partial x}\right)^2 + \left(\frac{\partial z}{\partial y}\right)^2 + 1} \ \ dA$$

$$= \sqrt{(2x)^2 + (1)^2 + 1} \ \ dA$$

$$= \sqrt{4x^2 + 2} \ \ dA$$

$$= \sqrt{2} \ \sqrt{2x^2 + 1} \ \ dA$$

Therefore our surface integral becomes

$$\iint_\sigma y\sqrt{z-y} \ dS = \iint_R xy\sqrt{2} \ \sqrt{2x^2 + 1} \ \ dA$$

where R is the projection of σ on the xy-plane.

Notice that <u>we have reduced the computation of a surface integral to the computation</u> <u>of a double integral!</u> However, it still remains to determine the limits of integration for this double integral. We know that σ is that part of the graph $z = x^2 + y$ for which $0 \le z \le 1$ and $0 \le x \le 1$. Thus, on substituting $x^2 + y$ for z, we have

$$0 \le x^2 + y \le 1 \qquad \text{and} \qquad 0 \le x \le 1$$

$$\boxed{-x^2 \le y \le 1 - x^2 \qquad \text{and} \qquad 0 \le x \le 1}$$

Ah ha! These are a set of limits for a region R of type I (see §17.2 of <u>The Companion</u>) as shown to the right. Thus our integral becomes

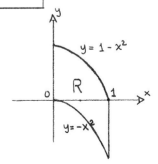

18.4.6

$$\iint_\sigma y\sqrt{z-y}\ dS = \sqrt{2}\ \iint_R xy\sqrt{2x^2+1}\ dA$$

$$= \sqrt{2}\ \int_0^1 \left[\int_{-x^2}^{1-x^2} xy\sqrt{2x^2+1}\ dy\right] dx$$

Inner integral:

$$\int_{-x^2}^{1-x^2} xy\sqrt{2x^2+1}\ dy = x\sqrt{2x^2+1}\ \int_{-x^2}^{1-x^2} y\ dy$$

$$= x\sqrt{2x^2+1}\ \left[y^2/2\right]\Big|_{-x^2}^{1-x^2}$$

$$= x\sqrt{2x^2+1}\ \left[\frac{\left(1-x^2\right)^2}{2} - \frac{\left(-x^2\right)^2}{2}\right]$$

$$= x\sqrt{2x^2+1}\ [(1-2x^2)/2]$$

$$= \tfrac{1}{2}\ x(1-2x^2)\sqrt{2x^2+1}$$

Full integral:

$$\sqrt{2}\ \int_0^1 [\ldots]\ dx = \frac{\sqrt{2}}{2}\ \int_0^1 x(1-2x^2)\sqrt{2x^2+1}\ dx$$

$$= \frac{\sqrt{2}}{2}\ \int_1^3 (2-u)\sqrt{u}\ (du/4)$$

$$= \frac{\sqrt{2}}{8}\ \int_1^3 (2u^{1/2} - u^{3/2})\ du$$

$$= \frac{\sqrt{2}}{8}\ \left[\frac{4u^{3/2}}{3} - \frac{2u^{5/2}}{5}\right]_1^3$$

$$
\boxed{
\begin{aligned}
&u = 2x^2+1\\
&\tfrac{1}{4}\ du = x\ dx\\
&2-u = 1-2x^2\\
&u=1\quad \text{when}\quad x=0\\
&u=3\quad \text{when}\quad x=1
\end{aligned}
}
$$

$$= \frac{\sqrt{2}}{8} \left[4\sqrt{3} - \frac{18\sqrt{3}}{5} - \frac{4}{3} + \frac{2}{5} \right]$$

$$\approx \boxed{-.0425}$$

$\square$

In addition to the case in which σ is a portion of the surface $z = f(x,y)$, Anton's Theorem 18.4.2 considers the analogous cases in which σ is a portion of surfaces of the forms $y = f(x,z)$ and $x = f(y,z)$. The formulas (and their uses) are all identical except that the roles of x, y and z are switched.

3.   Cylindrical and spherical surface integrals (optional).   Anton's Theorem 18.4.2 covers a wide range of surface integrals, but it is awkward to apply to certain very simple surfaces such as cylinders and spheres. For example, consider the unit sphere centered at the origin. To integrate a function $g(x,y,z)$ on this sphere by the methods of Theorem 18.4.2 we would need to break the sphere up into two hemispheres, σ_1 and σ_2, and integrate over each one separately. Moreover, the boundary of each hemisphere would have to be treated in the special way described in Anton's Example 6, and each of the two surface integrals would have numerous unpleasant square root terms.

There are, however, more general methods for computing surface integrals than those given in Theorem 18.4.2. Although a full discussion of these methods would take more space than we can devote to the subject here, we can introduce you to the computation of surface integrals in cylindrical and spherical coordinates. These techniques are the best ways to handle integrals on cylinders or on spheres, two types of surfaces which occur often in

scientific applications of surface integration.

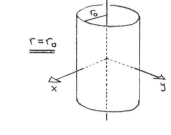

Surface integrals in cylindrical coordinates.

We need a formula for dS, the infinitesimal

change in surface area on the cylinder $r = r_0$ that

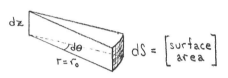

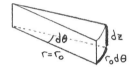

is produced by infinitesimal changes $d\theta$

and dz in the θ and z coordinates.

As we can see in the picture to the left, the

bottom edge of dS has length $r_0 d\theta$,

while the side edge has height dz. Thus

the infinitesimal surface area dS is the

product

$$dS = r_0 \, dz \, d\theta$$

Now we can use our three-step "infinitesimal method" of §18.4.1 of The Companion

to obtain a cylindrical formula for a surface integral:

Step 1. Let $d\sigma$ be an infinitesimal portion of the

cylindrical surface $r = r_0$ corresponding to

the infinitesimal changes $d\theta$ and dz. From

above the surface area of $d\sigma$ is

$$dS = r_0 \, dz \, d\theta$$

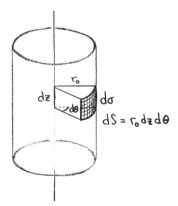

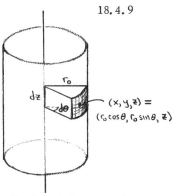

<u>Step 2.</u>   Pick a point $(x, y, z) = (r_0 \cos \theta, r_0 \sin \theta, z)$

   in $d\sigma$ and form the quantity

$$dM = g(x, y, z) \, dS$$

$$= g(r_0 \cos \theta, r_0 \sin \theta, z) \, r_0 \, dz \, d\theta$$

<u>Step 3.</u>   The surface integral of g over σ is then the integral ("sum") of the infinitesimal

   quantities dM:

| Surface integral on $r = r_0$ | $$\iint_\sigma g(x, y, z) \, dS = \iint_\sigma g(r_0 \cos \theta, r_0 \sin \theta, z) \, r_0 \, dz \, d\theta$$ | (A) |

The limits of integration
will be θ and z limits

Although this derivation is merely "intuitively plausible," it can be made rigorous by

using limits of Riemann Sums in place of infinitesimals.

<u>Example B.</u>   Evaluate $\displaystyle\iint_\sigma z\, x^2 \, dS$ where σ is that portion of the cylinder $x^2 + y^2 = 4$

which lies between $z = 0$ and $z = 1$.

<u>Solution.</u>   Although this integral can be evaluated in rectangular coordinates using

Theorem 18.4.2 (see Exercise 12 in Anton's Exercise Set 18.4), it is much easier if we

use our cylindrical surface integral formula:

$$\iint_\sigma z\, x^2 \, dS = \iint_\sigma z(2 \cos \theta)^2 \, (2 \, dz \, d\theta)$$

$x = 2 \cos \theta$
$dS = 2\, dz\, d\theta$

$$= 8 \iint_\sigma z \cos^2 \theta \, dz \, d\theta$$

The θ, z limits for σ
are $0 \le \theta \le 2\pi$,
$0 \le z \le 1$

$$= 8 \int_0^{2\pi} \left[\int_0^1 z \cos^2 \theta \, dz \right] d\theta$$

18.4.10

$$= 8 \int_0^{2\pi} \left[\frac{1}{2} z^2 \cos^2 \theta \right] \Bigg|_{z=0}^{z=1} d\theta$$

$$= 4 \int_0^{2\pi} \cos^2 \theta \, d\theta = 2 \int_0^{2\pi} (1 + \cos 2\theta) \, d\theta$$

$$= 2 \left[\theta + \frac{1}{2} \sin 2\theta \right] \Bigg|_{\theta=0}^{\theta=2\pi} = 2 [2\pi] = \boxed{4\pi} \qquad \square$$

$\rho = \rho_0$

Surface integrals in spherical coordinates

We need a formula for dS, the infinitesimal change in surface area on the sphere $\rho = \rho_0$ that

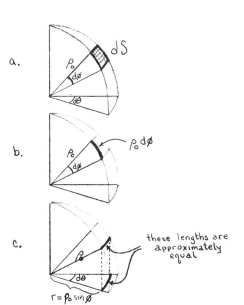

a.

b.

c. these lengths are approximately equal

$r = \rho_0 \sin \phi$

d.

$r \, d\theta = \rho_0 \sin \phi \, d\theta$

is produced by infinitesimal changes $d\theta$ and $d\phi$ in the θ and ϕ coordinates. We can see from Figure (b) to the left that the side edge of dS has length $\rho_0 \, d\phi$. The length of the bottom edge of dS is more complicated to obtain. Projecting it down onto the xy-plane as shown in Figure (c), we see that its length can be approximated by the curve in Figure (d) of length

$$r \, d\theta = \rho_0 \sin \phi \, d\theta$$

(This uses the formula $r = \rho \sin \phi$ of §15.9.) Thus we obtain the infinitesimal surface

area dS by multiplying the lengths of the side and bottom edges:

$$dS = \rho_0^2 \sin\phi \, d\phi \, d\theta$$

Using an "infinitesimal" approach, as we did for cylindrical coordinates, we can use

our expression for dS to obtain the following spherical surface integral formula:

| Surface integral on $\rho = \rho_0$ | $\iint_\sigma g(x,y,z)\, dS = \iint_\sigma g(\rho_0 \cos\theta \sin\phi, \rho_0 \sin\theta \sin\phi, \rho_0 \cos\phi) \rho_0^2 \sin\phi \, d\phi \, d\theta$ | (B) |

The limits of integration
will be θ and ϕ limits

This informal derivation can be made rigorous by the use of Riemann Sums. Here is an

example of the use of the formula:

Example C. Evaluate $\iint_\sigma \sqrt{x^2 + y^2}\, dS$ where σ is that portion of the sphere

$x^2 + y^2 + z^2 = a^2$, $a > 0$, which lies in the first octant and above the cone $x^2 + y^2 = z^2$.

Solution. Since our surface is part of a sphere $\rho = a$, the integral is best evaluated by

using the spherical surface integral formula:

$$\iint_\sigma \sqrt{x^2 + y^2}\, dS = \iint_\sigma r \, dS \qquad \text{since } r^2 = x^2 + y^2$$

$$= \iint_\sigma (a \sin\phi)(a^2 \sin\phi \, d\phi \, d\theta)$$

Using $r = a \sin\phi$
$dS = a^2 \sin\phi \, d\phi \, d\theta$

$$= \int_0^{\pi/2} \left[\int_0^{\pi/4} a^3 \sin^2\phi \, d\phi \right] d\theta$$

The ϕ and θ limits are easily computed
to be $0 \le \phi \le \pi/4$ and $0 \le \theta \le \pi/2$.

$$= a^3 \int_0^{\pi/2} \left[\int_0^{\pi/4} \frac{1 - \cos 2\phi}{2} \, d\phi \right] d\theta$$

$$= \frac{a^3}{2} \int_0^{\pi/2} \left[\phi - \frac{1}{2} \sin 2\phi \right] \Big|_0^{\pi/4} d\theta$$

$$= \frac{a^3}{2} \int_0^{\pi/2} \left[\frac{\pi}{4} - \frac{1}{2} \right] d\theta$$

$$= \frac{a^3}{2} \left[\frac{\pi}{4} - \frac{1}{2} \right] \cdot \frac{\pi}{2} = \boxed{\frac{\pi (\pi - 2)}{16} a^3}$$

If you wish to check your ability to use these cylindrical and spherical surface integral formulas, apply them to Exercises 3, 7, 8, 12, 14, 15, 20 and 22 in Anton's Exercise Set 18.4.

Section 18.5: Surface Integrals of Vector Functions

1. Oriented Surfaces. Loosely speaking, a surface σ is said to be orientable if it has two distinct "sides," i.e., you cannot get from one side of the surface to the other without going "through" the surface or "around an edge." For example, the sphere is orientable since its surface has both an outside and an inside: if we placed Howard Ant on one side, he would have to make a hole to get through to the

other side. The paraboloid shown to the left is also orientable. It has an upper and a lower side to its surface: if we placed Howard Ant on one side,

he would have to burrow through the surface or go around the upper edge to get to the other

side.

On the other hand, consider the M̈obius band,
as shown to the right (a band of paper with just one

"twist" in it). This surface has only one side!

Why? Well, place Howard Ant on any point on

the surface, and look at what happens:

If Howard Ant starts   ... and walks through   ... he'll end up on
   on one "side"...   the "twist"...   the "other" side !

Thus the M̈obius band is NOT an orientable surface.

Moving Howard Ant around a surface

is merely a playful way of visualizing a choice of

normal vectors moving continuously around the surface

(i.e., there is a normal vector pointing out from the

surface at every point on Howard's path). Thus, for

the M̈obius band, there is no way we can make a

continuous choice of normal vectors for every point, for we can always move $\overline{n}$ continuously

around the band and come back to $-\overline{n}$. Hence

the M̈obius band is not orientable.

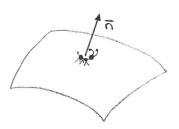

An oriented surface is merely an orientable surface (i.e., 2-sided surface) for which we have made a choice of which of the two sides of the surface we will place our normal

vectors on. Thus any orientable surface can be made into an oriented surface in exactly two

ways (one for each of the two sides of the surface).

When evaluating a vector surface integral over an oriented surface σ, <u>YOU MUST</u>

<u>TAKE CAREFUL NOTE OF THE CHOICE OF THE ORIENTATION</u>; if you don't, you risk

having a sign error occur in the evaluation of the integral.

2. <u>The definition of the vector surface integral.</u> Suppose we have a vector function

$$\overline{F}(x,y,z) = \langle f(x,y,z), g(x,y,z), h(x,y,z) \rangle$$

which has continuous components f, g and h on the oriented surface σ. If $\overline{n} = \overline{n}(x,y,z)$

is the unit normal vector of the orientation at (x,y,z), then the (vector) surface integral of

$\overline{F}$ over σ is defined to be

Definition of a Vector Surface Integral

$$\iint_\sigma \overline{F} \cdot \overline{n} \, dS$$

Since the dot product of two vectors is a scalar, the vector surface integral is merely

a scalar surface integral of the (scalar) function

$$g(x,y,z) = \overline{F}(x,y,z) \cdot \overline{n}(x,y,z)$$

Here is an example in which a vector surface integral is computed from this definition:

<u>Example A.</u> Let σ be the portion of the surface $z = x^2 + y$ for which $0 \le z \le 1$ and

$0 \le x \le 1$, oriented by upward normals. If $F(x,y,z) = \langle y\sqrt{2x^2+1}, xz, xz \rangle$,

evaluate the vector surface integral

$$\iint_{\sigma} \overline{F} \cdot \overline{n} \; dS$$

<u>Solution.</u> Anton gives general formulas for normals to surfaces in Table 18.5.2. According to that table, upward-pointing unit normals to the surface

$$z = x^2 + y$$

are given by

$$\overline{n} = \frac{-\dfrac{\partial z}{\partial x}\,\overline{i} - \dfrac{\partial z}{\partial y}\,\overline{j} + \overline{k}}{\sqrt{\left(\dfrac{\partial z}{\partial x}\right)^2 + \left(\dfrac{\partial z}{\partial y}\right)^2 + 1}}$$

Since $\dfrac{\partial z}{\partial x} = 2x$ and $\dfrac{\partial z}{\partial y} = 1$, we obtain

$$\overline{n} = \frac{-2x\,\overline{i} - \overline{j} + \overline{k}}{\sqrt{2}\,\sqrt{2x^2 + 1}} = \langle -2x, -1, 1 \rangle / \left(\sqrt{2}\,\sqrt{2x^2 + 1} \right)$$

Now computing $\overline{F} \cdot \overline{n}$ is easy:

$$\overline{F}(x, y, z) \cdot \overline{n}(x, y, z)$$

$$= \langle y\sqrt{2x^2 + 1}, \; xz, \; xz \rangle \cdot \langle -2x, -1, 1 \rangle / \left(\sqrt{2}\,\sqrt{2x^2 + 1} \right)$$

$$= \left(-2xy\sqrt{2x^2 + 1} - xz + xz \right) / \left(\sqrt{2}\,\sqrt{2x^2 + 1} \right)$$

$$= -\sqrt{2}\; xy$$

Thus

$$\iint_{\sigma} \overline{F} \cdot \overline{n} \; dS = -\sqrt{2} \iint_{\sigma} xy \; dS$$

our vector surface integral having been transformed into a (scalar) surface integral.

We use the techniques of Theorem 18.4.2(a) to evaluate this integral:

xy remains xy (i.e., there is no z to replace by $x^2 + y$)

dS converts to $\sqrt{\left(\dfrac{\partial z}{\partial x}\right)^2 + \left(\dfrac{\partial z}{\partial y}\right)^2 + 1}\ dA$

$$= \sqrt{(2x)^2 + (1)^2 + 1}\ dA \quad \text{since} \quad z = x^2 + y$$

$$= \sqrt{2}\ \sqrt{2x^2 + 1}\ dA$$

Therefore our surface integral becomes

$$-\sqrt{2}\ \iint_\sigma xy\ dS = -\sqrt{2}\ \iint_R xy\sqrt{2}\ \sqrt{2x^2 + 1}\ dA$$

where R is the projection of σ on the xy-plane.

Oh my, what a coincidence! Except for the $-\sqrt{2}$, the double integral on the right is the same as the double integral of Example A of the previous section. In that example we showed

$$\iint_R xy\sqrt{2}\ \sqrt{2x^2 + 1}\ dA \approx -.0425$$

Thus

$$\iint_\sigma \overline{F} \cdot \overline{n}\ dS = -\sqrt{2}\ \iint_R xy\ \sqrt{2}\ \sqrt{2x^2 + 1}\ dA$$

$$\approx -\sqrt{2}\ (-.0425) \approx \boxed{.0601} \qquad\qquad \square$$

3. Remembering the unit normals $\bar{n}$ from Table 18.5.2 is not really very hard. For each $\bar{n}$ in the table let $\bar{N}$ be the corresponding numerator term. Then we have:

equation for σ	dependent variable	$\pm \bar{N}$
$z = z(x,y)$	z	$\langle -\frac{\partial z}{\partial x}, \; -\frac{\partial z}{\partial y}, \; +1 \rangle$
$y = y(x,z)$	y	$\langle -\frac{\partial y}{\partial x}, \; +1, \; -\frac{\partial y}{\partial z} \rangle$
$x = x(y,z)$	x	$\langle +1, \; -\frac{\partial x}{\partial y}, \; -\frac{\partial x}{\partial z} \rangle$

Notice that the component of the dependent variable receives the $+1$, while the other components are the negatives of the corresponding partial derivatives.

The sign of $\bar{N}$ is also easy to remember:

upward, right, forward (all "positive" words)

have the $+1$ in the $\bar{N}$ formula

The denominator of $\bar{n}$ is merely the norm of

$\bar{N}$, so there is nothing new which

needs to be memorized.

4. Other computational formulas. Suppose σ is an oriented surface of one of the forms of Table 18.5.2. If we let $\bar{N}$ be the numerator of $\bar{n}$ as we did in the previous paragraph, then we have a very pleasant computational formula for vector surface integrals over σ:

Theorem 18.5.2	$$\iint_\sigma \overline{F} \cdot \overline{n} \, dS = \iint_R \overline{F} \cdot \overline{N} \, dA$$

> This is an ordinary double integral over the projection R of the surface σ onto the coordinate plane of the two independent variables.

This is an alternate way of stating Anton's Theorem 18.5.2. The significance of the result lies in eliminating a number of unnecessary square root calculations. To see this we'll verify the new formula for a surface of the form $z = z(x,y)$, oriented by upward normals:

$$\iint_\sigma \overline{F} \cdot \overline{n} \, dS = \iint_R \overline{F} \cdot \underbrace{\left(\frac{\overline{N}}{\sqrt{\left(\frac{\partial z}{\partial x}\right)^2 + \left(\frac{\partial z}{\partial y}\right)^2 + 1}} \right)}_{\overline{n} \text{ by Table 18.5.2}} \underbrace{\left(\sqrt{\left(\frac{\partial z}{\partial x}\right)^2 + \left(\frac{\partial z}{\partial y}\right)^2 + 1} \; dA \right)}_{dS \text{ as in Theorem 18.4.2}}$$

Hence by simple cancellation of the square root terms we obtain the desired expression

$$\iint_R \overline{F} \cdot \overline{N} \, dA .$$

Anton illustrates this formula in Example 5; here is an additional example:

Example B. Let σ be that portion of the paraboloid $y = x^2 + z^2$ for which $y \le 4$, oriented by right unit normals. If $\overline{F}(x,y,z) = z\,\overline{i} + y\,\overline{j} + x\,\overline{k}$, evaluate the vector surface integral

$$\iint_\sigma \overline{F} \cdot \overline{n} \, dS$$

Solution. Using our version of Theorem 18.5.2 we have

$$\iint_\sigma \overline{F} \cdot \overline{n} \, dS = \iint_R \overline{F} \cdot \overline{N} \, dA , \qquad \text{where } R \text{ is the projection of } \sigma \text{ on the } xz\text{-plane}$$

Now $\overline{F}(x,y,z) \cdot \overline{N}(x,y,z)$

$$= \underbrace{\langle z, y, x \rangle}_{\text{definition of } \overline{F}} \cdot \underbrace{\langle -\frac{\partial y}{\partial x}, \ 1, \ -\frac{\partial y}{\partial z} \rangle}_{\overline{N} \text{ from Table 18.5.2}}$$

$$= \langle z, x^2 + z^2, x \rangle \cdot \langle -2x, 1, -2z \rangle \qquad \text{since } y = x^2 + z^2$$

$$= -2xz + x^2 + z^2 - 2xz$$

$$= -4xz + x^2 + z^2$$

Hence

$$\iint_\sigma \overline{F} \cdot \overline{n} \ dS = \iint_R (-4xz + x^2 + z^2) \, dA$$

What remains is to find the x, z limits for the region R. However, we know

$$x^2 + z^2 = y \leq y$$

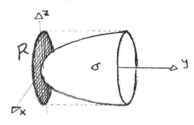

so that R is the disk

$$x^2 + z^2 \leq 4$$

The best way to integrate over such a disk is to use polar coordinates in the xz-plane:

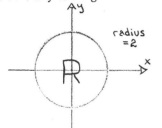

$$x = r \cos \theta$$

$$z = r \sin \theta \qquad\qquad dA = r \, dr \, d\theta$$

$$x^2 + z^2 = r^2$$

The limits for R are

$$0 \leq r \leq 2 \quad \text{and} \quad 0 \leq \theta \leq 2\pi$$

Thus

$$\iint_\sigma \overline{F} \cdot \overline{n} \; dS = \iint_R (-4xz + x^2 + z^2) \, dA$$

$$= \int_0^{2\pi} \left[\int_0^2 (-4r^2 \sin \theta \cos \theta + r^2) r \, dr \right] d\theta$$

Inner integral:

$$\int_0^2 (-4r^2 \sin \theta \cos \theta + r^2) r \, dr$$

$$= (-4 \sin \theta \cos \theta + 1) \int_0^2 r^3 \, dr$$

$$= (-2 \sin 2\theta + 1) \left[\frac{r^4}{4} \right] \Big|_0^2 \qquad \text{using} \quad \sin 2\theta = 2 \sin \theta \cos \theta$$

$$= (-2 \sin 2\theta + 1) [4]$$

Full integral:

$$\int_0^{2\pi} [\ldots] \, d\theta = 4 \int_0^{2\pi} (1 - 2 \sin 2\theta) \, d\theta$$

$$= 4 [\theta + \cos 2\theta] \Big|_0^{2\pi}$$

$$= 4 [2\pi + 1 - 0 - 1]$$

$$= \boxed{8\pi} \qquad\qquad\qquad \square$$

5. <u>The meaning of a vector surface integral: flux (optional).</u>* Suppose we have a fluid moving through a region of three-dimensional space. If there is a "penetrable" oriented surface σ in the path of this flow, we wish to obtain a formula for the <u>total volume of fluid passing through the surface σ in one unit of time.</u> This quantity is called the <u>flux of the fluid flow across σ.</u>

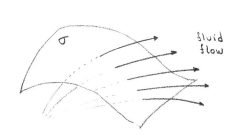

We need to be more precise in one aspect of this definition. Since σ is an oriented surface, it has a "positive side" (the side with the normal vector n̄) and a "negative side" (the side without the normal). Fluid volume passing from the negative side to the positive side of the surface will be assigned a positive sign; fluid volume passing from the positive side to the negative side of the surface will be assigned a negative sign. Thus, the flux is really the <u>net</u> total volume of fluid passing through the oriented surface σ in one unit of time, i.e.,

$$
\begin{bmatrix} \text{flux} \\ \text{across} \\ \sigma \end{bmatrix} = \begin{bmatrix} \text{total volume of} \\ \text{fluid passing from the} \\ \text{- to the +} \\ \text{side of } \sigma \\ \text{per unit time} \end{bmatrix} \text{ minus } \begin{bmatrix} \text{total volume of} \\ \text{fluid passing from the} \\ \text{+ to the -} \\ \text{side of } \sigma \\ \text{per unit time} \end{bmatrix}
$$

---

* This subsection will be necessary for §18.8.

Thus, if σ is a closed surface such
as a sphere, a zero flux across σ indicates
that the total volume of fluid contained inside
σ does not change. On the other hand
(assuming σ has an outward pointing

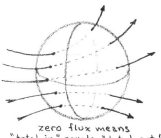

zero flux means
"total in" equals "total out"

normal), a positive flux across σ indicates that σ encloses a <u>source</u> of fluid, while a

negative flux indicates that σ encloses a <u>sink</u> for our fluid (i. e. , a spot where we are losing

fluid, as if it is "going down a sink drain").

What does all this have to do with vector surface integrals? The answer is that the

formula for flux will be given as a vector surface integral of the fluid flow over the surface

(Equation D below). Since this is the most basic application of vector surface integration,

we'll derive this formula.

We'll obtain our surface integral by using the infinitesimal approach of §18. 4. 1 of

<u>The Companion</u>. Let

$$\overline{F}(x,y,z) = \text{the velocity vector of the fluid}$$
$$\text{at the point } (x,y,z)$$

and consider $d\sigma$ as any "infinitesimal" piece of the surface σ with surface area dS.

Since $d\sigma$ is infinitesimally small, we can
assume that it is a flat piece of a plane with
normal vector $\overline{n}$; moreover, we can also
assume that $\overline{F}(x,y,z)$ is constant over

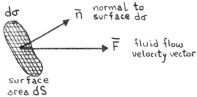

$d\sigma$
$\overline{n}$ normal to surface $d\sigma$

$\overline{F}$ fluid flow velocity vector

surface area dS

$d\sigma$, taking the constant value $\overline{F}$ everywhere

in $d\sigma$. (In what follows we assume that $\overline{n}$ and $\overline{F}$ both lie on the same side of $d\sigma$. The other case is handled similarly and also yields the key equation (C) below.)

To compute the flux through $d\sigma$ we must compute the volume of fluid which passes through $d\sigma$ in one unit of time. Since $\|\overline{F}\|$ is the speed of the fluid through $d\sigma$, it also equals the "straight line distance" that our fluid would travel* in <u>one</u> unit of time after it crosses

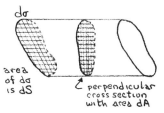

$d\sigma$ (distance equals rate times time). Hence the volume of fluid crossing $d\sigma$ in one unit of time can be "drawn" as the slanted cylinder shown to the right.

Thus a formula for dV is given by

$$dV = \|\overline{F}\| \, dA \qquad\qquad \text{(A)}$$

where dA is the area of a perpendicular cross-section of the cylinder.

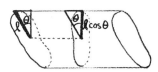

However, if θ is the angle of inclination of $d\sigma$ to the cylinder (as shown to the right), the area of the perpendicular cross-section is merely the area of the base $d\sigma$ multiplied by the factor $\cos\theta$. (The ratio of these areas equals the ratio of the line segment lengths $\ell \cos\theta$ and ℓ, as shown above.) Thus

$$\begin{bmatrix} \text{area of the} \\ \text{perpendicular} \\ \text{cross-section} \end{bmatrix} = \cos\theta \cdot \begin{bmatrix} \text{area of} \\ \text{the base} \\ d\sigma \end{bmatrix}$$

which is the same as

$$dA = \cos\theta \, dS$$

---

* We are not assuming that the fluid necessarily travels in a "straight line. " However, the volume of fluid crossing $d\sigma$ is the same whether it travels in a straight line or not, and we can <u>compute</u> this volume in the "straight line" case. This is what we proceed to do.

Thus our formula (A) for dV becomes

$$dV = \|\overline{F}\| \cos \theta \, dS \qquad\qquad (B)$$

We have one further change to make in this formula.

Notice that θ, the angle of inclination of $d\sigma$

to the cylinder, is also the angle between $\overline{F}$

and the unit normal $\overline{n}$ (see the figure to the

right). Thus

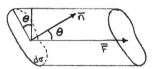

$$\|\overline{F}\| \cos \theta = \|\overline{F}\| \|\overline{n}\| \cos \theta = \overline{F} \cdot \overline{n}$$

| since $\|\overline{n}\| = 1$ | | definition of $\overline{F} \cdot \overline{n}$ |

Plugging this into (B) will yield our final formula for dV:

$$\boxed{dV = \overline{F} \cdot \overline{n} \, dS} \qquad\qquad (C)$$

(As we remarked earlier, if $\overline{F}$ and $\overline{n}$ are on

opposite sides of $d\sigma$, formula (C) still holds.

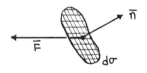

However, in this case the formula yields a

negative value for dV. This is fine since

we wish to assign a negative sign to the volume of fluid flow in this direction.)

Thus, the (net) total volume of fluid crossing σ per unit of time, i. e. , the flux of

$\overline{F}$ across σ, is

$$V = \iint_{\sigma} dV = \iint_{\sigma} \overline{F} \cdot \overline{n} \, dS \qquad \text{by Equation (C)}$$

Hence

$$\begin{bmatrix} \text{flux of } \overline{F} \\ \text{across } \sigma \end{bmatrix} = \iint_\sigma \overline{F} \cdot \overline{n} \, dS \qquad \text{(D)}$$

The "infinitesimal" arguments given above can be made rigorous by using Riemann Sums.
This is essentially what Anton will do in §18.8.

　　　We have derived a formula for volume flux; one can also derive similar formulas for
mass flux of a fluid flow, heat flux of a heat flow, and current flux of an electrical current
flow. These derivations are only slightly more complex than the one given here.

6.　Cylindrical and spherical vector surface integrals (optional).　In §18.4.3 of The Companion
we discussed the computation of surface integrals using cylindrical and spherical coordinates.
In this section we'll do the same for vector surface integrals.

Vector surface integrals in cylindrical coordinates.

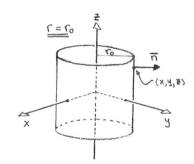

We let $\overline{n}(x,y,z)$ be the outward unit normal to
the cylinder $r = r_0$ at (x,y,z), and take
$g(x,y,z) = \overline{F}(x,y,z) \cdot \overline{n}(x,y,z)$ in Equation (A)
of §18.4.3 to obtain

$$\iint_\sigma \overline{F} \cdot \overline{n} \, dS = \iint_\sigma \underbrace{(\overline{F} \cdot \overline{n})(r_0 \cos\theta, r_0 \sin\theta, z)}_{\substack{\text{the function } \overline{F} \cdot \overline{n} \text{ evaluated} \\ \text{at } (r_0 \cos\theta, r_0 \sin\theta, z)}} r_0 \, dz \, d\theta \qquad \text{(A)}$$

The trick is to find a formula for $\overline{n}$ at the point (x,y,z).

But that's easy: in the figure to the right, observe that

letting $z = 0$ in the position vector $\overline{x} = \langle x, y, z \rangle$

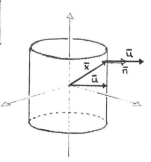

will yield an outward normal vector $\bar{u} = \langle x, y, 0 \rangle$. Hence the outward **unit** normal vector $\bar{n}$ at (x, y, z) is

$$\bar{n} = \bar{u} / \|\bar{u}\| = \langle x, y, 0 \rangle / \sqrt{x^2 + y^2} = \langle x, y, 0 \rangle / r_0$$

Substituting into Equation (A) yields

Vector surface integral on the cylinder $r = r_0$	$\displaystyle \iint_\sigma \overline{F} \cdot \bar{n} \, dS = \iint_\sigma \overline{F}(x, y, z) \cdot \langle x, y, 0 \rangle \, dz \, d\theta$ where x, y, z are expressed in cylindrical coordinates: $x = r_0 \cos\theta$, $y = r_0 \sin\theta$, $z = z$ The limits of integration will be θ and z limits

Example C. Evaluate $\displaystyle \iint_\sigma F \cdot n \, dS$ where σ is that portion of the cylinder $r = r_0$,

for which $y \geq 0$ and $z \geq 1$, oriented by outward unit normals, and $\overline{F}(x, y, z) = (y/z^2)\bar{j}$.

Solution. The half cylinder σ $(y \geq 0)$ extends infinitely far up $(z \geq 1)$, parallel to the z-axis. Thus, since $\overline{F}(x, y, z) = \langle 0, y/z^2, 0 \rangle$,

$$\iint_\sigma \overline{F} \cdot \bar{n} \, dS = \iint_\sigma \left\langle 0, \frac{r_0 \sin\theta}{z^2}, 0 \right\rangle \cdot \langle r_0 \cos\theta, r_0 \sin\theta, 0 \rangle \, dz \, d\theta$$

$$= \int_0^\pi \int_1^\infty \frac{r_0^2 \sin^2\theta}{z^2} \, dz \, d\theta \qquad \text{since} \quad 1 \leq z < \infty$$

$$\text{and} \quad 0 \leq \theta \leq \pi$$

Inner integral:

$$\int_1^\infty \frac{r_0^2 \sin^2 \theta}{z^2} \, dz = \lim_{a \to +\infty} \int_1^a \frac{r_0^2 \sin^2 \theta}{z^2} \, dz$$

$$= r_0^2 \sin^2 \theta \lim_{a \to +\infty} \int_1^a \frac{dz}{z^2} = r_0^2 \sin^2 \theta \lim_{a \to +\infty} \left(-\frac{1}{z} \right) \Bigg|_1^a$$

$$= r_0^2 \sin^2 \theta \lim_{a \to +\infty} \left(-\frac{1}{a} + 1 \right) = r_0^2 \sin^2 \theta$$

Full integral:

$$\int_0^\pi [\ldots] \, d\theta = \int_0^\pi r_0^2 \sin^2 \theta \, d\theta = r_0^2 \int_0^\pi \frac{1 - \cos 2\theta}{2} \, d\theta$$

$$= \frac{r_0^2}{2} \int_0^\pi (1 - \cos 2\theta) \, d\theta = \frac{r_0^2}{2} \left(\theta - \frac{\sin 2\theta}{2} \right) \Bigg|_0^\pi$$

$$= \frac{r_0^2}{2} (\pi) = \boxed{r_0^2 \pi / 2} \qquad \square$$

Vector surface integrals in spherical coordinates.

We let $\bar{n}(x, y, z)$ be the outward unit normal

to the sphere $\rho = \rho_0$ at (x, y, z), and take

$$g(x, y, z) = \bar{F}(x, y, z) \cdot \bar{n}(x, y, z)$$

in Equation (B) of §18. 4. 3 to obtain

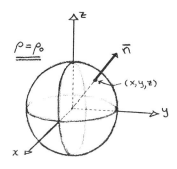

$$\iint_\sigma \bar{F} \cdot \bar{n} \, dS = \iint_\sigma (\bar{F} \cdot \bar{n}) \underbrace{(x, y, z)}_{} \rho_0^2 \sin \phi \, d\phi \, d\theta \qquad \text{(B)}$$

x, y, z expressed in
spherical coordinates

If (x,y,z) is a point on the sphere $\rho = \rho_0$, then the outward unit normal at (x,y,z) is given by

$$\bar{n} = \frac{\langle x,y,z \rangle}{\| \langle x,y,z \rangle \|} = \frac{\langle x,y,z \rangle}{\rho_0}$$

Substituting this into (B) yields

| Vector surface integral on the sphere $\rho = \rho_0$ | $$\iint_\sigma \bar{F} \cdot \bar{n} \, dS = \iint_\sigma \bar{F}(x,y,z) \cdot \langle x,y,z \rangle \, \rho_0 \sin \phi \, d\phi \, d\theta$$ where x,y,z are expressed in spherical coordinates $x = \rho_0 \sin \phi \cos \theta$ $y = \rho_0 \sin \phi \sin \theta$ $z = \rho_0 \cos \phi$ |

The limits of integration will be θ and ϕ limits

Example D. Evaluate $\displaystyle\iint_\sigma \bar{F} \cdot \bar{n} \, dS$ where σ is the sphere $\rho = 2$, oriented by outward normals, and $\bar{F}(x,y,z) = x\bar{i} + y\bar{j}$.

Solution. Since we are integrating over the full sphere, the limits of integration will be $0 \leq \theta \leq 2\pi$ and $0 \leq \phi \leq \pi$. Thus

$$\iint_\sigma \bar{F} \cdot \bar{n} \, dS = \iint_\sigma \langle x,y,0 \rangle \cdot \langle x,y,z \rangle \, 2 \sin \phi \, d\phi \, d\theta$$

$$= 2 \iint_\sigma (x^2 + y^2) \sin \phi \, d\phi \, d\theta$$

$$= 2 \iint_\sigma (4 \sin^2 \phi) \sin \phi \, d\phi \, d\theta$$

since $x^2 + y^2 = r^2 = 2^2 \sin^2 \phi$

$$= 8 \int_0^{2\pi} d\theta \int_0^{\pi} \sin^3 \phi \, d\phi$$

$$= 16\pi \int_0^{\pi} (1 - \cos^2 \phi) \sin \phi \, d\phi$$

$$= 16\pi \left(- \cos \phi + \frac{\cos^3 \phi}{3} \right) \Big|_0^{\pi}$$

$$= 16\pi \left(1 - \frac{1}{3} + 1 - \frac{1}{3} \right) = \boxed{64\pi / 3} \qquad \square$$

If you wish to check your ability to use the formulas in this subsection, apply them to Exercises 7, 9, 14, 15 and 21 in Anton's Exercise Set 18. 5. (Exercise 14 asks you to evaluate a surface integral over a cylinder which is parallel to the y-axis, not the z-axis. Such a case is best handled by rotating the xyz-axes into x' y' z'-axes by x = y', y = z' and z = x'. Then the cylinder is parallel to the z'-axis, and cylindrical coordinates in the x' y' z' coordinates can be used.)

Section 18. 6: The Divergence Theorem

1. The ∇ operator.   There is some notation which is useful in vector calculus. We define ∇ (pronounced "del") to be the ordered triple of the partial derivative operators in 3-space:

    $\boxed{\nabla \text{ operator}}$
    $$\boxed{\nabla = \langle \frac{\partial}{\partial x} , \frac{\partial}{\partial y} , \frac{\partial}{\partial z} \rangle}$$

This new notation corresponds to the notation for the gradient of a function f which we used in §16. 5:

gradient

$$\nabla f = \langle \frac{\partial f}{\partial x} , \frac{\partial f}{\partial y} , \frac{\partial f}{\partial z} \rangle$$

Thus the gradient of f is merely the application of the "del" operator to f, pronounced "del f."

Using this notation, the <u>divergence</u> of a vector function $\overline{F}$ can be expressed nicely as a "dot product" of ∇ with $\overline{F}$:

divergence

If $\overline{F}(x,y,z) = \langle f(x,y,z) , g(x,y,z) , h(x,y,z) \rangle$

then   div $\overline{F} = \nabla \cdot \overline{F} = \frac{\partial f}{\partial x} + \frac{\partial g}{\partial y} + \frac{\partial h}{\partial z}$

<u>Example A.</u>   If $\overline{F}(x,y,z) = \langle x^2 y , \cos(yz) , x^2 + y \rangle$,   find   div $\overline{F}$.

<u>Solution.</u>   div $F = \nabla \cdot \overline{F}$

$$= \langle \frac{\partial}{\partial x} , \frac{\partial}{\partial y} , \frac{\partial}{\partial z} \rangle \cdot \langle x^2 y , \cos(yz) , x^2 + y \rangle$$

$$= \frac{\partial}{\partial x} (x^2 y) + \frac{\partial}{\partial y} \cos(yz) + \frac{\partial}{\partial z} (x^2 + y^2)$$

$$= 2xy - z \sin(yz) \qquad\qquad\qquad\qquad \square$$

The ∇ notation makes remembering the definition of   div $\overline{F}$ very easy.

2.   <u>The Divergence Theorem.</u>   This important theorem applies to any solid   G   whose boundary surface σ is a finite collection of "smooth surfaces."   Intuitively a "smooth surface" is one

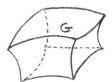

G is the solid, and
σ is the boundary,
or outer surface.

which has no sharp points, sharp bends,

or breaks. Thus the solid   G   shown to

the right has a boundary   σ   comprised

of six smooth surfaces.

a smooth surface

a non-smooth surface

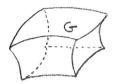

Divergence Theorem

Suppose (1)   G   is a solid whose boundary surface   σ
consists of a finite union of smooth
surfaces; and

(2) $\overline{F}$ is a vector function whose three component
functions have continuous first partial
derivatives on an open set containing   G.

If   σ   is oriented by outward-pointing normal vectors, then

$$\iint_\sigma \overline{F} \cdot \overline{n} \; dS = \iiint_G \operatorname{div} \overline{F} \; dV$$

In words this formula says

$$\left\{ \begin{array}{l} \text{the vector surface integral} \\ \text{of a vector function } \overline{F} \text{ over} \\ \text{the } \underline{\text{boundary}} \text{ of a solid} \end{array} \right\} \quad \text{equals} \quad \left\{ \begin{array}{l} \text{the triple integral} \\ \text{of the } \underline{\text{divergence}} \text{ of} \\ \overline{F} \text{ over the solid} \end{array} \right\}$$

The major computational value of the Divergence Theorem stems from the facts that:

- calculating   div $\overline{F}$ from $\overline{F}$ is generally quite easy, and

- calculating a triple integral is generally much easier than
  calculating a surface integral.

18.6.4

Thus:

> It is often easier to compute the triple
> integral obtained from the Divergence
> Theorem than it is to compute the
> original vector surface integral over
> the boundary of a solid.

Anton's Examples 2 through 5 use this procedure. In each example observe that

making the conversion from $\iint_\sigma \overline{F} \cdot \overline{n} \, dS$ to $\iiint_G$ div F dV is easy; the examples

are then solved by evaluating the triple integral, which is merely an application of the

techniques of §17.5.

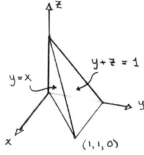

Example B. Suppose σ is the surface of the
tetrahedron bounded by $x = 0$, $z = 0$, $y = x$
and $y + z = 1$, and oriented by outward
normals. Compute $\iint_\sigma \overline{F} \cdot \overline{n} \, dS$, where

$\overline{F}(x, y, z) = \langle x^2 y/2, xz^2, xz + y \rangle$.

Solution. Since our surface has four "smooth" sides, a direct computation of the surface

integral would be unpleasant because it would entail evaluating four separate surface integrals.

Using the Divergence Theorem is definitely preferable:

$$\iint_\sigma \overline{F} \cdot \overline{n} \, dS = \iiint_G \operatorname{div} \overline{F} \, dV \qquad \text{by the Divergence Theorem, where } G \text{ is the solid enclosed by } \sigma$$

$$= \iiint_G \left[\frac{\partial}{\partial x}(x^2 y/2) + \frac{\partial}{\partial y}(xz^2) + \frac{\partial}{\partial z}(xz+y) \right] dV$$

$$= \iiint_G (xy + x) \, dV$$

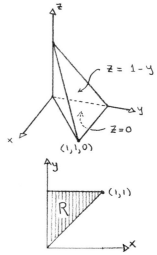

Now the problem is to compute the limits of

integration for G. Examining G we see

that it is the solid bounded above and below by

the graphs of $z = 1 - y$ and $z = 0$,

and that the projection of G onto the

xy-plane is the triangle R shown to

the right. Thus, by Theorem 17.5.2,

$$\iiint_G (xy + x) \, dV = \iint_R \left[\int_0^{1-y} (xy + x) \, dz \right] dA$$

The limits of integration for R are then

easily found by the techniques of §17.2 to

be $x \le y \le 1$ and $0 \le x \le 1$. This

gives

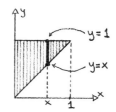

$$\iiint_G (xy + x) \, dV = \int_0^1 \int_x^1 \int_0^{1-y} (xy + x) \, dz \, dy \, dx$$

Inner integral:

$$\int_0^{1-y} (xy + x) \, dz = x(y + 1) z \Big|_{z=0}^{z=1-y}$$

$$= x(y + 1)(1 - y) = x(1 - y^2)$$

Middle integral:

$$\int_x^1 x(1 - y^2) \, dy = x(y - y^3/3) \Big|_{y=x}^{y=1}$$

$$= x(1 - \frac{1}{3}) - x(x - x^3/3)$$

$$= \frac{2x}{3} - x^2 + \frac{x^4}{3}$$

Outer integral:

$$\int_0^1 \left(\frac{2x}{3} - x^2 + \frac{x^4}{3} \right) dx = \left(\frac{x^2}{3} - \frac{x^3}{3} + \frac{x^5}{15} \right) \Big|_0^1$$

$$= \frac{1}{3} - \frac{1}{3} + \frac{1}{15} = \frac{1}{15}$$

Thus

$$\iint_\sigma \overline{F} \cdot \overline{n} \, dS = \boxed{1/15} \qquad \square$$

There are many other situations in which the Divergence Theorem can be applied. We will see some of these in subsequent sections.

Section 18. 7 :   Stokes' Theorem

1.   The ∇ operator.   In §18. 6. 1 we introduced the useful "del" operator, defined to be the

ordered triple of the partial derivative operators in   3-space :

$\boxed{\nabla \text{ operator}}$

$$\nabla = \langle \frac{\partial}{\partial x} , \frac{\partial}{\partial y} , \frac{\partial}{\partial z} \rangle$$

Using this notation, the curl of a vector function $\overline{F}$ can be expressed as a "cross-product"

of ∇ with $\overline{F}$:

$\boxed{\text{curl}}$

If $\overline{F}(x,y,z) = \langle f(x,y,z) , g(x,y,z) , h(x,y,z) \rangle$,

then   curl $\overline{F}$ = $\nabla \times \overline{F}$

$$= \begin{vmatrix} \overline{i} & \overline{j} & \overline{k} \\ \partial/\partial x & \partial/\partial y & \partial/\partial z \\ f & g & h \end{vmatrix}$$

$$= \langle \frac{\partial h}{\partial y} - \frac{\partial g}{\partial z} , \frac{\partial f}{\partial z} - \frac{\partial h}{\partial x} , \frac{\partial g}{\partial x} - \frac{\partial f}{\partial y} \rangle$$

(See §15. 4 if you need a review of the cross-product.)

Thus the concepts of gradient, divergence and curl are all easy to remember in terms of ∇ :

grad f = ∇f

div $\overline{F}$ = $\nabla \cdot \overline{F}$

curl $\overline{F}$ = $\nabla \times \overline{F}$

<u>Example A.</u>   If $\overline{F}(x,y,z) = \langle x^2 y , \cos(yz) , x^2 + y \rangle$,   find   curl $\overline{F}$.

Solution.

$$\text{curl } \overline{F} = \nabla \times \overline{F} = \begin{vmatrix} \overline{i} & \overline{j} & \overline{k} \\ \partial/\partial x & \partial/\partial y & \partial/\partial z \\ x^2 y & \cos(yz) & x^2 + y \end{vmatrix}$$

$$= \left\langle \frac{\partial}{\partial y}(x^2 + y) - \frac{\partial}{\partial z}\cos(yz) , \; \frac{\partial}{\partial z}(x^2 y) - \frac{\partial}{\partial x}(x^2 + y) , \right.$$

$$\left. \frac{\partial}{\partial x}\cos(yz) - \frac{\partial}{\partial y}(x^2 y) \right\rangle$$

$$= \left\langle 1 + y\sin(yz) , \; -2x , \; -x^2 \right\rangle \qquad \square$$

2. Stokes' Theorem.   This theorem applies to a "smooth"

   orientable surface σ (see §18.6.2 of The Companion)

   whose boundary curve C is a finite untion of smooth

   curves (see §14.3). Moreover, we give a direction to

   the boundary curve C   so that the curve C   and the

   oriented surface σ   obey the right-hand rule:

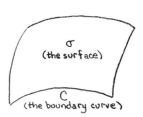

σ
(the surface)

C
(the boundary curve)

Right hand rule	If the thumb of your right hand
	points in the direction of the normal
	vectors of the oriented surface σ,
	then your fingers curl in the
	direction of the directed curve C

We now can state Stokes' Theorem as follows:

Stokes' Theorem	Suppose (1) σ is a smooth, oriented surface whose boundary curve C is comprised of a finite union of smooth curves; and

$\qquad\qquad$ (2) $\overline{F}$ is a vector function whose three component functions have continuous first partial derivatives on an open set containing σ.

If σ and C are oriented according to the right-hand rule, then

$$\int_C \overline{F} \cdot d\overline{r} = \iint_\sigma (\text{curl } \overline{F}) \cdot \overline{n} \, dS$$

In words this formula says

$$\left\{ \begin{array}{l} \text{the line integral of a} \\ \text{vector function } \overline{F} \text{ over} \\ \text{the } \underline{\text{boundary}} \text{ of a surface} \end{array} \right\} \quad \text{equals} \quad \left\{ \begin{array}{l} \text{the surface integral} \\ \text{of the } \underline{\text{curl}} \text{ of } \overline{F} \\ \text{over the surface} \end{array} \right\}$$

The major computational value of Stokes' Theorem stems from the facts that

- calculating curl $\overline{F}$ from $\overline{F}$ is generally quite easy, and

- calculating a surface integral can sometimes be easier than calculating a line integral, especially if you have some freedom in choosing a convenient surface.

Thus:

It is often easier to compute a surface integral obtained from Stokes' Theorem than it is to compute the line integral over the boundary of the surface.

This procedure can be particularly useful when the curve   C   is a union of curve

segments, and the surface   σ   can be chosen to be a very simple surface.   Anton's

Example 3   illustrates this technique; here is another illustration:

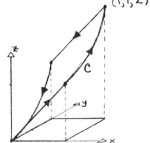

Example B.   Let   C   be that curve on the surface

$z = x + y^2$ which projects down onto the unit square

in the   xy-plane ,   and let

$$\overline{F}(x,y,z) = \langle -y^2 z ,\ 2xyz ,\ xy^2 \rangle$$

Find $\displaystyle\int_C \overline{F} \cdot d\overline{r}$,   if   C   is oriented as shown in the diagram.

Solution.   To evaluate this line integral directly would require the evaluation of four separate

line integrals, one over each piece of the curve.   To avoid doing this, we will apply Stokes'

Theorem, taking the surface   σ   to be that portion of the graph of $z = x + y^2$ which is

bounded by   C.   We orient   σ   by upward normals so that   σ   and   C   will obey the right-

hand rule.   (To see this is the correct choice of normals use your right-hand on the diagram

above.)

$$\int_C \overline{F} \cdot d\overline{r} = \iint_\sigma (\text{curl } \overline{F}) \cdot \overline{n} \ dS \qquad \text{by Stokes' Theorem}$$

But $\text{curl } \overline{F} = \nabla \times \overline{F} = \begin{vmatrix} \overline{i} & \overline{j} & \overline{k} \\ \partial/\partial x & \partial/\partial y & \partial/\partial z \\ -y^2 z & 2xyz & xy^2 \end{vmatrix}$

$$= \langle \frac{\partial}{\partial y}(xy^2) - \frac{\partial}{\partial z}(2xyz),\ \frac{\partial}{\partial z}(-y^2 z) - \frac{\partial}{\partial x}(xy^2),\ \frac{\partial}{\partial x}(2xyz) - \frac{\partial}{\partial y}(-y^2 z) \rangle$$

$$= \langle 2xy - 2xy,\ -y^2 - y^2,\ 2yz + 2yz \rangle$$

$$= \langle 0,\ -2y^2,\ 4yz \rangle$$

Thus

$$\int_C \overline{F} \cdot d\overline{r} = \iint_\sigma \langle 0, -2y^2, 4yz \rangle \cdot \overline{n} \, dS$$

$$= \iint_R \langle 0, -2y^2, 4yz \rangle \cdot \langle -\frac{\partial z}{\partial x}, -\frac{\partial z}{\partial y}, 1 \rangle \, dA$$

by Formula (6) of §18.5, where R is the unit square in the xy-plane (i.e., the projection of σ on the xy-plane).

$$= \iint_R \langle 0, -2y^2, 4y(x+y^2) \rangle \cdot \langle -1, -2y, 1 \rangle \, dA$$

by using $z = x + y^2$, the equation for the surface σ.

$$= \iint_R (4y^3 + 4xy + 4y^3) \, dA$$

$$= \iint_R (8y^3 + 4xy) \, dA$$

Thus we have reduced the integral to a simple double integral over the unit square (i.e., $0 \le x \le 1$, $0 \le y \le 1$) in the xy-plane. This integral is easily evaluated by the techniques of §17.1:

$$\int_C \overline{F} \cdot d\overline{r} = \int_0^1 \int_0^1 (8y^3 + 4xy) \, dy \, dx$$

$$= \int_0^1 (2y^4 + 2xy^2) \Big|_{y=0}^{y=1} \, dx$$

$$= \int_0^1 (2 + 2x) \, dx = (2x + x^2) \Big|_0^1 = \boxed{3}$$

Using Stokes' Theorem to compute line integrals is only the most elementary and pedestrian application of this remarkable result. In the next (optional) subsection we show how Stokes' Theorem leads to the 3-space generalization of the conservative vector field material of §18.2. Further applications of Stokes' Theorem will be given in §§18.8 and 18.9.

3. Conservative vector fields in 3-space. (Optional). In §18.2 of The Companion we stated the Conservative Vector Fields Theorem, essentially a list of four equivalent properties of a conservative vector field in the plane. Here is an abbreviated version of that result:

The Conservative Vector Fields Theorem	A vector function $\overline{F}(x,y) = f(x,y)\,\overline{i} + g(x,y)\,\overline{j}$ is conservative on an open set U if it possesses any one (hence all) of the following equivalent properties: 1. $\overline{F}$ is circulation-free 2. $\overline{F}$ is path-independent 3. $\overline{F}$ is a gradient field 4. $\overline{F}$ is irrotational (The last property implies the previous three only when U is simply-connected.)

The purpose of this section is to show that this theorem also applies to 3-space, and to demonstrate the crucial role Stokes' Theorem plays in this generalization.

The definitions of the first three properties in 3-space, and the proofs that they are equivalent, are the same as the corresponding definitions and proofs in 2-space. The definitions of irrotational and simply-connected must be changed, however:

| Irrotational Vector Field in 3-space | A vector function $\overline{F}(x,y,z)$ in 3-space is __irrotational__ on a set U if and only if curl $\overline{F}$ = 0 for all points of U. |

The use of the term "irrotational" will be explained and justified in §18. 8. 3 .

| Simply Connected Regions in 3-space | Suppose U is an open region in 3-space. Then U is said to be __simply-connected__ if 1. U is connected (i. e. , any two points in U can be joined by a smooth curve C lying entirely in U), and 2. every closed curve C in U can be "continuously shrunk to a point" in U. |

The statement that a closed curve C can be "continuously shrunk to a point" in U should

be intuitively clear: if we picture C as a rubber band then it can be shrunk in size until it is nothing but a point. Moreover, this can be done without breaking C (it shrinks continuously) and __without ever leaving the region U.__

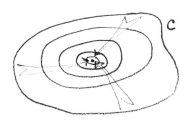

Consider what it means if a closed curve C in U __cannot__ be continuously shrunk to a point in U. It means that U has a rather major "hole" which passes through the interior of C in such a way that the curve

cannot "get around" it. For example, if

U_1 is the solid torus (doughnut) let C

be the curve shown to the right. C cannot

be shrunk to a point in the torus because

the "hole" gets in the way. Thus U_1 is not simply-connected.

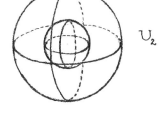

On the other hand, consider the set U_2 which is

the region between two concentric spheres. Although U_2

has a "hole," it is not a type of hole that gets in the way

of curves shrinking to a point in U_2. That's because

any closed curve can "slide off" a sphere as shown to

the right. Thus the region U_2 is simply connected. *

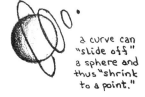

a curve can "slide off" a sphere and thus "shrink to a point."

$$* \qquad * \qquad * \qquad * \qquad *$$

We are now ready to state and prove the last part of the 3-space version of the

Conservative Vector Fields Theorem:

Suppose $\overline{F}(x,y,z)$ is a vector function on an open region U.

  (a) If $\overline{F}$ is conservative on U, then $\overline{F}$ is irrotational

      on U.

  (b) If $\overline{F}$ is irrotational on U, then $\overline{F}$ is conservative

      on U provided that U is simply-connected.

---

*
  We will discuss the hole in U_2 further, when we consider the concept of 2-simply

  connected sets in §18.9.5.

In this theorem the statement "$\overline{F}$ is conservative" means that $\overline{F}$ is circulation-free, path-independent, and a gradient field.

To prove statement (a), we assume $\overline{F}$ is conservative on U. Hence, since $\overline{F}$ is a gradient field on U, there exists a scalar function φ (a "potential function" for $\overline{F}$) such that $\overline{F} = \nabla \varphi$. Thus

$$\text{curl } \overline{F} = \text{curl } \nabla \varphi = 0$$

the last equality following from Anton's Exercise 21a in Exercise Set 18.7 (this is an elementary computation). This proves that $\overline{F}$ is irrotational, as desired.

To prove statement (b), we assume $\overline{F}$ is irrotational on U, and take C to be any simple closed curve in U which is the boundary of some smooth surface σ. We will show that $\int_\sigma \overline{F} \cdot d\overline{r} = 0$. To see this, use the fact that U is simply-connected to obtain a smooth surface σ' <u>contained in U</u> whose boundary is C (such a surface can be swept out by shrinking C to a point in U). Then Stokes' Theorem will apply to σ' and C (when σ' is properly oriented with respect to C):

$$\int_C \overline{F} \cdot d\overline{r} = \iint_{\sigma'} (\text{curl } \overline{F}) \cdot \overline{n} \, dS \qquad \text{by Stokes' Theorem}$$

$$= \iint_{\sigma'} o \, dS \qquad \begin{array}{l} \text{since } \sigma' \text{ is contained in } U \\ \text{and } \overline{F} \text{ is irrotational on } U \end{array}$$

$$= 0$$

Thus $\overline{F}$ is circulation-free on U, at least with respect to all simple closed curves in U which are boundaries for surfaces. Using this result one can then show that $\overline{F}$ is a gradient field on U, which in turn will show that $\overline{F}$ is circulation-free with respect to <u>all</u> closed curves in U (we'll omit the details). Thus $\overline{F}$ is conservative on U, as desired.

If you compare this result with the discussion in §18.3.4 of The Companion, you will see that Stokes' Theorem plays the same key role in the theory of conservative vector fields in 3-space that Green's Theorem plays in the 2-space theory.

Example C. Suppose $\overline{F}(x,y,z) = \langle z^2 + yz - x^2, xz, 2xz + xy \rangle$ and C is the curve which lies entirely on the unit sphere centered at the origin and joins $(0, -1, 0)$ and $(-1, 0, 0)$ as shown. Evaluate $\int_C \overline{F} \cdot d\overline{r}$.

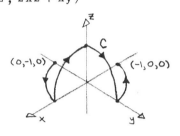

Solution. We'll first compute the curl of $\overline{F}$.

$$\text{curl } \overline{F} = \nabla \times \overline{F} = \begin{vmatrix} \overline{i} & \overline{j} & \overline{k} \\ \partial/\partial x & \partial/\partial y & \partial/\partial z \\ z^2 + yz - x^2 & xz & 2xz + xy \end{vmatrix}$$

$$= \langle x - x, \ 2z + y - 2z - y, \ z - z \rangle = \overline{0}$$

Thus, by the Conservative Vector Fields Theorem in 3-space, $\overline{F}$ is a conservative vector field on all of 3-space. Hence it is path-independent, and we can compute the line integral over a simpler curve C'. We'll chose C' to be the straight line segment joining $(0, -1, 0)$ to $(-1, 0, 0)$.

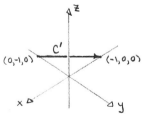

To parameterize C', note that a direction vector for the line is given by $\overline{u} = \langle -1, 0, 0 \rangle - \langle 0, -1, 0 \rangle = \langle -1, 1, 0 \rangle$. Then, using $(0, -1, 0)$ as the initial point, C' will be parameterized by

$$\overline{r}(t) = \langle 0, -1, 0 \rangle + t \langle -1, 1, 0 \rangle$$

$$= \langle -t, t - 1, 0 \rangle$$

where t is restricted to the interval $0 \le t \le 1$. We can now evaluate the line integral:

$$\int_C \overline{F} \cdot d\overline{r} = \int_{C'} \overline{F} \cdot d\overline{r} \qquad \text{since} \quad \overline{F} \quad \text{is path-independent}$$

$$= \int_C (z^2 + yz - x^2) \, dz + xz \, dy + (2xz + xy) \, dz$$

$$= \int_0^1 [0^2 + (t-1)(0) - (-t)^2] \frac{dx}{dt} + (-t)(0) \frac{dy}{dt} + (2(-t)(0) + (-t)(t-1)) \frac{dz}{dt}] \, dt$$

using $x = -t$, $y = t - 1$, $z = 0$
as in §18.1.4 of The Companion

$$= \int_0^1 [-t^2(-1) - t(t-1)(0)] \, dt = \int_0^1 t^2 \, dt = \frac{t^3}{3} \Big|_0^1 = \boxed{\frac{1}{3}}$$

Alternate Solution.   Since $\overline{F}$ is conservative (we proved that above), there is a potential

function $\varphi(x, y, z)$ such that $\overline{F} = \nabla\varphi$. We will determine φ by using the same techniques

employed in 2-space in §18.2.3. The vector equation $\nabla\varphi = \overline{F}$ is equivalent to three

scalar partial differential equations:

$$\frac{\partial\varphi}{\partial x} = z^2 + yz - x^2$$

$$\frac{\partial\varphi}{\partial y} = xz$$

$$\frac{\partial\varphi}{\partial z} = 2xz + xy$$

We integrate the first of these equations with respect to x to obtain

$$\varphi(x,y,z) = \int (z^2 + yz - x^2)\, dx$$

$$= xz^2 + xyz - x^3/3 + \psi(y,z)$$

where $\psi(y,z)$ is the "constant of integration," an unknown function of y and z. Now take the partial derivative with respect to y of this expression for $\varphi(x,y,z)$, and substitute into our 2-nd partial differential equation:

$$xz + \frac{\partial \psi}{\partial y}(y,z) = \frac{\partial \varphi}{\partial y} = xz$$

Thus $\frac{\partial \psi}{\partial y}(y,z) = 0$, which means that

$$\psi(y,z) = h(z)$$

where $h(z)$ is an unknown function of z alone (i.e., h has no dependence on y). Thus

$$\varphi(x,y,z) = xz^2 + xyz - x^3/3 + h(z)$$

and we take the partial derivative with respect to z of this expression for $\varphi(x,y,z)$ and substitute it into our 3-rd partial differential equation:

$$2xz + xy + h'(z) = \frac{\partial \varphi}{\partial z} = 2xz + xy$$

Thus $h'(z) = 0$, so $h(z)$ is a constant. Since we can choose $h(z)$ to be any constant, for convenience we'll choose $h(z) = 0$, which yields

$$\boxed{\varphi(x,y,z) = xz^2 + xyz - x^3/3}$$

It is easy to show that the Fundamental Theorem of Line Integrals (§18.2.3 of The Companion) holds in 3-space as well as in 2-space. Thus

$$\int_C \overline{F} \cdot d\overline{r} = \int_C \nabla\varphi \cdot d\overline{r}$$

$$= \varphi(-1,0,0) - \varphi(0,-1,0)$$

$$= -(-1)^3/3 = \boxed{1/3} \qquad\qquad \square$$

The advantage of the first solution to Example C is that it is faster when only one integral is being considered. The advantage of the second solution is that once the "potential function" φ is found, then all line integrals of $\overline{F}$ become trivial to compute.

Exercises.   These exercises cover the optional material of §18.7.3.

1.   Determine which of the following subsets of 3-space are simply connected:

        a.   U = all of 3-space minus the y-axis

        b.   V = all of 3-space minus the non-negative y-axis

        c.   W = all of 3-space minus the unit disk
             in the xy-plane $(x^2 + y^2 \le 1$, z = 0)

2.   Let $\overline{F}(x,y,z) = \langle 2xy + z, x^2 + z^2, 2yz + x\rangle$

        a.   On what set U is $\overline{F}$ irrotational?

        b.   Does this prove $\overline{F}$ is conservative on U?

        c.   Does $\overline{F}$ have a potential function? If so, determine one.

d. Evaluate

$$\int_C \overline{F} \cdot d\overline{r}$$

where C is the curve parameterized by

$$\overline{r}(t) = \langle \ln t, t^2 - (e+1)t + e, t \rangle, \quad 1 \le t \le e$$

3. Let $\overline{F}(x,y,z) = \langle 2xe^y, x^2 e^y, 2z \rangle$. Evaluate $\int_C \overline{F} \cdot d\overline{r}$ where C is

parameterized by $\overline{r}(t) = \langle t - t^2, t^5, t^3 \rangle, \quad 0 \le t \le 1.$

4. Let $\overline{F}(x,y,z) = \langle x, y, xz \rangle$. Does there exist a potential function

$\varphi(x,y,z)$ such that $\overline{F} = \nabla\varphi$?

5. Let $\overline{F}(x,y,z) = \langle -\dfrac{z}{x^2 + z^2}, y^2, \dfrac{x}{x^2 + z^2} \rangle$

a. On what set U is $\overline{F}$ irrotational?

b. Does this prove $\overline{F}$ is conservative on U ?

c. Evaluate

$$\int_C \overline{F} \cdot d\overline{r}$$

where C is the unit circle in the xz-plane, parameterized

counterclockwise.

d. Is $\overline{F}$ conservative on U ?

e. Evaluate

$$\int_{C_0} \overline{F} \cdot d\overline{r}$$

where C_0 is the ellipse $\dfrac{(x-1)^2}{9} + \dfrac{z^2}{4} = 1$ in the xz-plane,

parameterized counterclockwise.

Answers.

1. V and W are simply-connected; U and X are not.

2. a. $\overline{F}$ is irrotational on U = all of 3-space.

    b. $\overline{F}$ is therefore conservative on U since U is simply-connected.

    c. $\overline{F} = \nabla\varphi$ where $\varphi(x,y,z) = x^2 y + y z^2 + xz$

    d. $\int_C \overline{F} \cdot d\overline{r} = \varphi(\overline{r}(e)) - \varphi(\overline{r}(1)) = e$

3. Since curl $\overline{F} = 0$ on all of 3-space, then $\overline{F}$ is conservative. Hence

    $$\int_C \overline{F} \cdot d\overline{r} = \int_{C_0} \overline{F} \cdot d\overline{r} \text{ where } C_0 \text{ is the line segment connecting}$$

    $\overline{r}(0) = \langle 0,0,0 \rangle$ with $\overline{r}(1) = \langle 0,1,1 \rangle$. Using the parameterization

    $\overline{r}_0(t) = \langle 0,t,t \rangle$, $0 \leq t \leq 1$, yields $\int_C \overline{F} \cdot d\overline{r} = 1$.

4. Curl $\overline{F} = \langle 0, -z, 0 \rangle$. Since this is not zero on any open set in 3-space then $\overline{F}$ is not irrotational, and hence not a gradient field.

5. a. Curl $\overline{F} = 0$ at every (x,y,z) where $\overline{F}$ is defined. Thus $\overline{F}$ is irrotational on the set U = all 3-space minus the y-axis.

    b. This does not prove that $\overline{F}$ is conservative on U because U is not simply-connected.

    c. 2π.

    d. $\overline{F}$ is not conservative on U, since if it was conservative on U, then the answer to part (c) would be zero.

e. 2π. The answer is obtained by letting σ

be the surface in the   xz-plane   between

C   and C_0. Since   curl $\overline{F} = 0$ on σ,

then Stokes' Theorem will show that

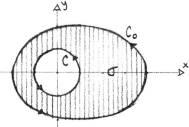

$$\int_{C_0} \overline{F} \cdot d\overline{r} = \int_C \overline{F} \cdot d\overline{r}, \text{ which equals } 2\pi \text{ by part (c).}$$

Section 18. 8: Applications

The techniques of vector calculus play a central role in those areas of physics and engineering in which "vector fields" are prominent, e. g. , fluid flows or electromagnetic theory. In this section we develop the important concepts of <u>flux</u> and <u>circulation</u> in a fluid flow, and analyze them using vector calculus techniques. (More will be said about electromagnetism in §18. 9. 8.)

The primary connections are those between flux and divergence, and between circulation and curl. Since these connections come about from the integral fromulations for divergence and curl, we will start with these formulas.

1. <u>Integral formulations for div $\overline{F}$ and curl $\overline{F}$.</u> Since both   div $\overline{F}$ and   curl $\overline{F}$ are defined in terms of partial derivatives, it is surprising that both can be expressed as (limits of) integrals.

Suppose $\overline{F}$ has continuous first partial derivatives in an open region   U   of   3-space. If   (x,y,z)   is any point of   U ,   then

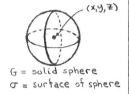

Integral formulation for div $\overline{F}$

$$\operatorname{div} \overline{F}(x,y,z) = \lim_{\text{vol}(G) \to 0} \frac{1}{\text{vol}(G)} \iint_{\sigma} \overline{F} \cdot \overline{n} \, dS$$

where G is a sphere centered on (x,y,z), with boundary surface σ.

G = solid sphere
σ = surface of sphere

Integral formulation for curl $\overline{F}$

$$\operatorname{curl} \overline{F}(x,y,z) \cdot \overline{n} = \lim_{\text{area}(D) \to 0} \frac{1}{\text{area}(D)} \int_{C} \overline{F} \cdot d\overline{r}$$

where $\overline{n}$ is any unit vector, D is a disk centered on (x,y,z) and perpendicular to $\overline{n}$, and C is the boundary curve of D, directed according to the right-hand rule with respect to D.

D = disk
C = circle

Anton gives the divergence formula after Equation (6) and the curl formula as Equation (9). The two formulas are based on the Divergence Theorem (Theorem 18.6.2) and Stokes' Theorem (Theorem 18.7.2), respectively. A discussion of the proofs will be given in our (optional) Subsection 5.

2. <u>Flux and divergence.</u> If fluid is flowing through a region of space, at every point (x,y,z) in the region let

$\overline{F}(x,y,z)$ = the velocity vector of the fluid at the point (x,y,z)

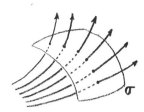

$\overline{F}$ is called the <u>flow field</u> for the moving fluid. Suppose σ is an oriented surface in the path of the fluid. Then

<table>
<tr><td>Flux
across
a surface</td><td>$\begin{bmatrix} \text{the } \underline{\text{flux}} \text{ of} \\ \overline{F} \quad \text{across} \quad \sigma \end{bmatrix}$ = $\begin{bmatrix} \text{the total (net) volume of fluid crossing from the} \\ \text{negative to positive side of } \sigma \text{ per unit time} \end{bmatrix}$

$= \displaystyle\iint_{\sigma} \overline{F} \cdot \overline{n} \; dS$</td></tr>
</table>

This surface integral formula for flux is the first result that Anton discusses in this section; an alternate derivation is given in §18.4.5 of <u>The Companion</u>. Flux is the fundamental, physical way in which a surface integral can be "visualized." In this sense, the interpretation

$$\iint_{\sigma} \overline{F} \cdot \overline{n} \; dS = \begin{bmatrix} \text{the } \underline{\text{flux}} \text{ of} \\ \overline{F} \quad \text{across} \quad \sigma \end{bmatrix}$$

is as basic as the area interpretation for ordinary integrals discussed in §5.7:

$$\int_{a}^{b} f(x) \, dx = \begin{bmatrix} \text{the } \underline{\text{area}} \text{ under the} \\ \text{graph of } f, \text{ over} \\ \text{the interval } [a, b] \\ (\text{when } f \geq 0) \end{bmatrix}$$

Anton's Example 1 illustrates the calculation of the flux of a flow field $\overline{F}$ across an oriented surface σ. Any of the methods for evaluating surface integrals given in §§18.5 - 18.6 may be used for such a calculation.

Often situations involving flux call for the conversion of "integral conditions" into "differential conditions." Here is a simple (yet important) application of this technique:

Example A. Suppose $\overline{F}$ represents a flow field of an <u>incompressible</u> fluid through an open

region U. Show that

$$\text{div } \overline{F} = 0 \qquad \text{at every point of U}$$

Solution. Let G be any spherical region in U, and let σ be its boundary surface.

Since the fluid is <u>incompressible</u>, then the volume of fluid entering σ per unit time must

equal the volume of fluid leaving σ per unit time. Thus the (net) flux of $\overline{F}$ across σ

must equal zero, i. e. ,

$$\iint_{\sigma} \overline{F} \cdot \overline{n} \; dS = 0$$

Now let (x, y, z) be any point in U. Then

$$\text{div } \overline{F}(x, y, z) = \lim_{\text{vol (G)} \to 0} \frac{1}{\text{vol (G)}} \iint_{\sigma} \overline{F} \cdot \overline{n} \; dS$$

$$\text{by the integral formulation for div } \overline{F}$$

$$= \lim_{\text{vol (G)} \to 0} 0 = 0$$

Thus $\text{div } \overline{F} = 0$ at every point of U   □

In view of Example A , we make the following definition:

Incompressible vector field	A vector function $\overline{F}$ is said to be an incompressible vector field on a region U if $\text{div } \overline{F} = 0$ at every point of U

This concept will play a central role in the study of "solenoidal vector fields" in §18.9.

Example A shows that flux and divergence are closely related. In fact, by using the integral formulation for $\text{div} \, \overline{F}(x,y,z)$, we can interpret $\text{div} \, \overline{F}(x,y,z)$ as the flux density of $\overline{F}$ at (x,y,z):

$$\text{div} \, \overline{F}(x,y,z) = \lim_{\text{vol}(G) \to 0} \frac{1}{\text{vol}(G)} \iint_\sigma \overline{F} \cdot \overline{n} \, dS$$

average flux of $\overline{F}$ per unit volume on G, i. e., average flux density of $\overline{F}$ on G

The limit $\text{vol}(G) \to 0$ changes the average flux density into an (instantaneous) flux density

3. Circulation and curl. Now we investigate the interpretation of line integrals as circulation. First we give a careful motivation for the definition of circulation.

Let C be a circular ring of radius a which is in a flow field $\overline{F}$. We are interested in measuring the tendency of the ring to turn under the influence of $\overline{F}$.

Assign a direction to C and assume C is parameterized by $\overline{r}(\theta)$, $0 \le \theta \le 2\pi$. Then $\overline{T}(\theta)$ denotes the unit tangent vector to C at $\overline{r}(\theta)$, and $(\overline{F} \cdot \overline{T})(\theta)$ is the component of the flow field at $\overline{r}(\theta)$ which is tangent to C.

There are three possibilities at the point of C   corresponding to θ :

(1) $(\overline{F} \cdot \overline{T})\, (\theta) > 0.$ Then $\overline{F}(\theta)$ and $\overline{T}(\theta)$

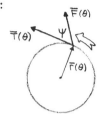

form an acute angle ψ,   and the fluid flow

at $\overline{r}(\theta)$ is trying to turn the ring counter-

clockwise;

(2) $(\overline{F} \cdot \overline{T})\, (\theta) < 0.$ Then $\overline{F}(\theta)$ and $\overline{T}(\theta)$

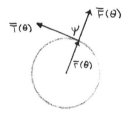

form an obtuse angle ψ,   and the fluid

flow at $\overline{r}(\theta)$ is trying to turn the ring

clockwise;

(3) $(\overline{F} \cdot \overline{T})\, (\theta) = 0.$ Then $\overline{F}(\theta)$ and $\overline{T}(\theta)$

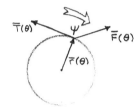

form a right angle ψ,   and the fluid flow

at $\overline{r}(\theta)$ is not trying to turn the ring

at all.

In order to assess the __full__ rotational effect of $\overline{F}$ (on all of   C)   we take the line integral

of $\overline{F}$ over   C:

Circulation around a circle	The circulation of $\overline{F}$ around C is defined to be the line integral $$\int_C \overline{F} \cdot d\overline{r} = \int_C \overline{F} \cdot \overline{T}\ ds$$	

Hence, if $\displaystyle\int_C \overline{F} \cdot \overline{T}\ ds > 0,$ then the forces trying to turn the ring counterclockwise

$(\overline{F} \cdot \overline{T}\, (\theta) > 0)$ outweigh the forces trying to turn the ring clockwise $(\overline{F} \cdot \overline{T}\, (\theta) < 0)$;

the net effect is that the ring is turned counterclockwise.

Similarly, if $\displaystyle\int_C \overline{F} \cdot \overline{T} \, ds < 0$, then the net effect is that the ring is turned clock-

wise, and if $\displaystyle\int_C \overline{F} \cdot \overline{T} \, ds = 0$ then the net effect is that the ring is not turned at all. Here

is an example:

Example B.   Suppose the ring   C   given by

$$\overline{r}(\theta) \; = \; \langle 2 \cos \theta \, , \, 0 \, , \, 2 \sin \theta \rangle$$

is in the flow field $\overline{F}(x,y,z) \; = \; \langle xz \, , \, x+y \, , \, x \rangle$.

Determine the circulation of $\overline{F}$ around   C.

Solution. $\displaystyle\int_C \overline{F} \cdot d\overline{r} \; = \; \int_C xz \, dx \; + \; (x+y) \, dy \; + \; x \, dz$

$$= \; \int_0^{2\pi} \; [\, xz \, \frac{dx}{d\theta} \; + \; (x+y) \, \frac{dy}{d\theta} \; + \; x \, \frac{dz}{d\theta} \,] \; d\theta$$

$$= \; \int_0^{2\pi} \; [\, 4 \cos \theta \sin \theta \, (- 2 \sin \theta) \; + \; 2 \cos \theta \, (2 \cos \theta) \,] \; d\theta$$

where we have used $x = 2 \cos \theta$, $y = 0$, $z = 2 \sin \theta$

$$= \; \int_0^{2\pi} \; - 8 \sin^2 \theta \cos \theta \; d\theta \; + \; \int_0^{2\pi} \; 4 \cos^2 \theta \; d\theta$$

$$= \; \int_0^0 \; - 8 u^2 \; du \; + \; 4 \int_0^{2\pi} \; \frac{1 \, + \, \cos 2\theta}{2} \; d\theta$$

where we have used $u = \sin \theta$

$$= \; 0 \; + \; 2 \, (\theta \, + \, \frac{1}{2} \sin 2\theta) \, \Big|_0^{2\pi} \quad = \quad \boxed{4\pi}$$

Since $4\pi > 0$, this means that the flow field $\overline{F}$ is turning C in the direction of its parameterization.   □

Any of the methods for evaluating line integrals given in §§18. 1 - 18. 3 and 18. 7 may be used to calculate circulation integrals.

As with flux in the previous subsection, often situations involving circulation call for the conversion of "integral conditions" into "differential conditions. " Here is a simple (yet important) application of this technique:

Example C.   Suppose $\overline{F}$ represents a flow field in a region U which is <u>irrotational</u> in the sense that any small circular ring placed in the flow will not rotate about its center. Show that

$$\operatorname{curl} \overline{F} = 0 \quad \text{at every point in} \quad U$$

<u>Solution.</u>   Since $\displaystyle\int_C \overline{F} \cdot d\overline{r}$ denotes the circulation of $\overline{F}$ about C, the irrotational condition translates into

$$\int_C \overline{F} \cdot d\overline{r} = 0 \quad \text{for all small rings} \quad C \quad \text{in} \quad U$$

Now let (x,y,z) be any point in U, and $\overline{n}$ any unit vector in 3-space. Then

$$\operatorname{curl} \overline{F}(x,y,z) \cdot \overline{n} = \lim_{\operatorname{area}(D) \to 0} \frac{1}{\operatorname{area}(D)} \int_C \overline{F} \cdot d\overline{r}$$

by the integral formulation for $\operatorname{curl} \overline{F}$, where D is a disk centered on (x,y,z), perpendicular to $\overline{n}$, with boundary curve C, directed according to the right-hand rule with respect to $\overline{n}$

$$= \lim_{\operatorname{area}(D) \to 0} 0 = 0$$

Thus, since this is true for all (x,y,z) in U, and for all unit vectors $\overline{n}$, this proves

$$\text{curl } \overline{F} = 0 \quad \text{at every point in } U \qquad\qquad \square$$

In view of Example C, we make the following definition:

Irrotational vector field	A vector function $\overline{F}$ is said to be an irrotational vector field on a region U if $\text{curl } \overline{F} = 0$ at every point of U

This is the same definition for irrotational that we encountered in §18.7.3, only now we see the reason for the name!

Example C shows that circulation and curl are closely related. In fact, by using the integral formulation for $\text{curl } \overline{F}(x,y,z) \cdot \overline{n}$, we can interpret $\text{curl } \overline{F}(x,y,z) \cdot \overline{n}$ as the rotation of $\overline{F}$ about $\overline{n}$ at (x,y,z):

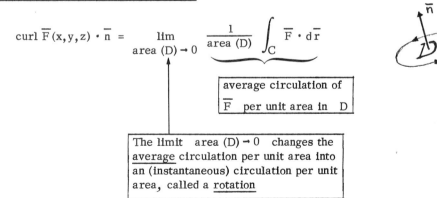

$$\text{curl } \overline{F}(x,y,z) \cdot \overline{n} = \lim_{\text{area }(D) \to 0} \underbrace{\frac{1}{\text{area }(D)} \int_C \overline{F} \cdot d\overline{r}}$$

average circulation of $\overline{F}$ per unit area in D

The limit area $(D) \to 0$ changes the average circulation per unit area into an (instantaneous) circulation per unit area, called a rotation

4. <u>All these similarities</u> ... ! No doubt you have noticed the startling similarities in our treatment of flux and circulation in this section. In case you missed them, here is a diagram of "corresponding concepts:"

flux	....	circulation
divergence	....	curl
Divergence Theorem	....	Stokes' Theorem
incompressible	....	irrotational

$$\iint_{\sigma} \overline{F} \cdot \overline{n} \ dS \qquad .... \qquad \int_{C} \overline{F} \cdot d\overline{r}$$

$$\iiint_{G} \operatorname{div} \overline{F} \ dV \qquad .... \qquad \iint_{\sigma} (\operatorname{curl} \overline{F}) \cdot \overline{n} \ dS$$

There are very deep and beautiful reasons for this correlation, and we will explore them in §§18. 9 - 18. 10 .

5. <u>More on the integral formulations for $\operatorname{div} \overline{F}$ and $\operatorname{curl} \overline{F}$ (optional)</u>. The proofs of the integral formulations for $\operatorname{div} \overline{F}$ and $\operatorname{curl} \overline{F}$, as well as many of the applications of these formulas, use generalizations of Theorem 5.9.4 , the Mean-Value Theorem (MVT) for Integrals:

MTV for Integrals	If $y = f(x)$ is continuous on a closed interval $[a, b]$ then there is at least one number x^* in $[a, b]$ such that $$\frac{1}{b - a} \int_{a}^{b} f(x) \ dx = f(x^*)$$

Intuitively this says that there is some point x^* in $[a,b]$ at which the value of f is the average value of f over $[a,b]$. This easy-to-believe result generalizes to triple integrals as follows:

MVT for Triple Integrals	If $w = f(x,y,z)$ is continuous on a closed sphere G, then there is at least one point (x^*, y^*, z^*) in G such that $$\frac{1}{\text{vol}(G)} \iiint_G f(x,y,z)\, dV = f(x^*, y^*, z^*)$$

We will use this result to prove the integral formulation for divergence. Suppose $\overline{F}$ has continuous first partial derivatives in an open region U of 3-space. If (x,y,z) is any point in U, and G is any small sphere centered on (x,y,z) with boundary surface σ, then

$$\frac{1}{\text{vol}(G)} \iint_\sigma \overline{F} \cdot \overline{n}\, dS = \frac{1}{\text{vol}(G)} \iiint_G \text{div}\, \overline{F}\, dV$$

G = solid sphere
σ = surface of sphere

by the Divergence Theorem (Theorem 18. 6. 2)

$$= \text{div}\, \overline{F}(x^*, y^*, z^*)$$

by the MVT for triple integrals,
where (x^*, y^*, z^*) is some point in G

Hence, if we let $\text{vol}(G)$ approach zero, then (x^*, y^*, z^*) must approach (x,y,z), the center of the sphere G. Thus

$$\lim_{\text{vol}(G) \to 0} \frac{1}{\text{vol}(G)} \iint_\sigma \overline{F} \cdot \overline{n}\, dS = \lim_{\substack{(x^*, y^*, z^*) \\ \to (x,y,z)}} \text{div}\, \overline{F}(x^*, y^*, z^*)$$

$$= \text{div}\, \overline{F}(x,y,z)$$

since $\overline{F}$ having continuous partial derivatives implies $\text{div}\, \overline{F}$ is continuous

This proves the integral formulation for divergence.

The proof of the integral formulation for curl is identical, with Stokes' Theorem replacing the Divergence Theorem, and a MVT for scalar surface integrals replacing the MVT for triple integrals.

The next example is a generalization of Example A , and represents the most common way the integral formulation for divergence is used in applications. It is another illustration of the conversion of "integral conditions" into "differential conditions" discussed in Subsection 2.

Example D. Suppose $\overline{F}$ is a vector function with continous partial derivatives, and f is a continuous scalar function on an open region U in 3-space. Suppose further that $\overline{F}$ and f are related by

$$\iint_\sigma \overline{F} \cdot \overline{n} \; dS = \iiint_G f \; dV$$

for every solid sphere G contained in U , with σ as its boundary surface (oriented with outward normals). Show that

$$\text{div } \overline{F} = f \qquad \text{at every point of U}$$

Solution. Let (x,y,z) be any point in U . Then

$$\text{div } \overline{F}(x,y,z) = \lim_{\text{vol}(G) \to 0} \frac{1}{\text{vol}(G)} \iint_\sigma \overline{F} \cdot \overline{n} \; dS \qquad \begin{array}{l} \text{by the integral} \\ \text{formulation for divergence} \end{array}$$

$$= \lim_{\text{vol}(G) \to 0} \frac{1}{\text{vol}(G)} \iiint_G f \; dV \qquad \begin{array}{l} \text{by the given} \\ \text{integral conditions} \end{array}$$

$$= \lim_{\text{vol}(G) \to 0} f(x^*, y^*, z^*) , \qquad \text{for } (x^*, y^*, z^*) \text{ some point in G}$$

$$\qquad\qquad\qquad \begin{array}{l} \text{The existence of } (x^*, y^*, z^*) \text{ is guaranteed} \\ \text{by the MVT for triple integrals} \end{array}$$

$$= f(x,y,z), \quad \text{since f is continuous}$$

Thus the corresponding set of differential conditions are

$$\text{div } \overline{F} = f \qquad \text{at every point of } U \qquad\qquad \square$$

The surface integral conditions in Example D lead to differential conditions in terms of the divergence; in the next example line integral conditions lead to differential conditions in terms of curl. The derivation, which we will omit, is similar to that of Example D.

Example E. Suppose $\overline{F}$ is a vector function with continuous partial derivatives, and $\overline{J}$ is a continuous vector function on an open region U in 3-space. Suppose further that $\overline{F}$ and $\overline{J}$ are related by

$$\int_C \overline{F} \cdot d\overline{r} = \iint_\sigma \overline{J} \cdot \overline{n} \; dS$$

for every smooth surface σ in U with a circle C as its boundary curve (oriented by the right-hand rule with respect to σ). Then the corresponding set of differential conditions on $\overline{F}$ are

$$\text{curl } \overline{F} = \overline{J} \qquad \text{at every point of } U \qquad\qquad \square$$

Section 18.9: Solenoidal Vector Fields (Optional)

1. Solenoidal vector fields. In §§18.2 and 18.7.3 of The Companion we studied a class of vector functions known as conservative vector fields, and verified four equivalent formulations for such functions under the title of the Conservative Vector Fields Theorem (CVFT). In this section we study another class of vector functions known as solenoidal vector fields, and verify a theorem which is very similar to the CVFT. This will explain the similarities

between the discussions of flux and circulation which we pointed out in §18. 8. 4 of <u>The</u>

<u>Companion.</u>

    Here is the theorem:

<table>
<tr>
<td>The
Solenoidal
Vector
Fields
Theorem</td>
<td>

Suppose $\overline{F}(x,y,z)$ is a vector function on an open region U.

Then $\overline{F}$ is <u>solenoidal</u> if it possesses any one (hence all of the

following equivalent properties:

    (1) $\overline{F}$ is <u>flux-free</u> on U, i.e., $\iint_\sigma \overline{F} \cdot \overline{n}\ dS = 0$ for

         every surface σ in U which bounds a solid in $\mathbb{R}^3$;

    (2) $\overline{F}$ has <u>span-independent</u> surface integrals for every

         surface σ in U, i.e., $\iint_\sigma \overline{F} \cdot \overline{n}\ dS$ depends

         only on the boundary curve of σ.

    (3) $\overline{F}$ is a <u>curl-field</u> on U, i.e., $\overline{F} = \operatorname{curl}\overline{H}$ for some

         vector function $\overline{H}$ defined on U.

Moreover, if $\overline{F}(x,y,z)$ has continuous first partial derivatives

on U, and U is an open, <u>2-simply connected</u> region, then

there is a fourth equivalence:

    (4) $\overline{F}$ is <u>incompressible</u> on U, i.e., $\operatorname{div}\overline{F} = 0$ at every

         point in U.

</td>
</tr>
</table>

    We will discuss these four properties in the order in which they are mentioned in

the theorem.

2. **Flux-free vector fields.** A vector function $\overline{F}(x,y,z)$ is said to be a <u>flux-free</u> vector

field on an open region U if

| Flux-free vector field | $\displaystyle\iint_\sigma \overline{F}\cdot\overline{n}\ dS = 0$ for every surface σ in U <br> which bounds a solid in $\mathbb{R}^3$ |

Since $\displaystyle\iint_\sigma \overline{F}\cdot\overline{n}\ dS$ measures flux across σ (as discussed in §§18.5.5 and 18.8.2

of <u>The Companion</u>) then the terminology "flux-free" is extremely natural.

<u>Example A.</u>   Evaluate $\displaystyle\iint_\sigma \overline{F}\cdot\overline{n}\ dS$ where σ is the surface of the unit sphere centered

on $(0,0,0)$ and

$$\overline{F}(x,y,z) = \langle\, 1\, ,\, -2z\,(x^2+y^2+z^2)^{-2}\, ,\, 1+2y\,(x^2+y^2+z^2)^{-2}\, \rangle$$

<u>Solution.</u>   Direct computation of this surface integral would be unpleasant, and the

Divergence Theorem (Theorem 18.6.2) does not apply since $\overline{F}$ is not defined at $(0,0,0)$,

a point in the interior of σ . However, as we will show in Examples C and F   below, $\overline{F}$

is <u>solenoidal</u> on all of 3-space minus the origin. Hence

$$\iint_\sigma \overline{F}\cdot\overline{n}\ dS = 0 \qquad\qquad\qquad \square$$

3. **Span-independent surface integrals.** A surface integral $\displaystyle\iint_\sigma \overline{F}\cdot\overline{n}\ dS$ is said to be

<u>independent of spanning surface</u>, or <u>span-independent</u>, in an open region U if

the value of $\displaystyle\iint_\sigma \overline{F} \cdot n \, dS$ depends only on

the boundary curve of σ and not on the

particular spanning surface in U which

has this boundary curve.

Thus, if σ_1 and σ_2 are surfaces in U with the

same boundary curve C, then if the surface integral

is span-independent, we have

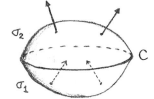

$$\iint_{\sigma_1} \overline{F} \cdot \overline{n} \, dS = \iint_{\sigma_2} \overline{F} \cdot \overline{n} \, dS$$

provided that the orientation of σ_1 and σ_2 induce the same direction for the boundary

curve C.

It is convenient to know when a surface integral is span-independent for the following

reason: if we must compute $\displaystyle\iint_\sigma \overline{F} \cdot \overline{n} \, dS$ for a complicated surface σ, we can replace

it with $\displaystyle\iint_{\sigma'} \overline{F} \cdot \overline{n} \, dS$ where σ' is a simpler surface which has the same boundary curve

as σ.

Example B.   Evaluate $\displaystyle\iint_\sigma \overline{F} \cdot \overline{n} \, dS$ where σ is the surface $x^2 + y^2 + z^3 = 4$, $z \geq 0$,

oriented by upward normals, and

$$\overline{F}(x,y,z) = \langle 1, \, -2z(x^2 + y^2 + z^2)^{-2}, \, 1 + 2y(x^2 + y^2 + z^2)^{-2} \rangle$$

Solution.   A direct computation would be awful. Instead we will verify in Examples C and F

below that the vector function $\overline{F}$ is solenoidal on all of 3-space minus the origin, and hence has span-independent surface integrals. Thus we can change the surface σ to make the surface integral more palatable. The boundary curve of σ is given when $z = 0$:

$$x^2 + y^2 = 4$$

the circle in the xy-plane with center $(0,0)$ and radius 2. It is tempting to replace σ by the flat disk in the xy-plane whose boundary is the given circle, but this will not work since the ill-behaved point $(0,0,0)$ is in this disk. Instead we will take σ' to be the (upper) hemisphere

$$x^2 + y^2 + z^2 = 4, \qquad z \geq 0$$

This seemingly minor change in σ will allow us to evaluate the surface integral

$\iint_{\sigma'} \overline{F} \cdot \overline{n} \; dS$ by the use of spherical coordinates as described in §18.5.6 of __The Companion__:

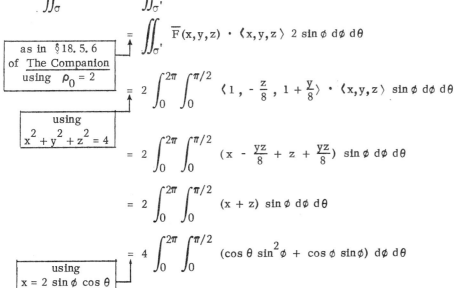

$$\iint_{\sigma} \overline{F} \cdot \overline{n} \; dS = \iint_{\sigma'} \overline{F} \cdot \overline{n} \; dS$$

as in §18.5.6 of The Companion using $\rho_0 = 2$

$$= \iint_{\sigma'} \overline{F}(x,y,z) \cdot \langle x,y,z \rangle \; 2 \sin\phi \; d\phi \; d\theta$$

$$= 2 \int_0^{2\pi} \int_0^{\pi/2} \langle 1, -\frac{z}{8}, 1+\frac{y}{8} \rangle \cdot \langle x,y,z \rangle \; \sin\phi \; d\phi \; d\theta$$

using $x^2 + y^2 + z^2 = 4$

$$= 2 \int_0^{2\pi} \int_0^{\pi/2} (x - \frac{yz}{8} + z + \frac{yz}{8}) \; \sin\phi \; d\phi \; d\theta$$

$$= 2 \int_0^{2\pi} \int_0^{\pi/2} (x + z) \; \sin\phi \; d\phi \; d\theta$$

$$= 4 \int_0^{2\pi} \int_0^{\pi/2} (\cos\theta \, \sin^2\phi + \cos\phi \, \sin\phi) \; d\phi \; d\theta$$

using $x = 2 \sin\phi \cos\theta$, $z = 2 \cos\phi$

$$= \dots = \boxed{4\pi} \qquad \text{(We'll leave the details to you)} \qquad \square$$

$$* \quad * \quad * \quad * \quad *$$

Only an elementary argument is needed to prove that a vector function $\overline{F}$ is flux-free if and only if it is span-independent. Such an argument can be constructed in the same way as we did in §18. 2. 2 of The Companion when we proved that a vector function is circulation-free if and only if it is path-independent. We'll leave the details to you.

This establishes the first two equivalences in the Solenoidal Vector Fields Theorem.

4. Curl fields. A vector function $\overline{F}(x, y, z)$ is said to be a curl field on an open region

U if

curl field		$\overline{F} = \text{curl } \overline{H}$ for some vector function $\overline{H}$ defined on U

The function $\overline{H}$ is called a vector potential function for $\overline{F}$

If we know that $\overline{F} = \text{curl } \overline{H}$, then evaluation of surface integrals for $\overline{F}$ is greatly simplified because of Stokes' Theorem (Theorem 18. 7. 2):

$$\iint_{\sigma} (\text{curl } \overline{H}) \cdot \overline{n} \; dS = \int_{C} \overline{H} \cdot d\overline{r}$$

where C is the boundary curve of σ, oriented by the right-hand rule with respect to σ. Stokes' Theorem also shows immediately that a curl field has span-independent surface integrals since the line integral $\int_{C} \overline{H} \cdot d\overline{r}$ depends only on the bounding curve C, not on the spanning surface σ itself. Thus condition (3) in the Solenoidal Vector Fields Theorem implies condition (2) (and hence condition (1) by the argument at the end of §18. 9. 3).

Conversely, assume conditions (1) and (2) of the Solenoidal Vector Fields Theorem hold for a vector function $\overline{F}$ on an open region U. Then it can be shown that $\overline{F}$ is indeed a curl field, although it requires some subtle analysis to construct a vector potential function $\overline{H}$ by using span-independence.

This establishes the equivalence of the first three conditions in the Solenoidal Vector Fields Theorem.

*   *   *   *   *

The process of determining a vector potential $\overline{H}$ for a curl-field $\overline{F}$ is not as easy as the corresponding process of determining a potential φ for a gradient field $\overline{F}$ (as done in §18.2.3 of The Companion). Although some general formulas for $\overline{H}$ do exist, they can be tedious to use, so we will not pursue them. However, here is an example showing how a vector potential may be found in a specific case:

Example C. Show that $\overline{F}(x,y,z) = \langle 1, -2z(x^2+y^2+z^2)^{-2}, 1+2y(x^2+y^2+z^2)^{-2} \rangle$ is a solenoidal vector field on the whole plane minus the origin.

Solution. Our only recourse at this stage is to prove that $\overline{F}$ is a curl field everywhere but at the origin. If $\overline{F} = \text{curl } \overline{H}$, and $\overline{H} = \langle f, g, h \rangle$, then, by Definition 18.7.1,

$$
\begin{cases}
\dfrac{\partial h}{\partial y} - \dfrac{\partial g}{\partial z} = 1 \\[2ex]
\dfrac{\partial f}{\partial z} - \dfrac{\partial h}{\partial x} = -2z(x^2+y^2+z^2)^{-2} \\[2ex]
\dfrac{\partial g}{\partial x} - \dfrac{\partial f}{\partial y} = 1 + 2y(x^2+y^2+z^2)^{-2}
\end{cases}
$$

We want to find three functions f, g and h which satisfy these equations; since these functions are not uniquely determined by the equations, we will have some freedom to make convenient choices. We'll start by choosing

$$h(x,y,z) = y + \psi_1(x,z), \quad \text{where} \quad \psi_1 \quad \text{is}$$

some yet-to-be-determined
function of x and z,

$$g(x,y,z) = \psi_2(x,y), \quad \text{where} \quad \psi_2 \quad \text{is some}$$

yet-to-be-determined
function of x and y

These choices satisfy the first of our three equations since $\dfrac{\partial h}{\partial y} = 1$ and $\dfrac{\partial g}{\partial z} = 0$. The

second equation then becomes

$$\frac{\partial f}{\partial z}(x,y,z) - \frac{\partial \psi_1}{\partial x}(x,z) = -2z(x^2 + y^2 + z^2)^{-2}$$

Convenient choices here are

$$f(x,y,z) = (x^2 + y^2 + z^2)^{-2}$$

$$\psi_1(x,z) = \psi_3(z), \quad \text{where} \quad \psi_3 \quad \text{is some yet-to-be-}$$

determined function of z

(Check for yourself that these choices satisfy the second equation.) The third equation now

gives

$$\frac{\partial \psi_2}{\partial x}(x,y) + 2y(x^2 + y^2 + z^2)^{-2} = 1 + 2y(x^2 + y^2 + z^2)^{-2}$$

so we can choose $\psi_2(x,y) = x$, and all our equations will be satisfied. Collecting these

results we have

$$f(x,y,z) = (x^2 + y^2 + z^2)^{-1}$$

$$g(x,y,z) = x$$

$$h(x,y,z) = y + \psi_3(z)$$

For convenience we choose $\psi_3(z) = 0$, so our final answer becomes

$$\overline{F} = \text{curl } \overline{H}, \quad \text{where} \quad \overline{H}(x,y,z) = \langle (x^2 + y^2 + z^2)^{-1}, x, y \rangle \qquad \square$$

<u>Example D.</u> Evaluate $\iint_\sigma \overline{F} \cdot \overline{n} \, dS$ where σ is the surface $x^2 + y^2 + z^3 = 4$, $z \geq 0$,

oriented by upward normals, and $\overline{F}(x,y,z) = \langle 1, -2z(x^2 + y^2 + z^2)^{-2},$

$1 + 2y(x^2 + y^2 + z^2)^{-2} \rangle$.

<u>Solution.</u> We evaluated this surface integral in Example B by changing the spanning surface.
In this solution we will make use of the vector potential $\overline{H}$ that we found for $\overline{F}$ in
Example C:

$$\iint_\sigma \overline{F} \cdot \overline{n} \, dS = \iint_\sigma (\text{curl } \overline{H}) \cdot \overline{n} \, dS \qquad \text{as in Example C}$$

$$= \int_C \overline{H} \cdot d\overline{r} \qquad \text{by Stokes' Theorem (Theorem 18.7.2),}$$

where C is the circle $x^2 + y^2 = 4$
in the xy-plane with the usual
counterclockwise orientation

$$= \int_C (x^2 + y^2 + z^2)^{-1} \, dx + x \, dy + y \, dz \qquad \text{from the}$$

formula for H in Example C.
Now parameterize C by

$$\overline{r}(\theta) = \langle 2 \cos \theta, 2 \sin \theta, 0 \rangle, \quad 0 \leq \theta \leq 2\pi$$

$$= \int_0^{2\pi} [\, 2^{-1}(-2 \sin \theta) + 2 \cos \theta (2 \cos \theta) + 0 \,] \, d\theta$$

$$= \int_0^{2\pi} [\, -\sin \theta + 4 \cos^2 \theta \,] \, d\theta$$

$$= \int_0^{2\pi} [\, -\sin\theta \; d\theta \; + \; 2(1 + \cos 2\theta)\,]\; d\theta$$

$$= (\cos\theta \; + \; 2\theta \; + \; \sin 2\theta) \Big|_0^{2\pi}$$

$$= \boxed{4\pi} \qquad \text{This agrees with Example B} \qquad\qquad \square$$

5. <u>2-Simply connected regions.</u> In §18.7.3 of <u>The Companion</u> we said that an open set   U

in   3-space   is <u>simply connected</u> if   U   is a connected set such that every closed curve   C

in   U   can be continuously shrunk to a point in   U. (Closed curves can be thought of as

stretched and twisted rubber bands, i.e., <u>continuous images of (deformed) circles.</u>)

Let's go up a dimension (from   2-space   to   3-space)   and replace rubber bands by

(hollow) rubber balls. Picturing rubber balls stretched and twisted into all sorts of weird

shapes gives you an intuitive feel for <u>continuous images of (deformed) spheres.</u> The concept

of   2-simple connectedness   is obtained merely by replacing deformed circles (closed curves)

in the definition of simply connected with deformed spheres:

<table>
<tr><td>2-Simply
Connected
Regions
in   3-space</td><td>Suppose   U   is an open region in   3-space.

Then   U   is said to be   <u>2-simply connected</u>   if

1.   U   is connected, and

2.   every continuous image of a sphere   σ   in   U

    can be continuously shrunk to a point in   U.</td></tr>
</table>

If   σ   continuously shrinks to a point in   U, then   σ   can be pictured as a (deformed) rubber

ball, confined to the region   U, shrinking in size until it is nothing but a point. In this process we do not allow the shrinking balls to rip or to tear in any way.

What does it mean if some continuous image of a sphere σ in U <u>cannot</u> be continuously shrunk to a point in U? It means that U has a fundamental "hole," a hole contained in the interior of σ. For example, take U_2 to be the region between two concentric spheres with radii 1 and 3, and let σ be the sphere with radius 2. (Note that σ is therefore inside of U_2). Then the "hole" in U_2 will prevent σ from shrinking to a point in U_2, and thus the region U_2 will not be 2-simply connected.

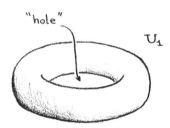

U_2 = region between spheres of radii 1 & 3

σ = sphere of radius 2

hole... a sphere of radius 1

On the other hand, consider the solid torus U_1. Although there is a "hole" in the torus, it is not possible to put that hole inside of any "rubber ball" contained in the torus! Thus U_1 is 2-simply connected!

"hole"

U_1

Notice the significant differences between simply connected sets as studied in §18.7.3 of <u>The Companion</u>, and 2-simply connected sets as studied in this subsection:

	simply connected?	2-simply connected?
U_1 (solid torus)	NO	YES
U_2 (region between two spheres)	YES	NO

Simple connectedness and 2-simple connectedness are sensitive to different types of "holes" in a region U.

6.   Incompressible vector fields.   We need a simple way to determine when a vector field

$\overline{F}(x,y,z)$ is solenoidal (since trying to compute a vector potential $\overline{H}$ for $\overline{F}$ can often

be extremely tedious and sometimes impossible).   The preferred method (at least when $\overline{F}$

has continous partial derivatives and   U   is an open,   2-simply connected   region) lies

with our fourth equivalent condition for solenoidal vector fields.   A vector function $\overline{F}(x,y,z)$

is said to be an incompressible vector field on an open region   U   if

Incompressible vector field	$\operatorname{div} \overline{F} = 0$ at every point in U

This definition was first given in §18. 8. 2 of The Companion,   where the

reason for the name "incompressible" was explained.

If $\overline{F}$ is a solenoidal vector field (i. e. , flux-free; span-independent; curl field)

and if the partial derivatives of $\overline{F}$ are continuous, then $\overline{F}$ is easily seen to be

incompressible, whether or not   U   is   2-simply connected.   This is Exercise 19 in

Anton's Exercise set 18. 7.

The converse of this result (i. e. , incompressible implies solenoidal) is also true

provided the region   U   is   2-simply connected.   The proof proceeds as follows: Suppose

σ is any continuous image of a sphere in   U.   Since   U   is   2-simply connected ,   the

interior solid   G   of σ is contained in   U,   and thus we can apply the Divergence Theorem

(Theorem 18. 6. 2):

$$\iint_{\sigma} \overline{F} \cdot \overline{n} \, dS = \iiint_{G} \operatorname{div} \overline{F} \, dV = 0$$

$$\underbrace{}_{\boxed{\text{Divergence Theorem}}} \qquad \underbrace{}_{\boxed{\overline{F} \text{ incompressible}}}$$

However, if every surface integral of $\overline{F}$ over a continuous image of a sphere in   U   equals zero, then we can construct a vector function $\overline{H}$ such that $\overline{F} = \text{curl } \overline{H}$ on U (as we discussed in Subsection 4 above). Thus $\overline{F}$ is a curl field, and hence it is solenoidal by the arguments given in §18.9.4.

This (finally!) completes the proof of the Solenoidal Vector Fields Theorem.

Example E.   Without computing a vector potential function, show that

$$\overline{F}(x,y,z) = \langle 1, -2z(x^2+y^2+z^2)^{-2}, 1 + 2y(x^2+y^2+z^2)^{-2} \rangle$$

is solenoidal on the region U which is all of 3-space minus some fixed infinite ray extending from the origin.

Solution.   We first check to see where $\overline{F}$ is incompressible:

$$\text{div } \overline{F}(x,y,z) = \frac{\partial}{\partial x}(1) + \frac{\partial}{\partial y}(-2z(x^2+y^2+z^2)^{-2}) + \frac{\partial}{\partial z}(1 + 2y(x^2+y^2+z^2)^{-2})$$

$$= 0 - 2z(-2)(x^2+y^2+z^2)^{-3}(2y) + 2y(-2)(x^2+y^2+z^2)^{-3}(2z)$$

$$= 0 \quad \text{if} \quad (x,y,z) \neq (0,0,0)$$

Thus $\overline{F}$ is incompressible everywhere except at the origin; in particular it is incompressible on U. Since U is 2-simply connected, then $\overline{F}$ is solenoidal on U by the Solenoidal Vector Fields Theorem.   □

7.   What if a region is not 2-simply connected?   Here the question is: if a region is not 2-simply connected, does checking that $\text{div } \overline{F} = 0$ help us in determining whether $\overline{F}$ is solenoidal?

If a region U is not 2-simply connected, then there are "holes" in U about which we can place "deformed spheres." Let $\sigma_1, \sigma_2, \ldots$ be a collection of such spheres covering all the "holes," with each σ_k containing only one "hole" in its interior. The

general principle is:

> If $\overline{F}$ is incompressible on U (i. e., $\operatorname{div} \overline{F} = 0$),
>
> and if $\displaystyle\iint_{\sigma_k} \overline{F} \cdot \overline{n} \, dS = 0$ for all the σ_k,
>
> then $\overline{F}$ is solenoidal on U.

Although we won't prove this result, here is a typical example of its use:

Example F.   Without computing a vector potential function, show that

$$\overline{F}(x,y,z) = \langle 1 , -2z(x^2 + y^2 + z^2)^{-2} , 1 + 2y(x^2 + y^2 + z^2)^{-2} \rangle$$

is solenoidal on the region U which is all of 3-space minus the origin.

Solution.   In the solution to Example E we showed that $\overline{F}$ is incompressible on U.

However, since U is not 2-simply connected , we cannot immediately use the Solenoidal

Vector Fields Theorem to conclude that $\overline{F}$ is solenoidal on U.

Instead, we proceed as follows. The only "hole" in U is at the origin, so let σ

denote the unit sphere centered on the origin. Using the "general principle" above, to

conclude that $\overline{F}$ is solenoidal on U we have only to prove $\displaystyle\iint_{\sigma} \overline{F} \cdot \overline{n} \, dS = 0$. We'll

do this by using spherical coordinates as described in §18. 5. 6 :

$$\iint_\sigma \overline{F} \cdot \overline{n} \; dS \; = \; \iint_\sigma \overline{F}(x,y,z) \cdot \langle x,y,z \rangle \; \sin \phi \; d\phi \; d\theta$$

$$= \int_0^{2\pi} \int_0^\pi \langle 1, -2z, 1+2y \rangle \cdot \langle x,y,z \rangle \; \sin \phi \; d\phi \; d\theta$$
$$\text{by using} \quad x^2 + y^2 + z^2 = 1$$

$$= \int_0^{2\pi} \int_0^\pi (x - 2yz + z + 2yz) \; \sin \phi \; d\phi \; d\theta$$

$$= \int_0^{2\pi} \int_0^\pi (x + z) \; \sin \phi \; d\phi \; d\theta$$

$$= \int_0^{2\pi} \int_0^\pi (\cos \theta \sin^2 \phi + \cos \phi \sin \phi) \; d\phi \; d\theta$$
$$\text{by using} \quad x = \sin \phi \cos \theta \quad \text{and} \quad z = \cos \phi$$

In this integral it is easiest to do the θ-integration first:

Inner integral: $\quad \displaystyle\int_0^{2\pi} (\cos \theta \sin^2 \phi + \cos \phi \sin \phi) \; d\theta$

$$= \sin^2 \phi \int_0^{2\pi} \cos \theta \; d\theta + \cos \phi \sin \phi \int_0^{2\pi} d\theta$$

$$= \sin^2 \phi \; (\sin \theta \Big|_0^{2\pi}) + \cos \phi \sin \phi \; (\theta \Big|_0^{2\pi})$$

$$= 2\pi \; \cos \phi \; \sin \phi$$

$$= \pi \; \sin 2\phi$$

Full integral: $\quad \displaystyle\int_0^\pi \pi \sin 2\phi \; d\phi = \frac{\pi}{2} \left(- \cos 2\phi \; \Big|_0^\pi \right) = 0$

Thus $\quad \displaystyle\iint_\sigma \overline{F} \cdot \overline{n} \; dS = 0$, as desired, proving that $\overline{F}$ is solenoidal on U. $\qquad \square$

8. Applications to electromagnetism.   Solenoidal and conservative vector fields arise time and time again in physics, especially in fluid mechanics and electromagnetism.   Here are a few comments on how they arise in electromagnetism.

Electromagnetism, at its elementary level, can be thought of as the interaction between two vector fields: $\overline{E}$,   the electric field

$\overline{B}$,   the magnetic field

The most basic laws governing these fields are Maxwell's Equations of Electromagnetism. These laws are essentially statments about how far away $\overline{E}$ and $\overline{B}$ are from being conservative or solenoidal! If we let t be time, and

$$\rho(t, x, y, z) = \text{charge density at time } t$$
$$\text{at the point } (x, y, z)$$

$$\overline{j}(t, x, y, z) = \text{current density at time } t$$
$$\text{at the point } (x, y, z) \quad ,$$

then Maxwell's Equations are

(1) $\operatorname{div} \overline{E} = \rho$ (Gauss' law) Thus $\overline{E}$ is incompressible in any region of space not containing charge

(2) $\operatorname{div} \overline{B} = 0$ Thus $\overline{B}$ is incompressible in all space, and is thus solenoidal

(3) $\operatorname{curl} \overline{E} = -\dfrac{\partial \overline{B}}{\partial t}$ (Faraday's law) Thus $\overline{E}$ cannot be conservative unless $\overline{B}$ remains constant with time

(4) $\operatorname{curl} \overline{B} = \overline{j} + \dfrac{\partial \overline{E}}{\partial t}$ (Ampère's law) Thus $\overline{B}$ will in general not be conservative

All four laws are first formulated in integral form (the underlying theory leads naturally to integrals, not derivatives), and are then converted to the more convenient derivative form by using the integral-to-derivative conversion techniques discussed in §18.8 of The Companion.

For example, Gauss' law is first formulated as

$$\iint_\sigma \overline{E} \cdot \overline{n} \; dS = \iiint_G \rho \; dV$$

for every solid sphere G contained in the region under consideration, with the boundary surface σ of G oriented with outward normals. Then Example D of §18.8.5

$$\operatorname{div} \overline{E} = \rho$$

the differential version of Gauss' law which is listed above.

Similarly, the law which states that $\overline{B}$ is incompressible is obtained from integral conditions by Example A of §18.8.2, while Faraday's law and Ampère's law are both obtained from integral conditions by Example E of §18.8.5.

Thus, practically all of the material of Chapter 18 is necessary to obtain Maxwell's equations!

Exercises.

1.   Determine which of the following subsets of 3-space are 2-simply connected:

    a.   U = all of 3-space minus the y-axis

    b.   V = all of 3-space minus the non-negative y-axis

c.   W = all of 3-space minus the unit disk in

the xy-plane $(x^2 + y^2 \le 1, z = 0)$

d.   X = all of 3-space minus the unit circle in

the xy-plane $(x^2 + y^2 = 1, z = 0)$

Compare your answers to those of Exercise 1 in §18.7 of The Companion. What does this say about the relationships between simply connected sets and 2-simply connected sets ?

2.   Consider the surface integral $\iint_\sigma \overline{F} \cdot \overline{n} \, dS$ where σ is the surface

$x^2 + y^2 + z^6 = 1$, $z \ge 0$, oriented by upward normals, and

$\overline{F}(x, y, z) = \langle -x^2 z, 2xyz, x^2 \rangle$.

a.   What is the largest open set U on which $\overline{F}$ is incompressible?

b.   Is $\overline{F}$ solenoidal on U ?

c.   Evaluate $\iint_\sigma \overline{F} \cdot \overline{n} \, dS$

d.   Does there exist a vector function $\overline{H}$ such that $\overline{F} = \operatorname{curl} \overline{H}$? If so, find such a function.

e.   If the answer to (d) was positive, then evaluate

$\iint_\sigma \overline{F} \cdot \overline{n} \, dS$ using $\overline{H}$.

3.   Consider the vector field

$\overline{F}(x, y, z) = \langle 2y(x^2 + y^2 + z^2)^{-2}, -2z(x^2 + y^2 + z^2)^{-2} - 1, 0 \rangle$

a.   What is the largest open set U on which $\overline{F}$ is incompressible?

  b.   Is U 2-simply connected ?

  c.   Prove that $\overline{F}$ is solenoidal on U without calculating

        a vector potential function

4.   Show that the following fact is true:

        If $\overline{F}$ is solenoidal on a simply connected open region U,

        then any two vector potential functions for $\overline{F}$ differ

        by at most a gradient field on U.

Answers.

1.   U and V are 2-simply connected; W and X are not. Thus, when

      combined with the results of Exercise 1 in §18.7 of The Companion,

      we have all four possible combinations of simply connected and 2-simply connected:

	simply connected?	2-simply connected?
U	NO	YES
V	YES	YES
W	YES	NO
X	NO	NO

2.   a.   U = all of 3-space

      b.   Since U is 2-simply connected and $\overline{F}$ is incompressible

            on U, then $\overline{F}$ is solenoidal on U.

c.   Replace σ with σ^*, the unit disk in the xy-plane,

oriented with the upward normals. Then

$$\iint_\sigma \overline{F} \cdot \overline{n}\, dS = \iint_{\sigma^*} \overline{F} \cdot \overline{n}\, dS = \iint_{\sigma^*} x^2\, dS = \pi/4$$

where the last evaluation is best done with polar coordinates

d.   Numerous choices are possible; one such choice is

$$\overline{H}(x,y,z) = \langle - x^2 y, \ 0, \ - x^2 yz \rangle$$

e. $$\iint_\sigma \overline{F} \cdot \overline{n}\, dS = \int_C \overline{H} \cdot d\overline{r} = \pi/4 \, ,$$

where C is the unit circle in the xy-plane,

oriented in a counterclockwise direction

3.   a. U = all of 3-space minus the origin

       b. U is not 2-simply connected

       c.   Use the techniques of Example F to prove $\overline{F}$ is solenoidal on U.

If σ is the unit sphere, then

$$\iint_\sigma \overline{F} \cdot \overline{n}\, dS = \int_0^{2\pi} \int_0^\pi (2xy - 2yz - y)\, \sin\phi\, d\phi\, d\theta$$

and this integral is easily seen to be zero if you first perform

the θ - integration.

4.   Suppose $\overline{F} = \operatorname{curl} \overline{H}_1 = \operatorname{curl} \overline{H}_2$. Then $\operatorname{curl}(\overline{H}_1 - \overline{H}_2) = 0$, so that $H_1 - H_2$

is irrotational on U. However, U is simply connected, so that $H_1 - H_2$

is conservative on U, and hence $H_1 - H_2 = \nabla\varphi$ for some function φ on U.

Section 18. 10 :   An Introduction to Differential Forms (Optional)

The similarity of the Conservative Vector Fields Theorem (§§18. 2 and 18. 7. 3)

and the Solenoidal Vector Fields Theorem (§18. 9) should make you suspect there is a

unifying principle for the topics we have so far discussed in vector calculus.   There is,

and the key to it is the use of differential forms.   Although in one section we can only give

a superficial treatment of differential forms, we can indicate how this concept is used to

unify and generalize the Conservative and Solenoidal Vector Fields Theorems.

1.   Differential forms.   The general definition for a (differential) k-form is beyond the

scope of this book.   However, in 3-space we can achieve our purposes by defining

k-forms (for k = 0 , 1 , 2 , 3) to be certain combinations of the differentials dx, dy

and dz :

Differential forms on 3-space

0-forms: expressions of the form f ,

where f is any function

1-forms: expressions of the form f dx + g dy + h dz ,

where $\overline{F} = \langle f, g, h \rangle$ is any vector function

2-forms: expressions of the form f dy dz + g dz dx + h dx dy ,

where $\overline{F} = \langle f, g, h \rangle$ is any vector function

3-forms: expressions of the form f dx dy dz

where f is any function

(We use dz dx rather than dx dz in the definition of 2-forms to correspond accurately

with the general definition of 2-forms.   This "backwards" notation has no effect on the

development of k-forms which we follow.)

Thus, for our purposes, 0-forms and 3-forms are just scalar functions denoted in funny ways, while 1-forms and 2-forms are just vector functions denoted in funny ways.

There is an operator d, called the exterior derivative, which takes a k-form to a $(k+1)$-form $(k = 0, 1, 2)$, essentially by differentiation. For k-forms in 3-space the d operator is a collection of familiar operators:

The exterior derivative on 3-space	d from 0-forms to 1-forms is the gradient
	d from 1-forms to 2-forms is the curl
	d from 2-forms to 3-forms is the divergence

Notice how this definition is based on interpreting 0-forms and 3-forms as scalar definitions, and 1-forms and 2-forms as vector functions. To be specific,

$$d(f) = \frac{\partial f}{\partial x}\, dx + \frac{\partial f}{\partial y}\, dy + \frac{\partial f}{\partial z}\, dz$$

$$d(f\, dx + g\, dy + h\, dz)$$

$$= \left(\frac{\partial h}{\partial y} - \frac{\partial g}{\partial z}\right) dy\, dz + \left(\frac{\partial f}{\partial z} - \frac{\partial h}{\partial x}\right) dz\, dx + \left(\frac{\partial g}{\partial x} - \frac{\partial f}{\partial y}\right) dx\, dy$$

$$d(f\, dy\, dz + g\, dz\, dx + h\, dx\, dy)$$

$$= \left(\frac{\partial f}{\partial x} + \frac{\partial g}{\partial y} + \frac{\partial h}{\partial z}\right) dx\, dy\, dz$$

The exterior derivative is defined in general to be that unique mapping from k-forms to $(k+1)$-forms that obeys a certain collection of rules. We will not list these rules since to do so would require developing more structure for the set of k-forms than is possible at our level. (An introduction to some of these rules is given in Exercise 5 at the end of this section.)

Example A. Suppose $\omega_1 = xz\,dx + x^2 z\,dy + (y+z)\,dz$,

and $\omega_2 = xz\,dy\,dz + x^2 z\,dz\,dx + (y+z)\,dx\,dy$

Compute $d\omega_1$ and $d\omega_2$.

Solution. $d\omega_1 = d(xz\,dx + x^2 z\,dy + (y+z)\,dz)$

$$= \left(\frac{\partial(y+z)}{\partial y} - \frac{\partial(x^2 z)}{\partial z}\right)dy\,dz + \left(\frac{\partial(xz)}{\partial z} - \frac{\partial(y+z)}{\partial z}\right)dz\,dx$$

$$+ \left(\frac{\partial(x^2 z)}{\partial x} - \frac{\partial(xz)}{\partial y}\right)dx\,dy$$

$$= (1 - x^2)\,dy\,dz + x\,dz\,dx + 2xz\,dx\,dy$$

$d\omega_2 = d(xz\,dy\,dz + x^2 z\,dz\,dx + (y+z)\,dx\,dy)$

$$= \left(\frac{\partial(xz)}{\partial x} + \frac{\partial(x^2 z)}{\partial y} + \frac{\partial(y+z)}{\partial z}\right)dx\,dy\,dz$$

$$= (z+1)\,dx\,dy\,dz \qquad\qquad \square$$

Before we can attach any meaning to k-forms other than "funny ways to write scalar and vector functions" we need some additional definitions...

2. <u>Oriented k-manifolds with boundary.</u> In 3-space this is just a new name for a collection of objects you are already familiar with:

M   is an oriented   k-manifold   with boundary means

k = 0:   M   is a two point set $\{\overline{X}_0, \overline{X}_1\}$,   where

$\overline{X}_0$ is designated as the initial point,

and X_1 is designated as the terminal

point. The boundary of   M   is the empty set.

k = 1:   M   is a smooth, directed curve   C.   The

boundary of   M   consists of the initial

and terminal points of   C.

k = 2:   M   is a smooth, oriented surface σ,   whose

boundary is assumed to be a closed

piecewise smooth curve   C,   directed

so that   C   and σ obey the right-

hand rule ($\S$ 18. 7. 2   of   The Companion).

k = 3:   M   is a solid whose boundary σ is assumed

to be a closed surface comprised of a finite

union of smooth surfaces. σ is oriented

by outward normals.

Thus the boundary of a   k-manifold   (k = 1, 2, 3)   is a finite union of   (k - 1) - manifolds ;

such a union will also be called a   (k - 1) - manifold.   In this way we obtain a natural mapping

∂ from   k-manifolds   to   (k - 1) - manifolds ,   known as the boundary operator:

Boundary
Operator

> If M is an oriented k-manifold with boundary,
>
> (k = 1, 2, 3), then ∂M is the boundary of M.
>
> ∂M is an oriented (k - 1)-manifold.

3. <u>Integrating k-forms on k-manifolds</u>. We now define the integral of a k-form on a k-manifold. We then will have a meaning for a k-form ω: it is a mapping that assigns a number $\displaystyle\int_M \omega$ to each k-manifold M:

<u>k = 0</u>: If $\omega = \varphi$ is a 0-form and $M = \{\overline{X}_0, \overline{X}_1\}$ is a 0-manifold, then

$$\int_M \omega = \varphi(\overline{X}_1) - \varphi(\overline{X}_1)$$

<u>k = 1</u>: If $\omega = f\,dx + g\,dy + h\,dz$ is a 1-form, and

M = C is a 1-manifold (a curve), then

$$\int_M \omega = \int_C f\,dx + g\,dy + h\,dz$$

$$= \int_C \overline{F} \cdot d\overline{r}$$

an ordinary <u>line integral</u> of $\overline{F} = \langle f, g, h \rangle$ over the directed curve C

$\underline{k = 2}$:   If ω = f dy dz + g dz dx + h dx dy is a 2-form, and

M = σ is a 2-manifold (a surface), then

$$\int_M \omega = \int_\sigma f \, dy \, dz + g \, dz \, dx + h \, dx \, dy$$

$$= \iint_\sigma \overline{F} \cdot \overline{n} \, dS$$

an ordinary <u>surface integral</u> of $\overline{F}$ = $\langle f, g, h \rangle$ over

the oriented surface σ

$\underline{k = 3}$:   If ω = f dx dy dz is a 3-form, and

M = G is a 3-manifold (a solid), then

$$\int_M \omega = \int_G f \, dx \, dy \, dz$$

$$= \iiint_G f \, dV$$

an ordinary <u>triple integral</u> of f

over the solid G.

As with the exterior derivative, the definition of integration of forms on manifolds is dictated by certain properties we want integration to have. Once again, however, the details are beyond the scope of this book.

<u>Example B.</u>   Suppose ω = z dy dz + y dz dx + x dx dy, and M = σ is that portion of the paraboloid y = $x^2 + z^2$ for which y $\leq$ 4, oriented by right unit normals. Compute

$$\int_M \omega .$$

Solution. By the definition above

$$\int_M \omega = \iint_\sigma \overline{F} \cdot \overline{n} \, dS$$

where $\overline{F}(x,y,z) = \langle z,y,x \rangle$ This surface integral was evaluated in Example B of §18.5.4, and gave the answer 8π. □

4. The Generalized Stokes' Theorem. Now we can put some of our earlier results into more general settings. We'll start by looking at the various integral theorems obtained earlier in this chapter:

The Fundamental Theorem of Line Integrals for 3-space* (see §18.2.3)

$$\int_C \nabla \varphi \cdot d\overline{r} = \varphi(\overline{X}_1) - \varphi(\overline{X}_0)$$

where $\overline{X}_0$ and $\overline{X}_1$ are the initial
and terminal points for C respectively

In terms of differential forms this becomes

$$\int_M d\omega = \int_{\partial M} \omega$$

where $\omega = \varphi$ is a 0-form
and M = C is a 1-manifold (curve)

---

* We proved this result only for 2-space, but the proof carries over verbatim to 3-space.

Stokes' Theorem (Theorem 18.7.2)

$$\iint_\sigma (\text{curl } \overline{F}) \cdot \overline{n} \; dS = \int_C \overline{F} \cdot d\overline{r}$$

where C is the boundary
curve for the surface σ

In terms of differential forms this becomes

$$\int_M d\omega = \int_{\partial M} \omega$$

where ω is the 1-form given by $\overline{F}$,

and $M = \sigma$ is a 2-manifold (surface)

Divergence Theorem (Theorem 18.6.2)

$$\iiint_G \text{div } \overline{F} \; dV = \iint_\sigma \overline{F} \cdot \overline{n} \; dS$$

where σ is the boundary
surface for the solid G

In terms of differential forms this becomes

$$\int_M d\omega = \int_{\partial M} \omega$$

where ω is the 2-form given by $\overline{F}$,

and $M = G$ is a 3-manifold (solid)

As you can see, when these three theorms are written in terms of differential forms, they all look the same! Here is the one statement that encompasses all three:

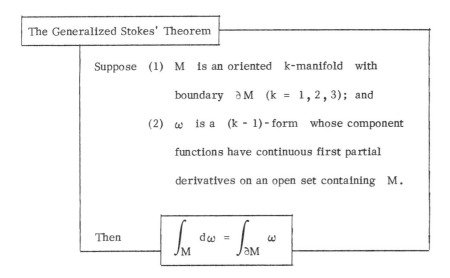

The Generalized Stokes' Theorem

Suppose (1) M is an oriented k-manifold with boundary ∂M (k = 1, 2, 3); and

(2) ω is a (k - 1)-form whose component functions have continuous first partial derivatives on an open set containing M.

Then $$\int_M d\omega = \int_{\partial M} \omega$$

(The component functions of a differentiable form are simply the component functions of the scalar or vector function which corresponds to the form.)

The real power of Stokes' Theorem is that it is true not simply in 3-space but in "n-space," for n any positive integer, at least when all the concepts involved are suitably generalized to n-space. For instance, here is what happens in the cases when n = 1 and n = 2:

For n = 1 we only allow k = 1. Then the Generalized Stokes' Theorem becomes just the Fundamental Theorem of Calculus (Theorem 5. 8. 1)!

For n = 2 we only allow k = 1 or 2. Then the Generalized Stokes' Theorem becomes the Fundamental Theorem of Line Integrals for 2-space (k = 1; §18. 2. 3) and Green' Theorem (k = 2; Theorem 18. 3. 1). This latter result is a consequence of the 2-space definition

$$d(f\ dx + g\ dy) = \left(\frac{\partial g}{\partial x} - \frac{\partial f}{\partial y}\right) dx\ dy$$

5.   <u>Exact and closed k-forms</u>.   In §§18.2, 18.7.3, and 18.9, we discussed the concepts

of conservative and solenoidal vector fields. Using the gradient field and curl field

formulations for conservative and solenoidal, those definitions can be generalized as follows:

$\overline{F}$ is a <u>conservative vector field</u> if it is a gradient field, i.e.,

$$\overline{F} = \nabla\varphi \quad \text{for some function } \varphi$$

In terms of differential forms this becomes

ω is a <u>conservative   1-form</u>

if $\omega = d\eta$ for some   0-form η

$\overline{F}$ is a <u>solenoidal vector field</u> if it is a curl field, i.e.,

$$\overline{F} = \text{curl } \overline{H} \quad \text{for some vector field } \overline{H} \quad \text{on   U}$$

In terms of differential forms this becomes

ω is a <u>solenoidal   2-form</u>

if $\omega = d\eta$ for some   1-form η

As you can see, conservative is a term that applies to   1-forms,   while solenoidal applies

to   2-forms. Moreover, the key ingredient in both is that

a   k-form ω is conservative or solenoidal

if and only if it is the exterior derivative

of some $(k-1)$-form η, i.e., $\omega = d\eta$

For this reason we single out the property $\omega = d\eta$:

Exact Differential Forms	A k-form ω is <u>exact</u> on an open set U if there exists a $(k-1)$-form η such that $$\omega = d\eta \quad \text{on all} \quad U$$

We will generalize the various equivalences in the Conservative and Solenoidal Vector Fields Theorems ($\S\S 18.2$, $18.7.3$ and 18.9) by using exact differential forms. It is best to start by generalizing condition (4) of these theorems:

$$\overline{F} \text{ is } \underline{\text{irrotational}} \text{ if } \quad \text{curl } \overline{F} = 0$$

$$\overline{F} \text{ is } \underline{\text{incompressible}} \text{ if } \quad \text{div } \overline{F} = 0$$

In terms of differential forms these definitions become

$$\omega \text{ is an } \underline{\text{irrotational } \text{1-form}} \text{ if } \quad d\omega = 0$$

$$\omega \text{ is an } \underline{\text{incompressible } \text{2-form}} \text{ if } \quad d\omega = 0$$

This leads us to single out the property $d\omega = 0$:

Closed Differential Forms	A k-form ω is <u>closed</u> on an open set U if $d\omega = 0$ on all of U

From the Conservative and Solenoidal Vector Fields Theorems we know that:

if $\overline{F}$ is a gradient field, then it is irrotational,

i.e., $\text{curl (grad } \varphi) = 0$ for all functions φ

if $\overline{F}$ is a curl field, then it is incompressible,

i.e., $\text{div (curl } \overline{H}) = 0$ for all vector functions $\overline{H}$

In terms of differential forms these results become:

> If ω is <u>exact</u>, then ω is <u>closed</u>,
>
> i. e., $d(d\eta) = 0$ for all k-forms η

In shorthand we simply write this as

$$d^2 = 0$$

(This is one of the general defining rules for the exterior derivative operator that we referred to earlier in Subsection 1.)

The converse of this result is what we are really interested in, however:

if ω is <u>closed</u> on an open set U,

then under what circumstances can

we conclude that ω is <u>exact</u> on U?

For conservative vector fields we need to assume U is simply connected; for solenoidal vector fields we need to assume U is 2-simply connected. The most general answer is beyond the scope of this section, but here is one which is pretty good:

Star-
Shaped
Sets

> An open set U is said to be star-shaped
> if there is some point P_0 in U such that
> any other point P in U can be
> joined to P_0 by a straight line
> segment lying entirely in U.

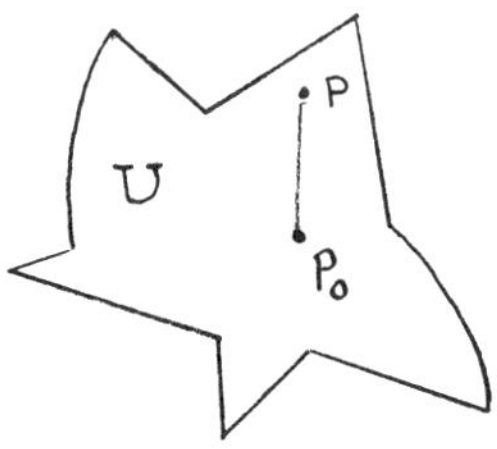

A star-shaped region is always simply connected and 2-simply connected.

The Exact Differential Forms Theorem

Suppose ω is a k-form on an open region U

Then the following properties are all equivalent:

(1) For every oriented k-manifold M in U such

that M is the boundary of some oriented

$(k+1)$-manifold N <u>in 3-space</u>, we have

$$\int_M \omega = 0$$

(2) Suppose M_1 and M_2 are two oriented k-manifolds

in U with the same oriented boundary. Then

$$\int_{M_1} \omega = \int_{M_2} \omega$$

(3) ω is <u>exact</u> on U, i.e., there is a $(k-1)$-form

η such that $\omega = d\eta$ on all U.

Moreover, if the component functions of ω have continuous first

partial derivatives on U, and U is an open, star-shaped region,

then there is a fourth equivalence:

(4) ω is <u>closed</u> on U, i.e., $d\omega = 0$ on all of U.

Conditions (1), (2) and (3) imply condition (4) for all types of open regions U;

it is the converse of this result that can fail if U has "inconvenient" types of holes.

We have already established this theorem in 3-space: for $k = 1$ it is the

Conservative Vector Fields Theorem, while for $k = 2$ it is the Solenoidal Vector Fields

Theorem.

The proof of the theorem in the general case follows the patterns we used in proving the Conservative and Solenoidal Vector Fields Theorems. Although we will not prove the theorem (we couldn't prove all the implications without the general definitions of k-forms, manifolds, etc.), we will give the argument that shows (3) implies (2), since it demonstrates the use of the Generalized Stokes' Theorem.

Suppose $\omega = d\eta$ on U, and M_1 and M_2 are two oriented k-manifolds with the same boundary $\partial M_1 = \partial M_2$. Then

$$\int_{M_1} \omega = \int_{M_1} d\eta \qquad \text{since } \omega = d\eta$$

$$= \int_{\partial M_1} \eta \qquad \text{by the Generalized Stokes' Theorem}$$

$$= \int_{\partial M_2} \eta \qquad \text{since } \partial M_1 = \partial M_2$$

$$= \int_{M_2} d\eta \qquad \text{by the Generalized Stokes' Theorem}$$

$$= \int_{M_2} \omega \qquad \text{since } \omega = d\eta$$

This proves the (3) implies (2).

6. A final comment. In vector calculus it is easy to get lost in terminology, definitions and strange formulas. That is why we think it helpful to place the material into the context of differential forms. The amazing unification that is achieved makes it much easier to remember individual results and to grasp the overall direction of the subject.

And that's not to say anything about the importance of differential forms in physics, engineering, geometry and topology. But those extensions are for courses beyond calculus....

Exercises.

1. Find $d\omega$ for the following k-forms ω in 3-space:

    a. $\omega_1 = x^2 y/z$

    b. $\omega_2 = xy\ dx + xe^z\ dy + y^2 z\ dz$

    c. $\omega_3 = x^2 y\ dy\ dz + x\ \ln y\ dz\ dx + xyz\ dx\ dy$

    d. $\omega_4 = 2xy^3 z\ dx + 3x^2 y^2 z\ dy + x^2 y^3\ dz$

    e. $\omega_5 = (x^2 - x)\ dy\ dz - (2xy + x/z^2)\ dz\ dx + z\ dx\ dy$

2. a. Which of the k-forms in Exercise 1 are closed? (Specify the largest open region on which the form is closed.)

    b. Which of the k-forms in Exercise 1 are exact? (Specify the largest open region -- or regions -- on which the form is exact.) In each case determine an η such that $\omega = d\eta$.

3. In 2-space the exterior derivative of a 1-form is defined by

$$d(f\ dx + g\ dy) = \left(\frac{\partial g}{\partial x} - \frac{\partial f}{\partial y}\right) dx\ dy$$

Is this consistent with the 3-space definition of

$$d(f\ dx + g\ dy + h\ dz)$$

when $h = 0$ and f and g are functions only of x and y (i. e. , do not depend on z)?

4.   Determine the value that the k-form ω assigns to the k-manifold M

in the following cases:

  a. $\omega = 3x^2 dx + 3y^2 dy + (4x + 2z) dz$

        M = the curve parameterized by $\overline{r}(t) = \langle t, t^2, t^3 \rangle$

                from t = -1 to t = 1

  b. $\omega = 3x^2 dx + 3y^2 dy + (4x + 2z) dz$

        M = the curve parameterized by $\overline{r}(t) = \langle t^2, -t, t \rangle$

                from t = 0 to t = 1

  c. $\omega = x^2 z \, dy \, dz + xy \, dz \, dx + xyz \, dx \, dy$

        M = the portion of the graph of z = xy over the
             unit square in the xy-plane, oriented with
             upward-pointing normals

5.   In xyzw-space (4-space), k-forms can be defined as certain

combinations of the differentials dx, dy, dz, dw $(f, f_1$, etc.,

denote real-valued functions on 4-space):

        0-forms: f

        1-forms: $f_1 dx + f_2 dy + f_3 dz + f_4 dw$

        2-forms: $f_{12} dz \, dw + f_{13} dy \, dw + f_{14} dy \, dz$

                        $+ f_{23} dx \, dw + f_{24} dx \, dz + f_{34} dx \, dy$

        3-forms: $f_1 dy \, dz \, dw + f_2 dz \, dx \, dw$

                        $+ f_3 dx \, dy \, dw + f_4 dy \, dx \, dz$

        4-forms: f dx dy dz dw

Assume we can "multiply" k-forms with ℓ-forms (e.g.,

dx multiplied by dy dz is dx dy dz), and that this

multiplication is associative and distributive; it is <u>not</u>

commutative, however. Instead,

$$du\ dv\ =\ -\ dv\ du \qquad (u, v\ =\ x, y, z\ \text{ or }\ w)$$

However, $\omega(f\eta)\ =\ (f\omega)\eta\ =\ f(\omega\eta)$ for f any 0-form,

and ω and η any k- and ℓ-forms

(a) Prove that du du = 0 for u = x, y, z, w

(b) Prove that dx dy dz = dz dx dy

The exterior derivative d maps k-forms to (k + 1) - forms and

is assumed to have the following properties

(i) $df\ =\ \left(\dfrac{\partial f}{\partial x}\right) dx\ +\ \left(\dfrac{\partial f}{\partial y}\right) dy\ +\ \left(\dfrac{\partial f}{\partial z}\right) dz\ +\ \left(\dfrac{\partial f}{\partial w}\right) dw$

(ii) $d(\eta + \omega)\ =\ d\eta\ +\ d\omega$

(iii) $d(f\ du\ dv\dots)\ =\ df\ du\ dv\dots$

(c) Determine the formulas for d as a map from k-forms

to (k + 1) - forms, for k = 0, 1, 2, 3.

(d) Evaluate $\displaystyle\int_{\partial M} \omega$, where M is the unit cube

in 4-space, and

$$\omega\ =\ x^2 w\ dy\ dz\ dw\ +\ x z^2 w\ dz\ dx\ dw$$

$$+\ x^2 y w\ dx\ dy\ dw\ +\ y w\ dy\ dx\ dz$$

Answers.

1. a. $d\omega_1 = (2xy/z)\,dx + (x^2/z)\,dy - (x^2y/z^2)\,dz$

    b. $d\omega_2 = (2yz - xe^z)\,dy\,dz - y^2\,dz\,dx + (e^z - x)\,dx\,dy$

    c. $d\omega_3 = (3xy + (x/y))\,dx\,dy\,dz$

    d. $d\omega_4 = 0 \cdot dy\,dz + 0 \cdot dz\,dx + 0 \cdot dx\,dy = 0$

    e. $d\omega_5 = 0 \cdot dx\,dy\,dz = 0$

2. a. The 1-form ω_4 is closed on all of 3-space

        The 2-form ω_5 is closed on all of 3-space

        except for the xy-plane (i. e., $z = 0$)

    b. Since $U = $ (all of 3-space) is simply connected, then

        the 1-form ω_4 is exact (conservative) on U.

        $\omega_4 = d\eta$ where $\eta = x^2y^3z$

        Since $U = $ (all of 3-space minus the xy-plane) consists

        of two disjoint, 2-simply connected regions $(z > 0$

        and $z < 0)$, then the 2-form ω_5 is exact

        (solenoidal) on both of these regions.

        $\omega_5 = d\eta$ where $\eta = (x/z)\,dx + xz\,dy + x^2y\,dz$ (Other η's
        are possible)

3. Yes, since $d(f\,dx + g\,dy + h\,dz)$

    $$= \left(\frac{\partial h}{\partial y} - \frac{\partial g}{\partial z}\right) dy\,dz + \left(\frac{\partial f}{\partial z} - \frac{\partial h}{\partial x}\right) dz\,dx + \left(\frac{\partial g}{\partial x} - \frac{\partial f}{\partial y}\right) dx\,dy$$

    If $h = 0$, then $\dfrac{\partial h}{\partial y} = \dfrac{\partial h}{\partial x} = 0$

Moreover, since f and g do not depend on z , then

$$\frac{\partial f}{\partial z} = \frac{\partial g}{\partial z} = 0$$

Thus

$$d(f\ dx\ +\ g\ dy\ +\ h\ dz) = \left(\frac{\partial g}{\partial x} - \frac{\partial f}{\partial y}\right) dx\ dy\ ,\quad \text{as desired.}$$

4.   a.   2   b.   7/3   c. $-5/36$

5.   a.   du du = - du du   from the anti-commutative rule.   Thus   du du = 0.

      b.   dx dy dz = dx (- dz dy) = (- dx dz) dy = dz dx dy

      c.   Let $f_x = \frac{\partial f}{\partial x}$, $f_y = \frac{\partial f}{\partial y}$, $f_z = \frac{\partial f}{\partial z}$, $f_w = \frac{\partial f}{\partial w}$

$\underline{k = 0}:$ $df = f_x\ dx + f_y\ dy + f_z\ dz + f_w\ dw$

$\underline{k = 1}:$ $d(f_1\ dx + f_2\ dy + f_3\ dz + f_4\ dw)$

$$= (f_{4z} - f_{3w})\ dz\ dw + (f_{4y} - f_{2w})\ dy\ dw$$
$$+ (f_{3y} - f_{2z})\ dy\ dz + (f_{4x} - f_{1w})\ dx\ dw$$
$$+ (f_{3x} - f_{1z})\ dx\ dz + (f_{2x} - f_{1y})\ dx\ dy$$

$\underline{k = 2}:$ $d(f_{12}\ dz\ dw + f_{13}\ dy\ dw + f_{14}\ dy\ dz$
$$+ f_{23}\ dx\ dw + f_{24}\ dx\ dz + f_{34}\ dx\ dy)$$

$$= (f_{12y} - f_{13z} + f_{14w})\ dy\ dz\ dw$$
$$+ (-f_{12x} + f_{23z} - f_{24w})\ dz\ dx\ dw$$
$$+ (f_{13x} - f_{23y} + f_{34w})\ dx\ dy\ dw$$
$$+ (-f_{14x} + f_{24y} - f_{34z})\ dy\ dx\ dz$$

$\underline{k = 3}:$ $d(f_1\ dy\ dz\ dw + f_2\ dz\ dx\ dw + f_3\ dx\ dy\ dw + f_4\ dy\ dx\ dz)$
$$= (f_{1x} + f_{2y} + f_{3z} + f_{4w})\ dx\ dy\ dz\ dw$$

Thus, d for $k = 0$ looks like a "gradient" and d for $k = 3$

looks like a "divergence;" for $k = 1$ and 2 we get strange

looking variants of the "curl"

d. $\displaystyle\int_{\partial M} \omega \; = \; \int_M \; d\omega$ by the Generalized Stokes' Theorem

$$= \int_M \left(\frac{\partial}{\partial x} \, (x^2 w) + \frac{\partial}{\partial y} \, (x \, z^2 w) + \frac{\partial}{\partial z} \, (x^2 y \, w) + \frac{\partial}{\partial w} \, (y \, w) \right) \, dx \; dy \; dz \; dw$$

$$= \int_0^1 \int_0^1 \int_0^1 \int_0^1 \; (2 \, x \, w \; + \; y) \; dx \; dy \; dz \; dw \; = \; 1$$

INDEX TO VOLUMES I AND II